Deep Learning Innovations in MRI Reconstruction and Analysis

Soumick Chatterjee

Deep Learning Innovations in MRI Reconstruction and Analysis

Enhancing Image Quality for Robust Image Processing and Clinical Decision Making

Springer Vieweg

Soumick Chatterjee
Human Technopole
Milan, Italy

ISBN 978-3-658-50740-4 ISBN 978-3-658-50741-1 (eBook)
https://doi.org/10.1007/978-3-658-50741-1

This Springer Vieweg imprint is published by the registered company Springer Fachmedien Wiesbaden GmbH, part of Springer Nature.
The registered company address is: Abraham-Lincoln-Str. 46, 65189 Wiesbaden, Germany

Pour toi papa, Prof. Swapan Kr. Chatterjee

Foreword by Prof Andreas Nürnberger

In the ever-evolving field of medical imaging, few challenges have proven as persistent and complex as the reduction of artefacts in magnetic resonance imaging (MRI). These artefacts—whether originating from undersampling during high-speed imaging or from motion during patient scanning—pose significant obstacles to the accuracy and reliability of MR images. Yet, with the rapid development of deep learning techniques, new possibilities have emerged to address these issues and enhance the performance of MRI. Soumick Chatterjee's *Deep Learning Innovations in MRI Reconstruction and Analysis: Enhancing Image Quality for Robust Image Processing and Clinical Decision Making* (Original title of the PhD thesis: *Reducing Artefacts in MRI Using Deep Learning: Enhancing Automatic Image Processing*) *Pipelines* stands as an exceptional contribution to this frontier, offering both a comprehensive review and innovative advancements aimed at mitigating artefacts in MR imaging.

The book is a rich exploration of how deep learning can be integrated into MRI image reconstruction and processing pipelines, specifically designed to reduce common artefacts. Soumick's work stands out for its systematic approach to this complex problem. By focusing on two key sources of artefacts—undersampling and patient motion—he presents a compelling case for the application of deep learning techniques in improving the quality of MR images, making the technology more effective in clinical settings. Through a careful selection of research questions, the author evaluates the feasibility of these models across different tasks, from improving image quality to aiding in specific medical applications such as brain tumour classification and small vessel segmentation.

This book is more than just a technical deep dive into MRI and deep learning; it bridges the gap between theory and practice. It offers an invaluable resource for researchers and professionals seeking to understand the interplay between cutting-edge AI models and the highly specialised domain of medical imaging. The author's ability to seamlessly integrate deep learning techniques with practical applications in the medical field is perhaps one of the most remarkable aspects of this work. Furthermore, the book's broad scope—covering everything from undersampled MRI reconstruction to motion artefact reduction—ensures that it is not only a deep dive into one aspect of MRI but rather a holistic view of the challenges and solutions in the domain.

Soumick's work draws on an impressive body of research, all of which has been published in top-tier journals and presented at major conferences in the fields of both computer science and medical imaging. This publication record speaks to the quality and significance of his contributions, as well as the meticulousness with which he approaches his research. The book is organised into six parts, each addressing different facets of the problem at hand. The first part of the book lays a solid foundation, introducing the essential principles of MRI technology, image processing, and neural networks, setting the stage for the innovative methods that follow. Although the chapters assume a certain level of technical background, they provide a comprehensive overview of these areas, making the book accessible to readers with a foundational understanding of medical imaging and machine learning. Part II provides a thorough review of existing methods for undersampled MRI reconstruction, while Part III introduces novel deep learning techniques designed to enhance these approaches. In Parts IV and V, Soumick takes a more applied approach, demonstrating how his deep learning methods can be incorporated into real-world medical imaging pipelines to address specific clinical tasks. Part V, focusing on motion artefact reduction, offers an additional layer of insight, demonstrating the versatility of the proposed methods beyond the problem of undersampling. Part IV of the book, which addresses medical tasks like brain tumour classification and small vessel segmentation, highlights the direct impact of these innovations on real-world clinical challenges. By embedding deep learning models in the image processing pipeline, Soumick shows how these approaches can improve not only image quality but also clinical decision-making. In Part VI, Soumick concludes with a forward-looking discussion on potential future research directions, demonstrating a deep understanding of the broader landscape of MRI image processing and deep learning. His insights into where the field might go next—such as exploring more efficient models for real-time processing and expanding applications to other imaging modalities—showcase his vision for continued progress. The thoroughness of the research, coupled with a clear articulation of future avenues for exploration, solidifies this work as an important reference point for anyone looking to push the boundaries of what is possible with deep learning in medical imaging.

One of the strengths of the book is Soumick's ability to present highly technical content in a manner that is both rigorous and accessible. Although some sections are dense with technical detail, the author takes care to explain complex concepts without overwhelming the reader. His clear explanations and methodical approach make the book an ideal resource for graduate students, researchers, and professionals in both computer science and medical imaging who are seeking to deepen their understanding of how AI can transform healthcare.

In summary, Soumick Chatterjee's *Deep Learning Innovations in MRI Reconstruction andAnalysis* is an impressive achievement in the intersection of machine learning and medical imaging. The work is not only a thorough examination of the state-of-the-art in MRI artefact reduction but also a demonstration of how deep learning can be practically applied to improve the quality and speed of MRI scans. Soumick's meticulous research, his ability to integrate deep learning with medical applications, and his forward-thinking approach to the future of MRI technology make this book an essential resource for anyone involved in medical imaging research or clinical practice. It is an indispensable guide for those who wish to understand the power of deep learning in solving some of the most pressing challenges in modern medical diagnostics.

Prof. Dr.-Ing. Andreas Nürnberger
Data & Knowledge Engineering Group
Faculty of Computer Science
Otto von Guericke University Magdeburg
Magdeburg, Germany

Foreword by Prof Oliver Speck

This book is based on the PhD thesis of Dr. Soumick Chatterjee. Soumick began his work in our lab in 2018, at a time when artificial intelligence and deep neural networks—now so omnipresent—had only just begun to enter broader scientific fields beyond their origins in computer science. Within our lab and the local imaging community, he was among the first to recognize the potential of these methods and to focus fully on the application of convolutional neural networks in the field of MR imaging.

In an era in which AI methods advance at a breathtaking pace, seven years may appear to be a very long time. Nevertheless, this book provides an overview of the fundamentals of MR imaging and image processing, together with an introduction to neural networks, that remains valid. These sections go far beyond the usual methodological introductions found in PhD theses, making the book a valuable resource for anyone seeking to enter the field.

The methods described are applied to a number of "real-world" MR acquisition challenges, such as the reconstruction of highly under-sampled MR data that violate traditional assumptions, including the Nyquist sampling theorem or the stationarity of the measured object. Furthermore, the book addresses several image-processing challenges, such as the segmentation of structures and pathologies in MR imaging data and the delineation of vessels in angiographic MR acquisitions. Finally, the correction of motion-induced image artefacts is explored as another important application.

The book thus presents a comprehensive overview of MR imaging, deep learning methods, and their application to a variety of imaging challenges. Throughout, Soumick's passion for Italy and Italian cuisine, as well as for *Star Trek*, becomes apparent in the naming of several approaches. I wish all readers insight and inspiration.

Prof. Dr. rer. nat. habil. Oliver Speck
Department of Biomedical Magnetic
Resonance, Institute for Physics,
Faculty of Natural Sciences
Otto von Guericke University
Magdeburg
Magdeburg, Germany

Acknowledgements

I would like to thank my supervisors, Prof. Andreas Nürnberger and Prof. Oliver Speck, for their guidance and support. When I started my PhD, I did not know anything about MRI, and I learnt everything from Oliver. I started learning about MR physics from the books he recommended and then by attending his lectures on the basics of MRI. The last time I had physics was in school, and it took some time for me to get the hang of it. Thank you, Oliver for being so patient with me. On the other hand, Andreas guided me from the machine learning / deep learning front. Our long meetings not only helped me to improve my work or fix problems, but also helped me reduce stress. Our discussions going beyond the PhD thesis topic helped me be a better human being. Thank you for being such a great supervisor. I feel lucky to have them and could not have asked for better supervisors than them.

Being from a computer science background, I never knew anything about MRI, let alone how to use one of these million-euro machines. On one hand, Oliver taught me from the theoretical side, and Cindy Lübeck taught me hands-on. She taught me how to use the scanner—from how to place a phantom or subject inside the scanner to how to acquire an image, from which sequence to choose for a specific requirement to how to improve/optimise the sequence to get the best image quality. As I had no background in radiology, my start was shaky at the beginning. But thanks to Cindy, I not only learnt how to use the system but also started loving doing the scans. Because of this, I would like to acknowledge her contribution and send her a special thanks. Starting a PhD is always stressful—from understanding the topic to trying to have an idea of how to proceed to the goal. On top of that, my topic was interdisciplinary—computer science and medical physics, and initially, I did not know anything about the latter. Dr Hendrik Mattern, one of my senior colleagues, explained to me the topics, gave me some initial ideas to work on and helped me find datasets to work with. Later on, we collaborated on a vessel segmentation project, which turned out to be one of the most popular topics I worked on during my PhD and which is now being extended in various new directions. Another senior colleague I would like to mention is Alessandro Sciarra. Since the very early days of my PhD, he has been my go-to person for any clarification I needed regarding MR physics. He always answered my questions with patience—helped me understand the complex topics of physics with ease. He also helped me cultivate an interest in MRI in general, helping me extend my original thesis topic of dealing with undersampling to dealing with MR artefacts in general. Together, we collaborated on numerous projects and co-supervised many students. Our collaborative work on motion correction became a major part of both of our PhD theses. Thank you so much for being such a nice colleague and, even more, the best big brother. Starting from discussing pasta to complaining about mensa food, one of the most important colleagues I had during my PhD was Domenico Iuso. Professionally, together we reverse-engineered the fake StarVIBE sequence and discovered its inaccuracies—which led to me reach out to Oliver, who then contacted Siemens to obtain the real StarVIBE. Even though he was in a different group, working with CT, we still discussed a lot regarding various aspects of our work. From a colleague, he became my best friend. Unfortunately, he had to leave the MEMoRIAL programme in the very first year of my PhD, but even then, we are still collaborating on topics such as anomaly detection using DL. The first three years of my work were in room 408 of G29. Directly after joining, the MEMoRIAL fellow I started working together was Mario Breitkopf (or Brietkopf). We had nearly identical thesis topics, and we started exchanging knowledge and ideas—he, from the MR physics side and me from deep learning.

Throughout our PhDs, we regularly discussed several overlapping topics. Another MEMoRIAL colleague from the same group (BMMR) who I collaborated with was Chompunuch Sarasaen. What initially started as discussions about abdominal imaging to model-based reconstruction, later grew into two co-first-authored papers regarding using deep learning to deal with the spatio-temporal trade-off of dynamic MRI. Moreover, her excellent anatomical and radiography-related knowledge, plus great artistic skills in figure creations, have been of great help in not only those two papers, but in many of my works.

The stress of a PhD, along with the gloomy weather of Magdeburg: meeting with friends and having a good time has not been a luxury, but a necessity. I have been lucky to have many great friends here. Alina was my first friend in this city right after I arrived, followed by Nirav. During this time, I met so many great people—made such nice friends, it's difficult to mention everyone. First, I would like to mention the best group I could imagine having: Alessandro, Irene, Daniele, Julia, Giacomo, Max. Here I would like to send my special regards to Irene, for being such a nice and supportive elder sister—for being with me through every problem. Another circle of four deep learning researchers who would go on weekly dinners every Wednesday: Max (again), Philipp, and Robert. Ever since meeting during the German A2 course, Francesca has become one of the treasured friends in my life. Her impromptu plans made my life ever so interesting here (and I have missed her tremendously ever since she moved to the UK). Talking about friends would be incomplete without mentioning spaghetti-breaker Elisa. After meeting for the first time in the LC meeting, she became one of my closest friends in no time. I believe all my friends here in Magdeburg, in Italy, and everywhere else definitely helped me continue during the tough times of this PhD. Hence, I would like to acknowledge their contributions.

When talking about my family, I would like to first mention my dad, Prof. Swapan Kumar Chatterjee. He has been my inspiration to come into research, and he has been the reason for my love for mathematics and computer science. Unfortunately, he could not see me doing this PhD, as he passed away in 2012. I'm sure he would have been very happy (however, he would have preferred me to do PhD either in maths or physics). I would like to thank him from the bottom of my heart. Next in line is my sister, Sunikkana. She is just 18 months younger than me, so basically, we grew up like pseudo-twins. Moving to Germany physically distanced us, and surely this is one of the negative aspects of coming here, but we are, as always, connected mentally. I hope she can move to Europe very soon! Another person who started as a friend but later became family—Rupali. She and her family have always been a major part of my support system. I would like to acknowledge all their contributions to my thesis for providing mental support throughout.

On a final note, I would like to thank the European Structural and Investment Funds (ESF) for funding my research within the context of the International Graduate School MEMoRIAL under the programme "Sachsen-Anhalt WISSENSCHAFT Internationalisierung" (project no. ZS/2016/08/80646), and our most-supportive coordinator Anke Ryll—for guiding and helping me with all the bureaucratic processes since the start of my PhD.

Magdeburg, Soumick Chatterjee
June 2022

This book is a publication of the PhD thesis titled “Reducing Artefacts in MRI using Deep Learning: Enhancing Automatic Image Processing Pipelines”, accepted by the Faculty of Computer Science at Otto von Guericke University Magdeburg and defended on 5 September 2022. The work was supervised by Prof. Andreas Nürnberger (Faculty of Computer Science) and Prof. Oliver Speck (Faculty of Natural Sciences). The research leading to these results was funded by the European Structural and Investment Funds (ESF) within the context of the International Graduate School MEMORIAL under the programme “Sachsen-Anhalt WISSENSCHAFT Internationalisierung” (Project No. ZS/2016/08/80646).

OTTO VON GUERICKE
UNIVERSITÄT
MAGDEBURG

Abstract

Magnetic resonance imaging (MRI) can provide high spatial resolution for detecting minute pathological changes in tissues. However, MRI is an inherently slow process due to consecutive data acquisition, which leads to a long scan time for high-resolution imaging. The acquisition speed can be increased by acquiring less data (undersampling). Consequently, this leads to the degradation of image quality, such as the loss of resolution or the introduction of image artefacts. On the other hand, longer scan times of the high-resolution images might result in the patient moving during the acquisition, leading to motion artefacts. The presence of artefacts in the image might hinder clinicians from a reliable diagnosis. Traditional techniques, such as compressed sensing for undersampled MRI reconstruction, can help improve the image quality of the undersampled MRIs, but their long execution time makes them unfit for realtime applications. Moreover, they face difficulties while reconstructing highly undersampled data, as the reconstructed image is heavily corrupted by artefacts. Prospective motion correction, one of the popular techniques to deal with motion artefacts in MRI, also performs poorly when it comes to a high level of corruption. Deep learning based methods have successfully been applied to image processing and reconstruction tasks. Given their tremendous potential that is yet to be explored, this thesis studies how these methods can be applied to correct MRIs with artefacts—focusing on the development of pipelines for reconstruction of undersampled Cartesian and radial MRIs, super-resolution of static and dynamic MRIs, and retrospective motion correction. Furthermore, what kind of available prior knowledge can be used to improve the performance of these methods, and how can they be incorporated—have also been studied here. Finally, this thesis explores the possibility of using complex-valued convolutions for these tasks.

MRIs are not only used by clinicians for performing diagnostics by visual inspections but also used in several automatic image processing pipelines, including deep learning based techniques, for further processing—for automated decision-making or as part of decision support systems—to perform classification, segmentation, anomaly detection, etc., or even to perform data standardisation—for example, image registration. Undersampling can also benefit these methods because it reduces scan time—making it possible to create larger datasets, and it also reduces the risk of motion artefacts due to shorter scan times—making it easier for the machine learning approaches to work with such data with fewer artefacts. Thus, this thesis additionally aims to propose methods for five different deep learning based automatic image processing pipelines—supervised classification, weakly- and semi-supervised segmentation, unsupervised anomaly detection, and image registration—initially trained and tested on high-resolution datasets, and then evaluate the performance of these methods on undersampled MRIs by employing the deep learning based reconstruction pipeline presented in this thesis. These pipelines were employed and evaluated for brain tumour classification and segmentation, vessel segmentation, and registration of brain MRIs.

Keywords: MRI, Deep Learning · Undersampling · Cartesian MRI · Radial MRI · Motion · Artefacts · Reconstruction · Retrospective Motion Correction

Zusammenfassung

Die Magnetresonanztomographie (MRT) bietet eine hohe räumliche Auflösung, um kleinste pathologische Veränderungen in Geweben zu erkennen. Allerdings ist die MRT aufgrund der sequentiellen Datenerfassung ein inhärent langsamer Prozess, was zu einer langen Scanzeit in der hochauflösenden Bildgebung führt. Die Aufnahmegeschwindigkeit kann durch die Erfassung von weniger Daten (undersampling) erhöht werden. Dies führt jedoch zu einer Verschlechterung der Bildqualität, z. B. zu einem Auflösungsverlust oder zur Entstehung von Bildartefakten. Andererseits kann eine längere Scanzeit bei hochauflösender Bildgebung ebenfalls zu Artefakten (Bewegungsartefakten) führen, da sich der Patient während langer Aufnahme häufiger bewegt. Das Vorhandensein von Artefakten im Bild kann den Kliniker an einer zuverlässigen Diagnose hindern. Herkömmliche Techniken wie die komprimierte Abtastung für die Rekonstruktion von MRT-Bildern mit Unterabtastung können dazu beitragen, die Bildqualität zu verbessern, sind aber aufgrund ihrer langen Ausführungszeit nicht für Echtzeitanwendungen geeignet. Außerdem haben sie Schwierigkeiten bei der Rekonstruktion stark unterabgetasteter Daten, die rekonstruierten Bilder sind durch Artefakte häufig stark beeinträchtigt. Die prospektive Bewegungskorrektur, eines der beliebtesten Verfahren zur Behandlung von Bewegungsartefakten in der MRT, ist bei einem hohen Grad an Verfälschung ebenfalls nicht sehr leistungsfähig. Auf Deep Learning basierende Methoden wurden bereits erfolgreich für Bildverarbeitungs-und Rekonstruktionsaufgaben eingesetzt. Angesichts ihres enormen Potenzials wird in dieser Arbeit untersucht, wie diese Methoden zur Korrektur von MRTs mit Artefakten eingesetzt werden können. Der Schwerpunkt liegt dabei auf der Enticklung von Verarbeitungspipelines zur Rekonstruktion von kartesischen und radialen MRTs mit Unterabtastung, der Superauflösung von statischen und dynamischen MRTs und der retrospektiven Bewegungskorrektur. Darüber hinaus wurde untersucht, welche Art von verfügbarem Vorwissen genutzt werden kann, um die Leistung dieser Methoden zu verbessern. Schließlich wird in dieser Arbeit die Möglichkeit der Verwendung komplexer Faltungen für diese Aufgaben untersucht.

MRTs werden nicht nur von Klinikern zur visuellen Diagnostik verwendet, sondern auch in verschiedenen automatischen Bildverarbeitungspipelines, einschließlich Deep-Learning-Verfahren, zur Weiterverarbeitung – für die automatische Entscheidungsfindung oder als Teil von Entscheidungsunterstützungssystemen – zur Durchführung von Klassifizierung, Segmentierung, Anomalieerkennung usw. oder sogar zur Datenstandardisierung – z. B. zur Bildregistrierung. Eine Reduzierung der Scanzeit durch Unterabtastung kann diesen Methoden ebenfalls zugute kommen, da hierdurch größere Datensätze erstellt werden können und durch die kürzere Scanzeit auch das Risiko von Bewegungsartefakten verringert wird. Hierdurch müssen maschinellen Lernverfahren weniger Artefakte berücksichtigen. Daher zielt diese Arbeit zusätzlich darauf ab, Methoden für fünf verschiedene Deep-Learning-basierte automatische Bildverarbeitungspipelines für hochauflösende Bilddaten vorzuschlagen – überwachte Klassifizierung, schwach- und halbüberwachte Segmentierung, unüberwachte Anomalieerkennung und Bildregistrierung – und dann die Leistung dieser Methoden auf unterabgetasteten MRTs zu bewerten, indem die in dieser Arbeit vorgestellte Deep-Learning-basierte Rekonstruktionspipeline eingesetzt wird. Diese Pipelines wurden für die Klassifizierung und Segmentierung von Hirntumoren, die Segmentierung von Gefäßen und die Registrierung von MRT-Bildern des Gehirns eingesetzt und bewertet.

Schlüsselwörter: MRT · Deep Learning · Unterabtastung · Kartesisches MRT · Radiales MRT · Bewegung · Artefakte · Rekonstruktion · Rückwirkende Bewegungskorrektur

Riassunto

La risonanza magnetica (RM) può offrire una risoluzione spaziale elevata per rilevare minime alterazioni patologiche nei tessuti. Tuttavia, il processo di acquisizione tramite risonanza magnetica è fondamentalmente lento a causa dell'acquisizione consecutiva dei dati, che comporta un lungo tempo di scansione per l'imaging ad alta risoluzione. La velocità di acquisizione può essere aumentata acquisendo meno dati (sottocampionamento). Ciò comporta un degrado della qualità dell'immagine, come la perdita di risoluzione o l'introduzione di artefatti. D'altra parte, tempi di scansione più lunghi necessari per le immagini ad alta risoluzione potrebbero aumentare il rischio di movimento del paziente durante l'acquisizione, con conseguenti artefatti da movimento. Ciò è particolarmente sfavorevole poichè la presenza di artefatti nell'immagine potrebbe impedire ai medici una diagnosi affidabile. Le tecniche tradizionali, come il compressed sensing per la ricostruzione di risonanze magnetiche sottocampionate, possono contribuire a migliorare la qualità delle immagini, ma il loro lungo tempo di esecuzione le rende inadatte alle applicazioni in tempo reale. Inoltre, la ricostruzione di dati fortemente sottocampionati incontra difficoltà, perché l'immagine ricostruita è fortemente corrotta da artefatti. Anche la correzione prospettica del movimento, una delle tecniche più popolari per trattare gli artefatti da movimento nella risonanza magnetica, si mostra poco efficace in presenza di un elevato livello di corruzione. I metodi basati sul deep learning sono stati applicati con successo all'elaborazione e alla ricostruzione delle immagini. Considerato il loro enorme potenziale ancora da esplorare, questa tesi viene dimostrato come questi metodi possano essere applicati per correggere le risonanze magnetiche con artefatti, focalizzando l'attenzione sullo sviluppo di pipeline per la ricostruzione di risonanze magnetiche cartesiane e radiali sottocampionate, la super-risoluzione di risonanze magnetiche statiche e dinamiche e la correzione retrospettiva del movimento. Inoltre, sono stati studiati anche i tipi di conoscenze pregresse disponibili che possono essere utilizzate per migliorare le prestazioni di questi metodi e come possono essere incorporate. Infine, in questa tesi viene esplorata la possibilità di utilizzare operazioni di convoluzioni a valore complesso per questi compiti. Le risonanze magnetiche non sono utilizzate solo dai medici per eseguire diagnosi tramite ispezioni, ma anche in diverse pipeline di elaborazione automatica delle immagini, comprese le tecniche di deep learning, per un'ulteriore elaborazione—per il processo decisionale automatizzato o come parte di sistemi di supporto alle decisioni—per eseguire la classificazione, la segmentazione, il rilevamento delle anomalie, ecc. Il sottocampionamento può essere vantaggioso anche per questi metodi, perché riduce i tempi di acquisizione, rendendo possibile la creazione di dataset più grandi, e riduce anche il rischio di artefatti da movimento grazie a tempi di acquisizione più brevi, rendendo più facile per gli approcci di Machine Learning lavorare con questi dati con meno artefatti. Quindi, in questa tesi vengono proposti metodi per cinque diverse pipeline di elaborazione automatica delle immagini basate sul deep learning—classificazione supervisionata, segmentazione debolmente e semi-supervisionata, rilevamento di anomalie non supervisionato e l'allineamento delle immagini. Queste reti sono state inizialmente addestrate e testate su dataset ad alta risoluzione, per poi valutare le prestazioni di questi metodi su risonanze magnetiche sottocampionate utilizzando la pipeline di ricostruzione basata sul deep learning presentata in questa tesi. Queste pipeline sono state utilizzate e valutate per la classificazione e la segmentazione dei tumori cerebrali, la segmentazione dei vasi e l'allineamento delle risonanze magnetiche cerebrali.

Keywords: RM · deep learning · Sottocampionamento · RM cartesiana · RM radiale · movimento · artefatti · ricostruzione · correzione retrospettiva del movimento

Contents

Acronyms

ACS	Autocalibration Signal
AD	Axial Diffusivity
ADMIR	Affine and Deformable Medical Image Registration
AE	Autoencoder
ANTs	Advanced Normalization Tools
BCE	Binary Cross Entropy
BHPF	Butterworth Highpass Filter
BLPF	Butterworth Lowpass Filter
BOLD	Blood Oxygenation Level Dependent
CAD	Computer-assisted Diagnosis
CAE	Convolutional Autoencoder
cceVAE	Compact Context-Encoding Variational Autoencoder
CE	Cross Entropy
ceVAE	Context-Encoding Variational Autoencoder
CG	Conjugate Gradient
CM	Classical Mechanics
CNN	Convolutional Neural Network
CReLU	Complex Rectified Linear Unit
CS	Compressed Sensing
CSF	Cerebrospinal fluid
CS-MRI	Compressed Sensing MRI
CSVD	Cerebral Small Vessel Disease
CT	Computed Tomography
CV-CNN	Complex-valued Convolutional Neural Network

DAE	Denoising Autoencoder
DCF	Density Compensation Function
DCT	Discrete Cosine Transform
DDoS	Dynamic Dual-channel of Super-resolution
DFT	Discrete Fourier Transform
DL	Deep Learning
DS6	Deformation-aware Segmentation, trained on 6 MR Volumes
DSC	Dice Similarity Coefficient
DTI	Diffusion Tensor Imaging
DVF	Deformation Vector Fields
DWI	Diffusion-Weighted Imaging
E2E-VN	End-to-End Variational Network
ELBO	Evidence Lower Bound
ELU	Exponential Linear Unit
FA	Fractional Anisotropy
FBP	Filtered Backprojection
FFT	Fast Fourier Transform
FID	Free Induction Decay
FLAIR	FLuid Attenuated Inversion Recovery
fMRI	Functional MRI
FOV	Field of View
FT	Fourier Transform
FTL	Focal Tversky Loss
GAN	Generative Adversarial Network
GBP	Guided Backpropagation
GCN	Graph Convolution Network
GE	Gradient Echo
GELU	Gaussian Error Linear Unit
GMVAE	Gaussian Mixture Variational Autoencoder
GPGPU	General-Purpose computing on Graphical Processing Unit
GPU	Graphical Processing Unit
GRAPPA	Generalized Auto-calibrating Partially Parallel Acquisitions
GRASP	Golden-angle RAdial Sparse Parallel
HGG	High-Grade Glioma
IC	Inverse Consistency
iFFT	Inverse Fast Fourier Transform
iFT	Inverse Fourier Transform
IHPF	Ideal Highpass Filter
ILPF	Ideal Lowpass Filter

IoU	Intersection over Union
IR	Inversion Recovery
LGG	Low-Grade Glioma
LSA	Lenticulostriate Arteries
LSR	Latent Space Representation
LSTM	Long short-term memory
MAE	Mean Absolute Error
MAP	Maximum A Posteriori
MD	Mean Diffusivity
MICDIR	Multi-scale Inverse-Consistent Deformable Image Registration
MIP	Maximum Intensity Projection
ML	Machine Learning
MLP	Multilayer perceptron
MoCo	Motion Correction
MOJITO	Multiple Overlapping k-space Junctions for Investigating Translating Objects
MPT	Moiré Phase Tracking
MR	Magnetic Resonance
MRA	Magnetic Resonance Angiography
MRI	Magnetic Resonance Imaging
MRT	Magnetic Resonance Tomography
MSCGUNet	Multi-scale UNet with Self-Constructing Graph Latent
MSE	Mean-Squared Error
MSFDF	MultiScale Frangi Diffusion Filter
MSS	Multi-Scale Supervision
NCC	Normalised Cross-correlation
NMI	Normalised Mutual Information
NMR	Nuclear Magnetic Resonance
NMRI	Nuclear Magnetic Resonance Imaging
NRMSE	Normalised Root Mean-Squared Error
NUFFT	Nonuniform Fast Fourier Transform
PC-MRI	Phase-Contrast MRI
PD	Primal-Dual
PELU	Parametric Exponential Linear Unit
PET	Positron Emission Tomography
PGSE	Pulsed Gradient Spin Echo
PI	Parallel Imaging
PLN	Perceptual Loss Network
PMC	Prospective Motion Correction

PR	Projection Reconstruction
PReLU	Parameterised Rectified Linear Unit
PROPELLER	Periodically Rotated Overlapping ParallEL Lines with Enhanced Reconstruction
PSNR	Peak Signal to Noise Ratio
QM	Quantum Mechanics
RAKI	Robust Artificial-neural-networks for k-space Interpolation
RAM	Random Access Memory
ReLU	Rectified Linear Unit
RF	Radio Frequency
RGB	Red Green Blue
RMC	Retrospective Motion Correction
RMSE	Root Mean-Squared Error
RNN	Recurrent Neural Network
ROI	Region Of Interest
RS	Resting-state
RSS	Root Sum of Squares
SCG	Self-Constructing Graph
SE	Spin Echo
SELU	Scaled Exponential Linear Unit
SENSE	SENSitivity Encoding
SiLU	Sigmoid-Weighted Linear Unit
SISR	Single-Image Super-Resolution
SME	Sensitivity Map Estimation
SNAILS	Self-navigated Interleaved Spiral
SNARC	Stochastic Neural Analogue Reinforcement Calculator
SNR	Signal-to-Noise Ratio
SOTA	State Of The Art
SPECT	Single Photon Emission Computed Tomography
SR	Super-Resolution
SSAE	Scale-Space Autoencoder
SSD	Single Shot MultiBox Detector
SSIM	Structural Similarity Index
StRegA	Segmentation Regularised Anomaly
SWI	Susceptibility Weighted Imaging
TanH	Hyperbolic Tangent
TE	Echo Time
TI	Inversion Time
ToF	Time of Flight

TP	Time-point
TR	Repetition Time
TV	Total Variation
UAD	Unsupervised Anomaly Detection
UQI	Universal Quality Index
VAE	Variational Autoencoder
Varden	Variable density
VED	Vesselness Enhancement Diffusion
VIPR	Vastly-undersampled Isotropic Projection Reconstruction
VN	Variational Network
VQ-VAE	Vector-Quantised Variational Autoencoder
wCE	Weighted Cross Entropy

List of Figures

List of Tables

1 Introduction

Magnetic Resonance Imaging (MRI) is one of the most significant innovations in the field of medical imaging, which uses the principles of Nuclear Magnetic Resonance (NMR) to image hydrogen nuclei [160]. It can provide high spatial resolution for detecting minute pathological changes in tissues—without using harmful ionising radiation. MR images can also offer better differentiation between fat, water and muscle than other imaging modalities—making it a preferred modality for differentiating tumours or other lesion tissues from the surroundings [110]. Therefore, image guidance based on MRI is a favourable tool for identifying and characterising tumours in interventions [110, 218]. In interventional scenarios, such as MR-guided liver biopsy, real-time or near real-time imaging is required to show excellent contrast between the target organ or structure and adjacent soft tissue while visualising the changes of internal organs during an examination. In such applications, dynamic MRI is used, which is obtained by acquiring the raw k-space data (in the frequency domain) continuously and reconstructing a sequence of images over time [69].

However, due to consecutive data acquisition, MRI is an inherently slow process [234]. Fast imaging can improve patient compliance, reduce patient motion, increase patient throughput etc. MRI data is not acquired in the spatial but the frequency domain, which is the 2D Fourier transform of the image, the so-called k-space. To get back to image space, a 2D inverse Fourier transform has to be applied. The speed of acquisition can be increased by ignoring parts of the k-space (undersampling). Taking the inverse Fourier transform from insufficiently densely sampled k-space frequency data might cause the resultant image to lose resolution and might also have artefacts due to the violation of the Nyquist-Shannon sampling theorem [4, 8]. Hence, while achieving high temporal resolution during dynamic MRIs, these acquisitions suffer from the restricted spatial resolution because only a limited part of the data can be measured—known as the spatio-temporal trade-off of MRI. Depending upon the type of undersampling and the amount of undersampling,

S. Chatterjee, *Deep Learning Innovations in MRI Reconstruction and Analysis*,
https://doi.org/10.1007/978-3-658-50741-1_1

the impact of undersampling can vary—from loss of resolution to introduction of different types of artefacts, such as aliasing, ghosting, streaking, etc.

Undersampling is not the only source of artefacts in MRI. There can be artefacts originating from the object that is being acquired—such as chemical shift, bounce point, susceptibility, metal, dielectric, etc., can also be technique-related, such as partial-volume, wrap-around, and Gibbs (truncation), as well as motion artefacts. Motion is one of the most common causes of artefacts in MR imaging—which can originate from patient motion [191, 197]. Physiological motion, such as cardiac or respiratory movement, can be controlled by gating or by specific sequence design [115]. Dealing with less-predictable voluntary or involuntary movement of the patient, often called bulk motion [31], which, e.g., in Parkinsonism, is more complicated and leads to ghosting and blurring [107]. It is noteworthy that if the scan time is very long, the chances of the patient moving during the acquisition are also high.

If the level of artefacts is too high, the scan must be repeated. Even a moderate level may lead to an incomplete or inaccurate diagnosis [256]. Many of the approaches currently available to deal with the undersampling artefacts by improving the reconstruction of undersampled data are very slow because they are very computationally heavy or iterative in nature. For example, compressed sensing techniques are successfully applied to accelerate MR image acquisition [352, 396]. However, it expands reconstruction time and may result in incoherent artefacts or unrealistic images [82, 95, 125]. Several approaches have also been proposed for the problem of motion artefacts—they can be typically grouped as Prospective Motion Correction (PMC) and Retrospective Motion Correction (RMC) [40]. PMC performs real-time adaptive updates of the data acquisition to avoid motion artefacts in the final reconstructed image, with the help of various tracking modalities [197], while RMC techniques can be considered as post-processing techniques where the MR image data is modified after the k-space data acquisition has been performed—during or after the reconstruction process [197, 372]. However, such approaches have limitations in terms of the amount of motion they can correct [pub:15].

1.1 Motivation

The incorporation of prior knowledge has been proven to be a useful mechanism for iterative reconstructions of undersampled MRIs, where it has been used as a penalty function [92]. Deep learning models have also been employed to introduce strong, complex priors in the reconstruction process [204, 269, 367]. Several deep learning based techniques have been shown to outperform the traditional iterative reconstruction techniques [267, 269, 271, 335]. However, there can be further unexplored network architectures that might be created to improve the reconstruction quality. Moreover, these methods focus on Cartesian MRI, while these Cartesian acquisitions might be corrupted by image artefacts caused by respiratory or patient motion if not handled explicitly by the acquisition sequence [153]. Nevertheless, there are many different ways to sample the k-space and non-Cartesian samplings, such as Radial sampling might be able to cope with this problem of motion artefacts by a repetitive sampling of the high-energy k-space centre [235]. Various techniques have been proposed to reconstruct undersampled radial k-space [92, 120]. Howbeit, the explorations of the deep learning based techniques for undersampled radial MRI reconstruction have been limited. Radial MRI and Computed Tomography (CT) are mathematically very similar, following the Fourier slice theorem. Even the earliest method to reconstruct radial MRIs was using Filtered Backprojection (FBP), which is essentially the technique used for CT reconstruction [234]. Similar to MRI, several deep learning based techniques have also been proposed for CT reconstruction [227, 249]. However, their applicability for the reconstruction of radial MRIs has not been explored.

During undersampling, if the k-space coverage is decreased (undersampling by taking only the centre of the k-space), then instead of coherent artefacts, the resulting image loses resolution. In such situations, a different direction of undersampled reconstruction of MRIs can be taken—by treating this as a Super-Resolution (SR) problem. This has been applied for both static [294, 353], as well as dynamic MRIs [279, 357]. One of the typical network architectures employed for this task is UNet. But it might result in smoothed output and chequerboard artefacts [215]. So there are still possibilities to find better network architectures for the task. On the other hand, most deep learning based methods need large training data sets and finding such training data—matching with the data of the real-time acquisition that needs to be reconstructed in terms of contrast and sequence—can be a challenging task, especially for dynamic MRIs. Using a training set significantly different from the test set can produce results of poor quality [290, 374]. Exploiting the available prior knowledge or incorporating the temporal relationship in the reconstruction process—might improve the performance without demanding large training datasets. One of the common approaches to deep learning based undersampled MRI reconstruction, including the super-resolution methods, is to perform filtering in the image space. Several approaches have also been proposed to work directly in k-space [301, 316, 384], or in a mix of image and k-space [267, 297, 368]. But they treat the complex-valued input as two separate real-valued data by splitting the complex data into real and imaginary parts and then concatenating them on the channel dimension to supply to the real-valued neural network models—discarding the richer algebraic structure of the complex data [382]. Complex-valued Convolutional Neural Network (CV-CNN) can preserve the geometric structure of the data and has been shown to perform better than ordinary convolutional networks (CNNs) [259, 376, 382]. However, a structured study is yet to be performed for MRI applications and has not yet been explored much in the field of undersampled MRI reconstruction.

MRIs are not only used by radiologists directly for diagnostics but also in several automatic processing pipelines—such as lesion classification and localisation—to serve as decision support systems, or even vessel segmentation or image registration—which can not only be used clinically but also used for several research purposes. With the advancements in deep learning, several such pipelines have been proposed. Undersampling can benefit such pipelines by saving acquisition time—more data can be acquired within the same period of time because of the increase in patient throughput, while by reducing acquisition time, the probability of the subject moving during the acquisition also reduces. However, those pipelines are typically trained and evaluated on fully-sampled data, and the effect of undersampling on such pipelines has not been explored.

Motion, on the other hand, is another source of artefacts in MRI, which can also be dealt with using deep learning [278, 282, 317, 320, 328, 372]. The motion correction process might also benefit from the incorporation of prior knowledge. Moreover, the network models proposed for undersampled reconstruction might also be applicable for correcting motion artefacts—an open field yet to be fully explored.

1.2 Research Questions

Based on the motivation, this thesis addresses research questions keeping the central focus on undersampled MRI reconstruction and the secondary focus on retrospective motion correction.

A) The following generic research question regarding both undersampled reconstruction and retrospective motion correction is addressed in this thesis:-

1. What type of prior knowledge can be incorporated in the reconstruction and motion correction process, and how? ***[RQ:1]***

    - Will the incorporation of prior knowledge improve the performance (e.g. accuracy, speed)? ***[RQ:1.1]***

B) The following research questions regarding the undersampled reconstruction of Cartesian MRIs are addressed here:-

2. What kind of deep learning model can improve the results of the artefact reduction methods working in the image space of the undersampled Cartesian MRIs? ***[RQ:2]***
3. What kind of deep learning model can improve the super-resolution reconstruction of undersampled MRIs? ***[RQ:3]***

    - How can the temporal information be incorporated in super-resolution reconstruction of undersampled dynamic MRIs and can it improve the super-resolution performance? ***[RQ:3.1]***

4. Is performing reconstruction in the k-space or in the hybrid space (mix of image and k-space) better than working purely in the image space? ***[RQ:4]***

    - Is complex-valued convolution (CV-Conv) better than real-valued convolution (Conv)? ***[RQ:4.1]***
    - What kind of deep learning model would be suitable for working in the k-space and/or hybrid space for undersampled MRI reconstruction? ***[RQ:4.2]***

C) Moreover, this thesis addresses the following research question regarding the undersampled reconstruction of radial MRIs:-

5. What kind of deep learning based method be applied for undersampled radial MRIs? ***[RQ:5]***

    - Can a unified deep learning based method be developed for the reconstruction of undersampled radial MRI and sparse CT? ***[RQ:5.1]***

D) A common research question for the undersampled reconstruction of both Cartesian and radial MRIs is also addressed here:-

6. What level of undersampling can the proposed model handle? ***[RQ:6]***

E) Furthermore, the following research question regarding retrospective motion correction is addressed here:-

7. Can the deep learning models for undersampled MRI reconstruction be used for dealing with motion artefacts? ***[RQ:7]***

F) Finally, this thesis addresses the following orthogonal question:-

8. How is the performance of deep learning based automatic MRI processing pipelines affected when they face undersampled MRIs while being trained on fully-sampled MRIs? ***[RQ:8]***

    - Can that performance be improved by incorporating undersampled reconstruction techniques as a precursor, and how much? ***[RQ:8.1]***

1.3 Thesis Outline

This thesis is structured into six main parts and an appendix. Part I provides the reader with a fundamental knowledge to understand the thesis, which is divided into three chapters. Chapter 2 introduces the fundamental concepts of MRI. This chapter starts with a bit of the history of MRI (Sect. 2.1). Then in Sect. 2.2, a brief overview of an MR scanner is provided—different components of the scanner are explained. Afterwards, a concise introduction to the physics behind MRI and how MR imaging is performed—are provided in Sect. 2.3. Finally, different types of MRI and different types of MR artefacts including their reasons of occurrence, are explained in Sects. 2.4 and 2.5, respectively. Chapter 3 provides the fundamentals of image processing and presents its connection to MRI. Finally, Chap. 4 provides a background on the neural networks—the main technology behind the methods used in this thesis and presents its connections to the earlier chapter on image processing. This chapter starts by providing a succinct history from machine learning to deep learning in Sect. 4.1. Then, Sect. 4.2 provides a short explanation of the basic building blocks of neural networks—perceptron and Multilayer perceptron (MLP) and Sect. 4.3 explains the learning mechanism of the neural networks in a nutshell. Sect. 4.4 afterwards provides different loss functions that can be used for the tasks related to this thesis, followed by Sect. 4.5—which provides information regarding the different evaluation methods that are relevant for this thesis. Finally, Sect. 4.6 introduces the concept of deep learning and Sect. 4.7 establishes the "black-box" problem of deep learning.

Part II introduces the different state-of-the-art (SOTA) technologies currently available for dealing with the artefacts in MRI (undersampling and motion) and presents the open questions. Moreover, this part also introduces different SOTA automatic MR image processing pipelines. Chapter 5 presents the techniques currently employed for undersampled MRI reconstruction, which is divided into two main sections—non-deep learning based and deep learning based techniques. Among the non-deep learning based techniques (Sect. 5.1), parallel imaging and iterative reconstruction techniques are discussed in Sects. 5.1.1 and 5.1.2, respectively. Section 5.2 presents the different deep learning techniques that have been proposed to deal with the problem of undersampled MRI. These methods are then grouped into three different groups: methods working in the image space (Sect. 5.2.1), methods working in the k-space (Sect. 5.2.2), and finally the methods working in a mix of image and k-space (Sect. 5.2.3). The methods which work in the image space, are further subdivided based on their problem statements: artefact reduction (Sect. 5.2.1.1) and super resolution (Sect. 5.2.1.2). Finally, Sect. 5.3 presents the open questions. Chapter 6 provides a concise view on the current motion correction techniques. Initially, Sect. 6.1 briefly explains different techniques that can be employed to track motion during MR acquisitions. Then Sects. 6.2 and 6.3 present the two main types of motion correction technologies—Prospective Motion Correction (PMC) and Retrospective Motion Correction (RMC). RMC techniques are further divided into non-deep learning based techniques (Sect. 6.3.1) and deep learning based techniques (Sect. 6.3.2). Finally, Sect. 6.4 presents a few of the open questions related to motion correction with deep learning. Chapter 7 introduces four different automatic image processing tasks. Section 7.1 presents the task of brain tumour classification and segmentation, within which three different types of SOTA models are presented - classification models, segmentation models, and unified models performing both in Sects. 7.1.1, 7.1.2, and 7.1.3, respectively. Afterwards, Sect. 7.2 explains the task of anomaly detection in brain MRIs and presents seven different deep learning based methods. Section 7.3 presents the task of vessel segmentation—focusing on small vessel segmentation and furnishes different non-deep learning based and deep learning techniques which have been employed so far for this task in Sects. 7.3.1 and 7.3.2, respectively. Finally, Sect. 7.4 introduces the topic of image registration and furnishes different SOTA methods and discusses three deep learning based pipelines in detail.

Part III proposes several methods to advance the field of undersampled MRI reconstruction. Based on their working principles, these methods are divided into three chapters: artefact reduction in the image space, undersampled reconstruction as a super-resolution problem, and methods working in a hybrid space (image + k-space). Chapter 8 proposes two different methods for artefact reduction in the image space. Section 8.1 proposes the first major contribution of this thesis: undersampled reconstruction in the image space by proposing the ReconResNet model and the NCC1701 pipeline for both Cartesian as well as radial MRIs. Section 8.2 proposes CT reconstruction-inspired undersampled radial MR reconstruction technique PDUNet, which optimises in both sinogram space (obtained by applying preprocessing on radial MRI k-space) and image space. Chapter 9 proposes super-resolution techniques to perform undersampled MRI reconstruction, which is divided into two sections based on the type of MRI used in the research: static and dynamic. Section 9.2 proposes the ShuffleUNet model for super-resolution and shows its applicability while working with static diffusion-weighted MRIs. Section 9.3 proposes two different techniques for the super-resolution of undersampled dynamic MRIs. Section 9.3.1 exploits the available prior knowledge to improve the quality of super-resolution, while Sect. 9.3.2 proposes a model to learn the spatio-temporal relationship to improve the super-resolution reconstruction. Chapter 10 proposes techniques working in the k-space and hybrid space. This chapter starts by presenting some preliminary comparisons of CNNs and CV-CNNs (Sect. 10.1.1). Section 10.1.2 proposes KSPResNet, which is a modified version of ReconResNet as a CV-CNN model working directly with the undersampled k-space data. Section 10.2 proposes two new models, Fourier-PD and Fourier-PDUNet, working in the hybrid space, by extending the research of PDUNet (Sect. 8.2) by modifying the original models with complex-valued functions. These models optimise in both k-space and image space.

Part IV proposes five different deep learning based automatic MRI processing pipelines and demonstrates their performance while working with undersampled MRIs. Furthermore, NCC1701 (Sect. 8.1) is employed as a precursor to those processing pipelines to reconstruct the undersampled MRIs, and then the change in performance is evaluated. This part presents five different deep learning based MRI processing pipelines for four different tasks - supervised spatiospatial models for the classification of brain tumours in Chap. 11, GP-models for supervised classification and weakly-supervised segmentation of brain tumours using inherently-explainable models in Chap. 12, unsupervised anomaly detection in brain MRIs using the StRegA pipeline in Chap. 13, deformation-aware semi-supervised learning (DS6) for small vessel segmentation in 7T MRAs using noisy training data in Chap. 14, and multi-scale inverse-consistent deformable image registration of intramodal brain MRIs using multi-scale UNet with self-constructing graph latent (MICDIR) in Chap. 15. Each chapter introduces the task, defines the aims and motivates, and proposes a new method for the given task—improving the performance of the available SOTA techniques for that task. Then those methods are evaluated on undersampled and NCC1701 reconstructed MRIs.

Part V presents the application of the methods proposed in Part III in the field of retrospective motion correction. Section 16.1 proposes a prior knowledge based motion correction in the image space using the ReconResNet model (Sect. 8.1). Section 16.2 demonstrates the applicability of the hybrid spaces models (Sect. 10.2) while performing deep learning based retrospective motion correction.

Finally, Part VI summarises the research presented in this thesis, provides conclusions drawn from the experiments, and finally presents the directions for future research.

Part I
Fundamentals

Summary
This part of the thesis presents some fundamental information regarding the three major aspects of this thesis: MRI, image processing, and neural networks. The chapter regarding MRI starts with a bit of the history, followed by a brief overview of an MR scanner, then a concise introduction to the physics behind MRI and how MR imaging is performed, and finally, different types of MRI and different types of MR artefacts including their reasons of occurrence are explained. The succeeding chapter provides the fundamentals of image processing and presents its connection to MRI. The final chapter provides a background on the neural networks—the main technology behind the methods used in this thesis and presents its connections to the earlier chapter on image processing. This chapter starts by providing a succinct history—from machine learning to deep learning. Then, a short explanation of the basic building blocks of neural networks—perceptron and Multilayer perceptron (MLP) is provided, and the learning mechanism of the neural networks is explained in a nutshell. Afterwards, the next section provides information regarding different loss functions that can be used for the tasks related to this thesis, followed by some information regarding the different evaluation methods that are relevant for this thesis. Finally, the concept of deep learning and the "black-box" problem of deep learning are explained.

2 Magnetic Resonance Imaging

Magnetic Resonance Imaging (MRI), also known as Nuclear Magnetic Resonance Imaging (NMRI) and Magnetic Resonance Tomography (MRT), is a medical imaging modality used for obtaining internal images of the body in a non-invasive manner. MRI can provide high spatial resolution for detecting minute pathological changes in tissues. Having great soft-tissue contrast and the absence of harmful ionising radiation make it one of the most popular imaging modalities in the field of radiology.

2.1 Radiology and MRI: a Brief History

Ever since the accidental discovery of x-rays by Röntgen on 8 November 1895 [3], and it was used under clinical conditions on 11 January 1896 by John Hall-Edwards, radiology has been an integral part of medical treatment. By exploiting the high concentration of calcium in bones, which can efficiently absorb x-rays, and the different absorption rates of the tissues, internal images of the human body can be obtained non-invasively in 2D (a single projection): known as radiographs, or simply, x-ray images. By acquiring projections from different angles—by creating the absorption profiles of the object from many directions, a 3D volume of the object can be reconstructed—which can enable a 3-dimensional view inside the human body. This is known as Computed Tomography, or CT [105]. In the year 1979, Hounsfield and Cormack won the Nobel Prize in Medicine and Physiology for their contribution to its creation.

Apart from x-ray and CT, the field of medical imaging has seen many more innovations. Kuhl, Chapman, and Edwards introduced the concept of emission and transmission tomography in the last 1950s, which later on became the building block of Single Photon Emission Computed Tomography (SPECT) [86] and Positron Emission Tomography (PET) [75]. PET uses radioactive substances, known as radio-

S. Chatterjee, *Deep Learning Innovations in MRI Reconstruction and Analysis*,
https://doi.org/10.1007/978-3-658-50741-1_2

tracers, to visualise and measure changes in metabolic processes and physiological activities. World War II has seen many scientific innovations; sonar is one of them. Another non-invasive imaging modality was developed in the 1950s based on sonar: the ultrasound. The non-invasive, ionising-radiation-free nature of ultrasound and its capabilities to image in real time while being sensitive to flow because of the Dopler effect, make this one of the most frequently used medical imaging modalities for cardiology, abdominal and vascular disease, real-time biopsy guidance, and many more.

The journey of yet another medical imaging modality, Magnetic Resonance Imaging or MRI, started even before the discovery of x-ray, with Nikola Tesla discovering the rotating magnetic field in 1882 [60]. In 1937, Isidor I. Rabi observed nuclear magnetic resonance [5], for which he received the Nobel Prize in Physics in 1944. In 1946, Bloch and Purcell discovered the magnetic resonance phenomenon, awarding them the Nobel Prize for Physics in 1952. In 1950, Erwin Hahn first detected spin echoes and free induction decay [9]. Two years later, in 1952, American physicist Herman Y. Carr, the pioneer of Magnetic Resonance Imaging, demonstrated the usage of gradients in magnetic fields for spatial localisation in his doctoral thesis, and he also created the first-ever 1D MRI. Jay Singer studied blood flow in 1959, with the help of Nuclear Magnetic Resonance (NMR) relaxation time measurements [12]. After a medley of various inventions over a decade, Raymond Damadian discovered in 1971 that the tumours and normal tissues could be distinguished based on their relaxation times [14]. This enabled the possibility of having better contrast between tissues and disease than x-ray and ultrasound. The first NMR image of a test tube was produced by Paul Lauterbur in 1973 [15], by using magnetic field gradients to localise the origin of the NMR signals and reconstructing them using projections—similar to CT, for which in 2003, he earned the Nobel Prize in Medicine, jointly with Sir Peter Mansfield. To call his invention, Lauterbur coined the term "zeugmatography": which means imaging by coupling together the two: magnetic field and gradients. In 1975, Richard Ernst proposed the usage of two-dimensional Fourier transform to obtain MR images - by using phase and frequency encoding. Raymond Damadian hand-built the world's first MRI scanner with the help of his post-doctoral researchers Michael Goldsmith and Larry Minkoff and acquired the first MRI of a healthy human body in 1977 and a human body with cancer in 1978. In the same year, Clow and Young published the first-ever human head MRI. In 1983, Ljunggren and Tweig introduced the concept of k-space—a representation of the acquired signal in the Fourier space. The first commercial system was called the "Neptune" and was installed at Hammersmith Hospital in London, which had a field strength of 0.15T. Around 1984, General Electric introduced the first high-field system with a field strength of 1.5T. MRI techniques developed rapidly in the late 1980s to 1990s and soon became the preferred imaging modality in the world of neurology.

2.2 An MR Scanner

The human body is full of hydrogen atoms—MRI exploits this to acquire the images. Let's have a closer look at the different components of an MR system. There are three components (or group of components) of an MR system: the main magnet, the gradients, and the Radio Frequency (RF) system (Fig. 2.1).

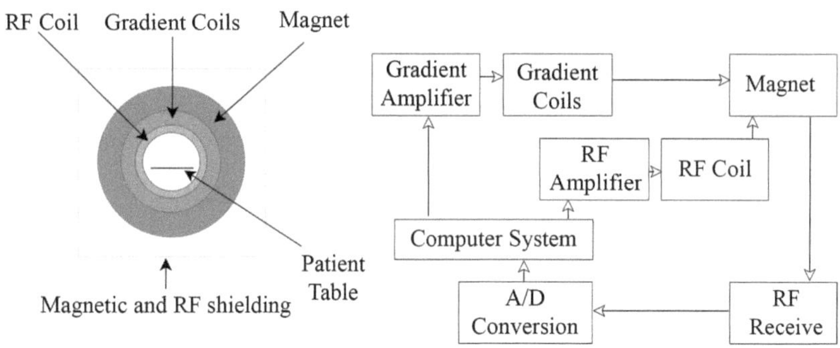

Fig. 2.1 Components of an MR scanner

2.2.1 Magnet

As suggested by its name, the main component of an MR scanner is the magnet. The magnet creates a static magnetic field, and the field strength or flux density of the magnet is measured in tesla (T). Before moving forward, to put things in perspective, the earth's magnetic field is about 30 μT at the equator and 70 μT at the poles (1 Microtesla [μT] = 0.000001 Tesla [T]). The field strength of a routine clinical scanner is between 0.2–3.0T, while human scanners for research can even have higher field strengths: 7.0T to 11.7T. Scanners meant for small animals are also available with 17.2T and NMR spectrometers with 23.5T [url:13]. Higher field strengths provide better Signal-to-Noise Ratio (SNR), which can help increase the spatial resolution or decrease the scan time; also, fat suppression and spectroscopy improve because of the increased chemical shift effects [234]. The magnetic field created by this main magnet is called the static magnetic field or B_0 field.

Early generation MR scanners used air-cored resistive magnets, which are now replaced by superconducting magnets. These magnets use materials that have zero electrical resistance in temperatures approaching absolute zero (−273.16°C, 0 K). Electric current is passed in a loop of superconducting wires and continues to circulate indefinitely as these wires are kept at such low temperatures. They are typically cylindrical in shape, and the patient is placed within the hollow centre of the magnet. The magnets use cryogenic coolers, which use compressed helium gas. These magnets do not require an external power supply for regular operation, but it is required to "ramped" up the magnet to the required field strength. In case of emergency, or even for an intentional "ramp down" of the magnet, the system is "quenched"—which results in a collapse of the magnetic field, as well as boils-off of the helium. Every magnet suffers from some level of inhomogeneity, which is more in higher field strengths. Fixed shimming is performed to improve the homogeneity of the magnet, which can be further improved by per-patient dynamic shimming [234].

2.2.2 Gradients

The next important set of components in an MR scanner is the magnetic field gradients, which are used for spatial localisation. Gradients are looped wires or thin metal sheets. When the current is passed, a secondary magnetic field is created, which distorts the static magnetic field in a controlled manner—causing the resonance frequency of the protons to vary as a function of position [url:13] (Fig. 2.2).

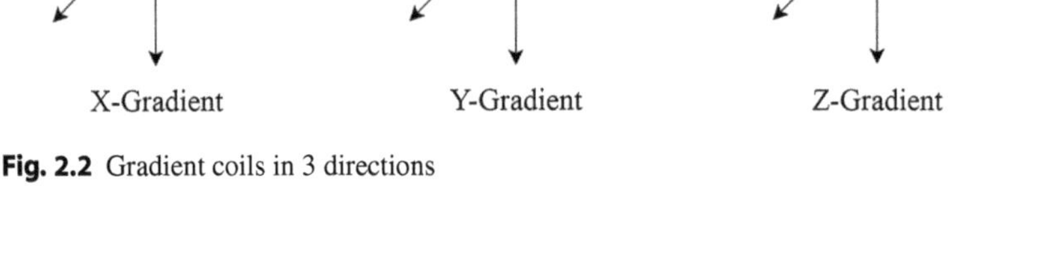

Fig. 2.2 Gradient coils in 3 directions

There are three sets of gradient coils in a typical MR scanner: one for each x, y and z direction, where the z-gradients have a circular (Maxwell) coil configuration, while the x–y (transverse) gradients have a saddle (Golay) coil configuration. Each coil set is driven by an independent power amplifier [url:13]. These magnetic fields are parallel to the B_0 field. The null point at the centre of the gradient coils, as well as of the main magnet, is called the isocentre.

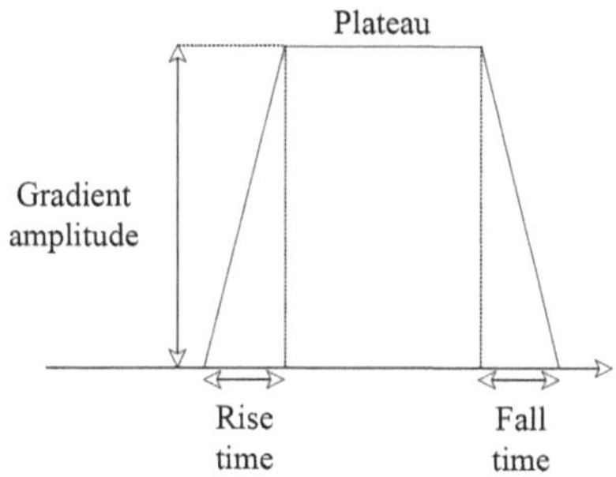

Fig. 2.3 Gradient pulse

The gradient pulse (shown in Fig. 2.3) in the conventional pulse sequences is trapezoidal in shape. The rise time of the pulse is typically between $200\mu s$ to $1000\mu s$. The strength of the gradient is determined by the rate of change of the field over distance, and it is expressed in millitesla per metre (mTm^{-1}). The maximum strength of the gradient is typically $10 - 50\ mTm^{-1}$ [234].

2.2.3 Radio Frequency System

There are three components that constitute the Radio Frequency (RF) system of an MR scanner: a transmitter, transmit coils and receiver coils. MR imaging relies on the absorption and emission of energy by the nuclei at a specific Larmor frequency, which is directly proportionate to the B_0 field strength. Larmor frequency at a 0.3T scanner is 12.8MHz, whereas, for 3T and 7T scanners, it is 127.8MHz and 298.2MHz, respectively. The Radio-frequency field, also known as the B_1 field, is time-varying and is transmitted near the Larmor frequency, applied perpendicular to the B_0 field, and it is generated by flowing electrical currents through the RF-transmit coils.

The transmitter, also known as the frequency synthesiser, is the first component of the MR scanner's RF system, which is responsible for generating a continuous sinusoidal carrier wave near the Larmor frequency for exciting the nuclei. The offset from the Larmor frequency is dependent on the slice position. The pulse modulator, the next component of this system, divides the continuous carrier wave into short bursts of 1–5 ms—known as the RF-pulse. The RF-pulse then goes to a high-power amplifier which generates the currents with power peaks between 10–30 kW required to drive the RF-coils—which results in a maximum B_1 field of 10–50 μT. Transmit attenuator is responsible for adjusting the gain of the power amplifiers, and the flip angle of the RF-pulse is changed by changing the degree of RF attenuation. Next in the pipeline is the quadrature hybrid coupler, which splits the output of the amplifier into two equal parts: in-phase and quadrature, commonly known as I and Q, respectively. They are 90° out of phase from each other and are supplied to the two ports of the quadrature transmit coil. The pre-final set of components of the RF system is the coils. With the received electrical currents, the RF-transmit coils are responsible for generating the B_1 field. These coils might also be used to receive MR signal, and in that case, a transmit/receive (T/R) switch is used to choose the appropriate mode. The receiver coil, or simply known as the coil, is responsible for detecting the MR signal. The final component of the RF system is the receiver. The MR signal of $\pm$16kHz is embedded into the Larmor frequency, and the receiver's job is to demodulate the combined signal to retrieve the frequency of the MR signal.

2.3 From Proton to Picture

2.3.1 MR Physics

The human body is full of micro-magnets: hydrogen atoms. An electrically neutral hydrogen atom consists of a negatively charged electron bound to the nucleus by the Coulomb force. The nucleus contains one positively charged proton, and has a spin angular momentum of $I = \frac{1}{2}\hbar$, where $\hbar$ is the reduced Planck's constant (h) by 2π: $\hbar = \frac{h}{2\pi}$.

The proton's spin is quantised in the presence of an external magnetic field, and it can be in one of the two quantum states: nearly aligned to the magnetic field, known as spin-up or parallel, and almost opposite to the field, known as spin-down or anti-parallel (Fig. 2.4 and 2.5). The state of the proton is based on the amount of energy it possesses, as the anti-parallel state requires more energy than the parallel state. Protons can swap their states by gaining or losing energy. The energy difference

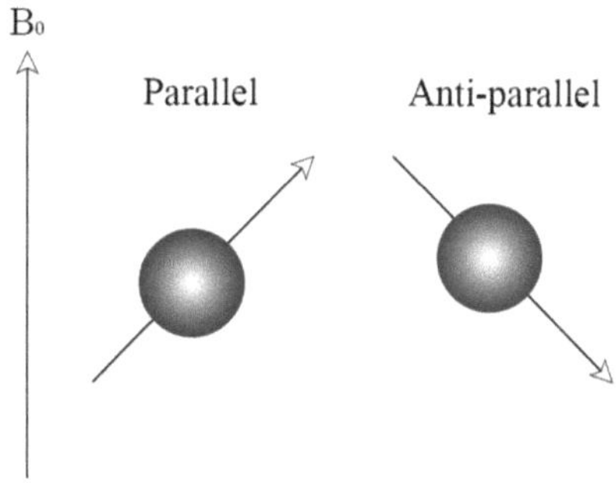

Fig. 2.4 Fundamentals: Two possible orientations of the proton in an external magnetic field B_0

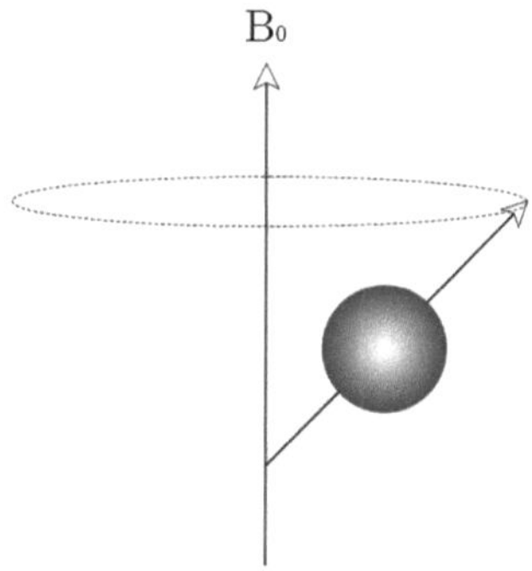

Fig. 2.5 Precession of the magnetic moment

between the two states is directly proportional to the external magnetic field, which can be defined by the Larmor frequency.

The magnetic moment of the proton can be derived as:

$$\boldsymbol{\mu} = \gamma \mathbf{J} \tag{2.1}$$

where γ is the gyromagnetic ratio and $\mathbf{J}$ is the angular momentum, which can further be expanded following Quantum Mechanics (QM) as:

$$\begin{aligned} \boldsymbol{\mu} &= \gamma \mathbf{J} \\ &= \gamma \hbar \mathbf{I} \end{aligned} \tag{2.2}$$

where, $\mathbf{J}$ is the quantised angular momentum, and I is the spin angular momentum quantum number. In the presence of an external field $\mathbf{B}$, the angular momentum can have $2I + 1$ number of possible values, protons can have two possible states: $\pm\frac{1}{2}$. The energy of each state ε can be defined as:

$$\begin{aligned} \varepsilon &= \boldsymbol{\mu} \cdot \mathbf{B} \\ &= \gamma \hbar I \cdot B \end{aligned} \tag{2.3}$$

and the energy difference of the two states is:

$$\begin{aligned} \Delta\varepsilon &= \left(\frac{1}{2} - -\frac{1}{2}\right) \gamma \hbar B \\ &= \gamma \hbar B \end{aligned} \tag{2.4}$$

De Broglie's wave function defines the frequency associated with this energy is:

$$\Delta\varepsilon = \hbar\omega \tag{2.5}$$

Hence, the precessional frequency can be derived as:

$$\hbar\omega_0 = \gamma \hbar B_0 \tag{2.6}$$

$$\Rightarrow \omega_0 = \gamma B_0 \tag{2.7}$$

This ω_0 is the Larmor frequency. This can also be derived using Classical Mechanics (CM). In the presence of an external magnetic field $\mathbf{B}$, the angular momentum changes following:

$$\begin{aligned} |\frac{d\mathbf{J}}{dt}| &= |\boldsymbol{\mu} \times \mathbf{B}| \\ &= |\gamma \mathbf{J} \times \mathbf{B}| \\ &= \gamma \mathbf{J}\mathbf{B} \sin\theta \end{aligned} \tag{2.8}$$

where θ is the angle between the $\boldsymbol{\mu}$ and $\mathbf{B}$, and

$$d\mathbf{J} = \mathbf{J} \sin\theta d\phi \tag{2.9}$$

From Eq. 2.8 and Eq. 2.9, the Larmor frequency can be derived as:

$$\begin{aligned}\omega &= \frac{d\phi}{dt} \\ &= -\frac{d\phi}{d\mathbf{J}} \cdot \frac{d\mathbf{J}}{dt} \\ &= \frac{1}{\mathbf{J}\sin\theta} \cdot -\gamma J B \sin\theta \\ &= \gamma \mathbf{B}\end{aligned} \tag{2.10}$$

$$\Rightarrow \omega_0 = \gamma \mathbf{B_0} \tag{2.11}$$

There are millions of hydrogen protons in the human body, and they are usually present in both states—while slightly more are in the low-energy parallel state. The ratio between the states is directly related to the field strength and inversely proportional to the body temperature. If a human being with a temperature of 37° C is inside a 1.5T MR scanner, this ratio would be $\sim$ 1.000004. Boltzmann distribution tells us the number of protons present in each energy state:

$$\frac{N_{\mathrm{up}}}{N_{\mathrm{down}}} = \exp\left(\frac{\Delta\varepsilon}{k_{\mathrm{B}}T}\right) \tag{2.12}$$

where $K_B = 1.39 \times 10^{23} J K^{-1}$ is the Boltzmann constant. At human body temperature, $\gamma\hbar B_0 \ll k_{\mathrm{B}}T$, Eq. 2.12 can be rewritten as:

$$\begin{aligned}\frac{N_{\mathrm{up}}}{N_{\mathrm{down}}} &= 1 + \frac{\gamma\hbar B_0}{k_{\mathrm{B}}T} \\ \Rightarrow N_{\mathrm{excess}} &= N_{\mathrm{up}} - N_{\mathrm{down}} \\ &= \frac{N_{\mathrm{total}}}{2} \cdot \frac{\gamma\hbar B_0}{k_{\mathrm{B}}T}\end{aligned}$$

By replacing N_{total} with the proton density ρ and by using the magnitude of the magnetic moment of the proton in the equation, the net magnetisation M_0 per unit volume can be obtained as:

$$M_0 = \frac{\rho\gamma^2\hbar^2 B_0}{4k_{\mathrm{B}}T} \tag{2.13}$$

It is to be noted that the M_0 is always perfectly aligned with the B_0.

The tips of the magnetic moments of the protons are spread out as shown in Fig. 2.6, because they are out of phase from each other. The average magnetic moment of a large group of protons that are precessing exactly at the same frequency can be represented using a vector—which is called the isochromat or the spin. As there are

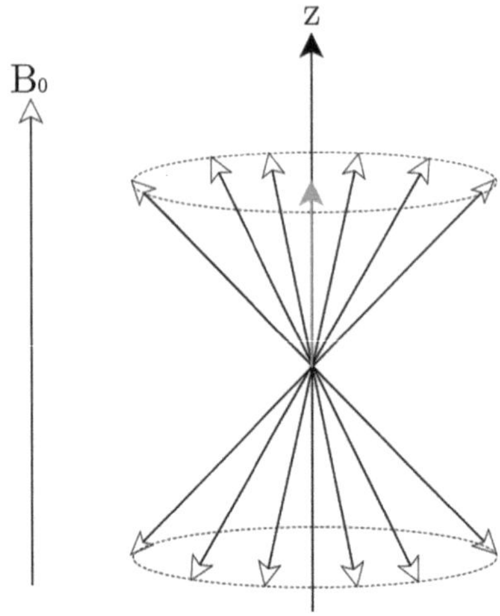

Fig. 2.6 Average spin of many protons produces the net magnetisation M_0

many protons in the body and they precess at different frequencies at a given point in time, there are also many such vectors. The vector sum of all these spins is the net magnetisation (M_0), and it is measured in μT. For a given volume V, the net magnetisation is:

$$\mathbf{M} = \frac{1}{V} \sum_{\substack{\text{protons} \\ \text{in V}}} \mu i \tag{2.14}$$

With the help of Eq. 2.8, the motion of the magnetisation **M** in an external filed **B** can be described as:

$$\frac{d\mathbf{M}}{dt} = \gamma \mathbf{M} \times \mathbf{B} \tag{2.15}$$

The net magnetisation of a body is very small—in order of microtesla, while the static magnetic field is in order of tesla. For example, say $M_0 = 1\mu T$ and $B_0 = 3T$, then the B_0 field is 3000000 times stronger than the M_0. This makes it virtually impossible to measure when the protons are in equilibrium in parallel to B_0. If they are 90° in the x–y plane, also known as the transverse plane, using a detector that only measures the magnetic field in this plane, it is possible to measure M_0. This is achieved using the RF-pulse, which creates a magnetic field perpendicular to B_0 at (or near) the Larmor frequency. To explain the effects of RF-pulse, a rotating frame of reference can be used, which rotates about the z-axis exactly at the Larmor frequency. Protons that are precessing exactly at the Larmor frequency appear stationary in such a rotating frame of reference (shown as Fig. 2.7). Protons that have lower or higher frequencies would gain or lose phase, respectively.

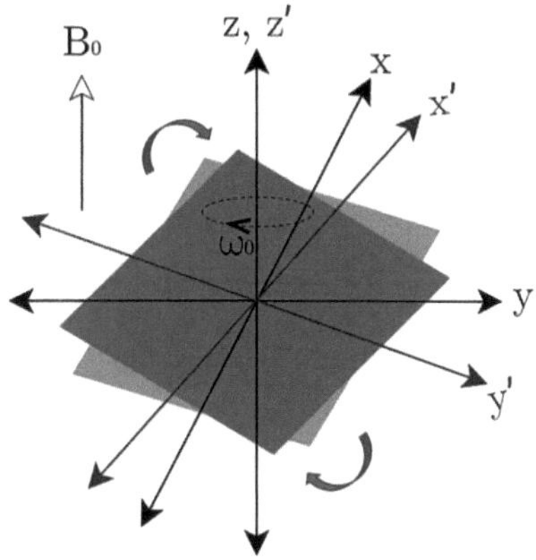

Fig. 2.7 Rotating frame of reference

$$\left(\frac{d\mathbf{M}}{dt}\right)_{rot} = \left(\frac{d\mathbf{M}}{dt}\right)_{fixed} - \boldsymbol{\omega} \times \mathbf{M} \quad (2.16)$$

where $\boldsymbol{\omega}$ is the frequency of the rotating frame.
With the help of Eq. 2.15 (as $(\frac{d\mathbf{M}}{dt})_{\text{fixed}} = \frac{d\mathbf{M}}{dt}$), Eq. 2.16 can be rewritten as:

$$\begin{aligned}\left(\frac{d\mathbf{M}}{dt}\right)_{\text{rot}} &= \gamma\mathbf{M} \times \mathbf{B} - \boldsymbol{\omega} \times \mathbf{M} \\ &= \gamma\mathbf{M} \times \mathbf{B} + \gamma\mathbf{M} \times \frac{\boldsymbol{\omega}}{\gamma} \\ &= \gamma\mathbf{M} \times \left(\mathbf{B} + \frac{\boldsymbol{\omega}}{\gamma}\right)\end{aligned} \quad (2.17)$$

and where $\frac{\boldsymbol{\omega}}{\gamma}$ is the magnetic field generated due to rotation and $\mathbf{B} + \frac{\boldsymbol{\omega}}{\gamma}$ is the total magnetic field experienced by the spins. In the rotating frame of reference, x and y axes are referred to as x' and y', respectively. As the rotating takes place around the z-axis, $z = z'$.

The motion of M_0 can be calculated as:

$$\begin{aligned}\left(\frac{d\mathbf{M}}{dt}\right)_{rot} &= \gamma\mathbf{M} \times \left(\mathbf{B_0} + \frac{\omega}{\gamma} + \mathbf{B_1}\right) \\ &= \gamma\mathbf{M} \times \mathbf{B_1}\end{aligned} \tag{2.18}$$

when the frame is rotating clockwise at the frequency ω_0, so $\omega = -\omega_0$.

In this rotating frame of reference, the radio-frequency field B_1 is is aligned along x' and M_0 is away from the z-axis for the duration of the RF-pulse. The flip angle α during an RF-pulse switched on-off cycle is:

$$\alpha = \gamma B_1 t_p \tag{2.19}$$

where t_p is the duration of the pulse, by changing the strength of the B_1 field, the required α over a certain t_p can be achieved. The required α can also be achieved by modifying the t_p. However, it is not desired as the acquisition time is a crucial factor in MRI. M_0 precessing in the transverse plane induces a voltage in the receiver coils, which the receiver senses to obtain the signal. As the protons dephase rapidly, the amplitude of this signal decays exponentially to zero, as shown in Fig. 2.8, known as the Free Induction Decay (FID).

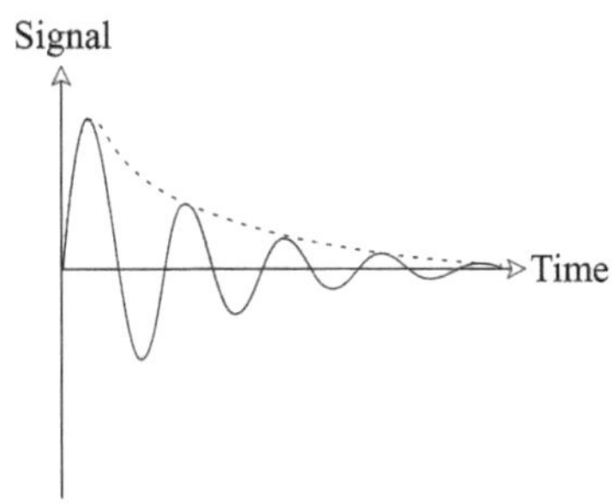

Fig. 2.8 Free Induction Decay (FID)

After the protons are excited by applying the RF-pulse, they begin to "relax" and go back to their equilibrium state as soon as the pulse is switched off. The time required to return to the equilibrium state is called the relaxation time. In the rotating frame of reference, rotating at the Larmor frequency, spins with higher frequency dephase clockwise and the ones with lower—anticlockwise.

Two effects can be observed for the FID: B_0 field inhomogeneity and spin-spin relaxation. Spin-spin relaxation is the interaction among the spins while moving within the tissue, which is completely independent of the field strength. This is characterised by the spin-spin relaxation time, commonly known as the T_2 relaxation time.

Spin-spin relaxation does not cause any net loss of energy, and the protons transfer the stored energy to the surrounding tissues, known as the lattice, to lose energy. This is known as the spin-lattice relaxation and is characterised using the spin-lattice relaxation time, commonly known as the T_1 relaxation time—which is inversely proportionate to the B_0 field strength.

To describe the behaviour of the magnetisation during excitation and relaxation for a sample placed in a static magnetic field, the Bloch equations can be used—which are based on classical mechanics. The magnetisation and its differential equation can be analysed, separating two of its components: parallel and perpendicular with respect to the static magnetic field B_0. The parallel component, also known as the longitudinal magnetisation, is:

$$M_{\parallel} = M_z \tag{2.20}$$

The perpendicular, also known as the transverse magnetisation, can be defined as:

$$\mathbf{M}_{\perp} = M_x \hat{x} + M_y \hat{y} \tag{2.21}$$

with the help of Eq. 2.21 and 2.20, for B to be an external magnetic field B_ext, Eq. 2.15 can be decoupled as:

$$\frac{dM_z}{dt} = 0 \tag{2.22}$$

and

$$\frac{d\mathbf{M}_{\perp}}{dt} = \gamma \mathbf{M}_{\perp} \times \mathbf{B}_{ext} \tag{2.23}$$

for protons not interacting with neighbouring tissues. To calculate the magnetisation for the protons which are interacting with their neighbourhood, additional terms are required in these equations. Two different terms are required for the parallel and perpendicular components, both dependent on decay parameters.

Longitudinal magnetisation is dependent on the spin-lattice interaction and is proportional to the difference $M_0 - M_z$, and the change can be defined by modifying Eq. 2.20:

$$\frac{dM_z}{dt} = \frac{1}{T_1}(M_0 - M_z) \quad \text{for } \mathbf{B}_{ext} \| \hat{z} \tag{2.24}$$

where T_1 is the spin-lattice relaxation time. The change in the magnetisation over time t can be calculated as:

$$M_z(t) = M_z(0)e^{-t/T_1} + M_0\left(1 - e^{-t/T_1}\right) \tag{2.25}$$

where $M_z(0)$ is the initial value of the magnetisation and M_0 is the equilibrium.

The change in the transverse magnetisation for the protons with spin-spin interaction can be defined by introducing the decay rate term in Eq. 2.21:

$$\frac{d\mathbf{M}_\perp}{dt} = \gamma\mathbf{M}_\perp \times \mathbf{B}_{ext} - \frac{1}{T_2}\mathbf{M}_\perp \tag{2.26}$$

where T_2 is the spin-spin relaxation time. In the rotating frame of reference, this can be written as:

$$\left(\frac{d\mathbf{M}_\perp}{dt}\right)_{rot} = -\frac{1}{T_2}\mathbf{M}_\perp \tag{2.27}$$

The exponential decay of the magnitude of $\mathbf{M}_\perp$ over time t of the transverse magnetisation can be calculated as:

$$\mathbf{M}_\perp(t) = \mathbf{M}_\perp(0)e^{-t/T_2} \tag{2.28}$$

where the initial value of the transverse magnetisation is $\mathbf{M}_\perp(0)$.

Eq. 2.24 and 2.26 can be combined into one vector equation:

$$\frac{d\mathbf{M}}{dt} = \gamma\mathbf{M} \times \mathbf{B}_{ext} + \frac{1}{T_1}(M_0 - M_z)\,\hat{z} - \frac{1}{T_2}\mathbf{M}_\perp \tag{2.29}$$

This is known as the Bloch Equation. For a constant field B_0, the components of the cross product in Eq. 2.29 produce three components:

$$\begin{aligned}
\frac{dM_x}{dt} &= \omega_0 M_y - \frac{M_x}{T_2} \\
\frac{dM_y}{dt} &= -\omega_0 M_x - \frac{M_y}{T_2} \\
\frac{dM_z}{dt} &= \frac{M_0 - M_z}{T_1}
\end{aligned} \tag{2.30}$$

If $B_x = B_1 cos\omega t$, $B_y = B_1 sin\omega t$, and $B_z = B_0$ By the introduction of integrating factors by replacing $M_x = m_x e^{-t/T_2}$ and $M_y = m_y e^{-t/T_2}$, the relaxation terms can be removed, and the change in magnetisation over time t for each of the three components can be defined as:

$$\begin{aligned} M_x(t) &= e^{-t/T_2}\left(M_x(0)\cos\omega_0 t + M_y(0)\sin\omega_0 t\right)\\ M_y(t) &= e^{-t/T_2}\left(M_y(0)\cos\omega_0 t - M_x(0)\sin\omega_0 t\right)\\ M_z(t) &= M_z(0)e^{-t/T_1} + M_0\left(1 - e^{-t/T_1}\right) \end{aligned} \tag{2.31}$$

2.3.2 MR Imaging

2.3.2.1 Echoes

During the acquisition of MRIs, Gradient Echo (GE) and Spin Echo (SE) are used instead of directly measuring the FID.

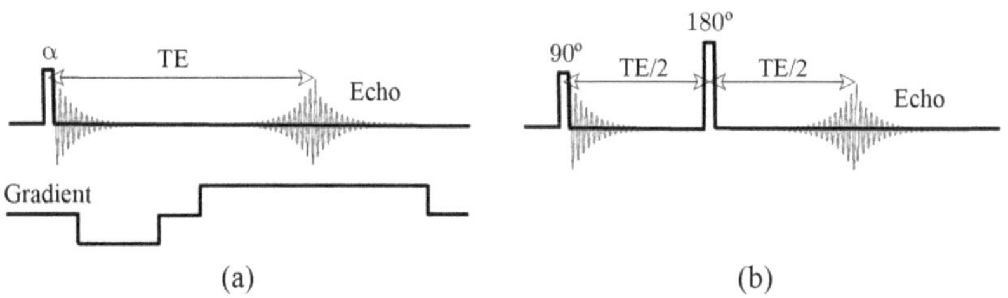

Fig. 2.9 (**a**) Gradient Echo (GE) sequence and (**b**) Spin Echo (SE) sequence

GE sequences (Fig. 2.9(a)) consist of two main components: negative gradient lobe and positive gradient. After the excitation phase, a negative gradient lobe is applied, which causes rapid (fast than FID) dephasing of the transverse magnetisation. When a positive gradient is applied afterwards, the magnetic field gradient is rephased, and those spins which were dephasing get rephased—in-phase along the y' axis—which forms the GE.

The amplitude of the echo, S_{GE}, is determined using the FID curve as:

$$S_{GE} = S_0 \exp\left(-\frac{TE}{T_2^*}\right) \tag{2.32}$$

where S_0 is the initial amplitude of the FID, TE is the echo time, and T_2* is a composite relaxation time: T_2 relaxation time + field inhomogeneities + tissue susceptibilities + diffusion of the protons.

In a SE sequence (Fig. 2.9(b)), two different RF-pulses are used: 90° and 180°. First, a 90° pulse is applied, and then the spins start dephasing. After a certain delay, a 180° pulse is applied on the y' axis, which flips all the spins by 180°—changing their phase angle while keeping their precessional frequency unchanged. This will cause all the spins to come in phase along with the y' axis after the same time as the delay between the two pulses—forming the SE. Due to the phase reversal, the amplitude of the echo will be totally dependent only on the T_2 relaxation time and the diffusion of the protons and will be unaffected by the field inhomogeneities or tissue susceptibilities. The amplitude can be calculated as:

$$S_{SE} = S_0 \exp\left(-\frac{TE}{T_2}\right) \tag{2.33}$$

where S_0 is the initial amplitude of the FID, TE is the echo time, and T_2 is the T_2 relaxation time.

2.3.2.2 Pulse Sequence

In an MR system, the static magnetic field B_0 is always present, but the gradients are applied in a controlled fashion based on requirements. A programmed set of gradients is known as a pulse sequence. Multiple such pulse sequences are grouped together to form an MRI protocol, which is used for imaging.

Two basic parameters are used to create any pulse sequence (along with many other parameters), and by changing these parameters, the MR contrast can be manipulated. They are (shown in Fig 2.10): Repetition Time (TR)—the time difference between any two consecutive points in a series of repeating pulses and echoes, and Echo Time (TE)—the time between the centre of the RF-pulse to the centre of the echo.

A basic GE sequence (shown in 2.11) consists of an RF-pulse and three different type of gradients: slice-selection gradient (G_{SS}), phase-encoding gradient (G_{PE}), and frequency-encoding gradient (G_{FE}). These three gradients can also be referred to as G_x, G_y, and G_z, and are defined mathematically as the derivative of the magnetic field (say, B_z) in space:

$$G_x = \frac{\partial B_z}{\partial x} \quad G_y = \frac{\partial B_z}{\partial y} \quad G_z = \frac{\partial B_z}{\partial z} \tag{2.34}$$

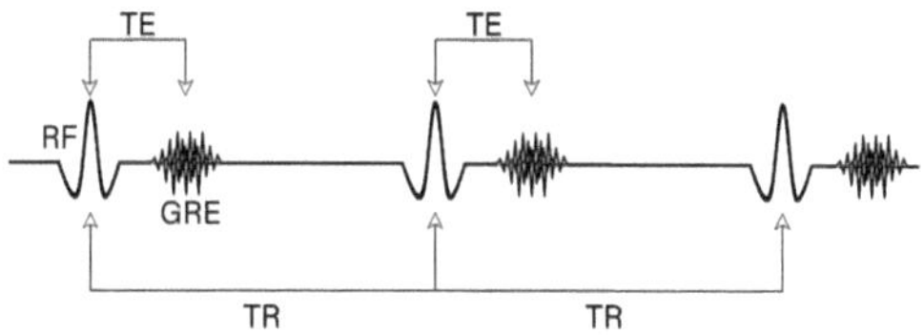

Fig. 2.10 Fundamental parameters of the pulse sequence: Repetition Time (TR) and Echo Time (TE)

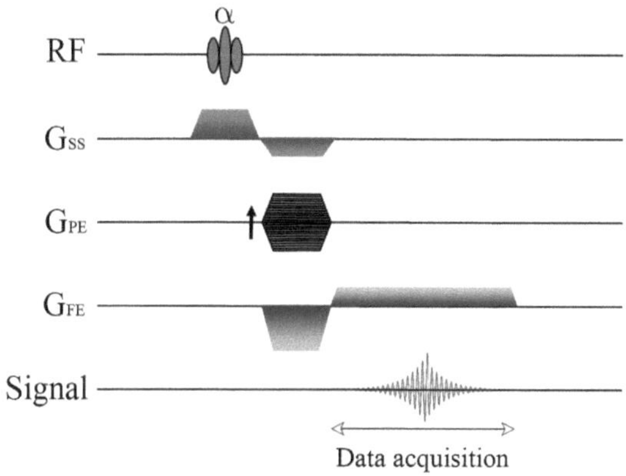

Fig. 2.11 Basic Gradient Echo (GE) sequence

RF-pulse excites the protons by applying the G_{SS} simultaneously, it restricts it within a 2D plane—within a slice. G_{PE} encodes the MR-signal in the phase-encoding direction and is applied in a direction that is orthogonal to the slice. The MR-signal is acquired when the G_{FE}, also known as the read-out gradient, is applied, and its effect on a discrete signal element ∂s can be defined as:

$$\partial s(t) = \rho(x) \cdot \exp\left(\frac{-t}{T_2^*}\right) \cdot \exp\left(i\gamma x G_{\text{FE}} t\right) \cdot \mathrm{d}x \tag{2.35}$$

where t is a certain time of acquisition of the MR-signal, $\rho(x)$ is the proton density along x-axis, T_2^* is a composite relaxation time (explained in Eq. 2.32), and $i = \sqrt{-1}$ is the complex notation. G_{FE} is applied continuously during the signal acquisition, known as sampling. G_{PE} is applied before sampling for τ amount of time, and the signal after applying both gradients is defined as:

$$\partial s(t) = \rho(x, y) \cdot \exp\left(\frac{-t}{T_2^*}\right) \cdot \exp\left(i\gamma x G_{\mathrm{FE}} t\right) \cdot \exp\left(i\gamma y G_{\mathrm{PE}} \tau\right) \cdot \mathrm{d}x \, \mathrm{d}y \quad (2.36)$$

This can be explained in a non-mathematical manner as:

Signal = Spin Density × T_2^* Decay × Phase Change due to G_{FE} × Phase Change due to G_{PE}

This complete sequence is repeated for acquiring every phase-encoding line present in the selected slice. By stacking all the phase-encoding lines, a 2D spatial frequency space can be obtained, known as the k-space, which is the 2D Fourier Transform of the measured MRI.

2.3.2.3 k-space and Fourier Transform

During an MR acquisition, the signal (Eq. 2.36) is sampled M times with an interval of Δt and the complete pulse sequence is repeated N times while increasing the amplitude of G_{PE} following:

$$G_{\mathrm{PE}}(n) = \Delta G \cdot n \quad (\text{ for } n = -\frac{\mathrm{N}}{2} \text{ to } \frac{\mathrm{N}}{2} - 1) \quad (2.37)$$

Now, the spatial frequency can be defined as:

$$\begin{aligned} k_{\mathrm{FE}} &= \bar{\gamma} \cdot G_{\mathrm{FE}} \cdot \Delta t \cdot m \\ k_{\mathrm{PE}} &= \bar{\gamma} \cdot \Delta G \cdot n \cdot \tau \end{aligned} \quad (2.38)$$

where $\bar{\gamma} = \gamma / 2\pi \approx 42\mathrm{MHzT}^{-1}$ (representation of the angular frequency of protons as scalar frequency), m is the number of samples in the FE direction, n is the number of samples in the PE direction. The total acquired MR-signal S in two dimensions: time t and pseudo-time $n \cdot \tau$, is the integral of the multiple samples of Eq. 2.36 with respect to x and y:

$$S(m, n) = \iint \rho(x, y) \cdot \exp\left(\frac{-t}{T_2^*}\right) \cdot \exp\left(i 2\pi x k_{\mathrm{FE}}\right) \cdot \exp\left(i 2\pi y k_{\mathrm{PE}}\right) \cdot \mathrm{d}x \, \mathrm{d}y \quad (2.39)$$

If the T_2^* is ignored, this is an Inverse Fourier Transform of the spin density $\rho(x, y)$:

$$S(m, n) = \rho\left(k_{\mathrm{FE}}, k_{\mathrm{PE}}\right) \quad (2.40)$$

Hence, the spin density distribution in two-dimensional space can be obtained by performing 2D Fourier Transform on the encoded signal, and a Fourier transformation pair is between with spatial frequency (k_{FE}, k_{PE}) and the positional coordinates x, y.

This raw data space $S(m, n)$, which is the discrete signal matrix in the world of MRI is known as the k-space. The "k" in the name of k-space is taken from the wavenumber, which is the number of waves per unit distance, calculated by the reciprocal of the wavelength ($k = 1/\lambda$)—the distance between two successive peaks in the signal (Fig. 2.12 and 2.13).

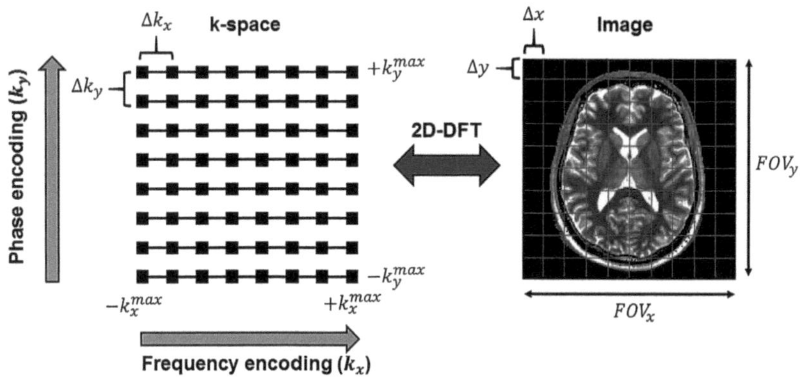

Fig. 2.12 Relationship between k-space and image-space. (Fig source: [225])

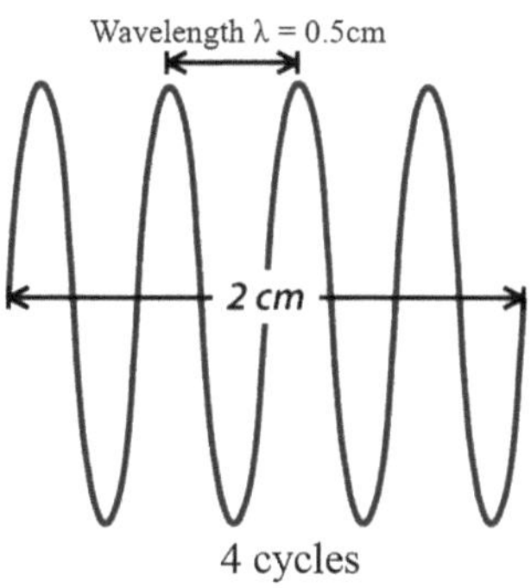

Fig. 2.13 Origin of the name "k-space": k is the wavenumber, which is the reciprocal of the wavelength ($k = 1/\lambda$). In this example, k = 1/0.5cm = 2 cycles/cm = $2cm_{-1}$

If a 2D object is represented by a function $f(x, y)$, it can be expressed in terms of the spatial frequencies as:

$$f(x, y) = \iint F\left(k_x, k_y\right) \cdot \exp\left(-i2\pi\left(xk_x + yk_y\right)\right) \cdot \mathrm{d}k_x\, \mathrm{d}k_y \tag{2.41}$$

where the spatial frequencies k_x, k_y are given by the function $F\left(k_x, k_y\right)$ as:

$$F\left(k_x, k_y\right) = \iint f(x, y) \cdot \exp\left(i2\pi\left(xk_x + yk_y\right)\right) \cdot \mathrm{d}x\, \mathrm{d}y \tag{2.42}$$

The functions f and F form a Fourier transformation pair. The function $f(x, y)$ denotes the image and the $F\left(k_x, k_y\right)$ is the k-space. This relationship and the Fourier Transform in general are further discussed in Sect. 3.2. The maximum resolution of the image is given by the pixel size as:

$$\Delta x = \frac{1}{\gamma G_{FE} M \Delta t} \quad \Delta y = \frac{1}{\gamma \Delta G N \tau} \tag{2.43}$$

and the size of the image, known as the Field of View (FOV), is derived as the inverse of the minimum spatial frequency step (Eq. 2.38):

$$FOV_{\mathrm{FE}} = \frac{1}{\bar{\gamma} G_{FE} \Delta t} \quad FOV_{\mathrm{PE}} = \frac{1}{\bar{\gamma} \Delta G_0 \tau} \tag{2.44}$$

In simpler terms, the smallest object detectable, the pixel size, is represented by the highest spatial frequency:

$$\begin{aligned} \Delta x &= \frac{1}{N \Delta k} \\ &= \frac{1}{FOV_k} \end{aligned}$$

The largest object detectable or the FOV, is:

$$FOV_x = \frac{1}{\Delta k_x} \tag{2.45}$$

where Δk_x is the sampling frequency,

To reconstruct the image, a 2D Inverse Fourier Transform is performed on the k-space, which results in a complex image comprising real and imaginary parts. MRIs used in the clinical routines are the magnitude of the reconstructed complex image. However, the phase image can also be used for various purposes, like in the case of Phase-Contrast MRI (PC-MRI). An example is shown in Fig. 2.14.

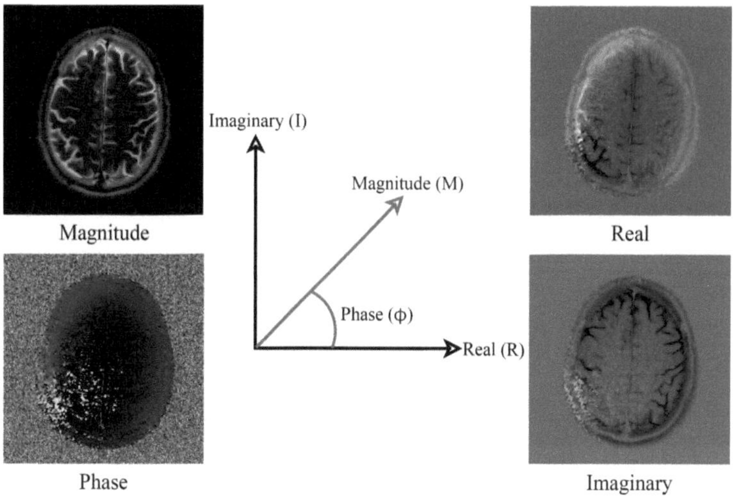

Fig. 2.14 Reconstruction of the k-space data. (source: fastMRI [293]) to obtain complex images and its relationship to the magnitude (M), phase (ϕ), real (R), and imaginary (I) images, where $M = \sqrt{R^2 + I^2}$ and $\phi = \tan^{-1}(I/R)$

The centre of the k-space corresponds to the low-frequency components—general shapes and contours, whereas the non-centre (or periphery) corresponds to the high-frequency components—edges, fine details, as shown in Fig. 2.15

On a final note, it should be said that even though here the explanation of 2D imaging is provided, a simple 1D imaging or 3D volumetric imaging is also possible with MR, using 1D and 3D Fourier transforms, respectively.

2.3.2.4 k-space Samplings

k-space can be sampled in various ways (known as trajectories): Cartesian, radial, and spiral, to name a few. Cartesian sampling is the most frequently used sampling strategy—which has been explained in the previous section. An interesting fact is that the first MR image was produced not using Cartesian sampling but using a non-Cartesian sampling pattern followed by a CT-like filtered back-projection reconstruction—known as Projection Reconstruction (PR). However, Cartesian acquisition followed by 2D Fourier Transform for the reconstruction of

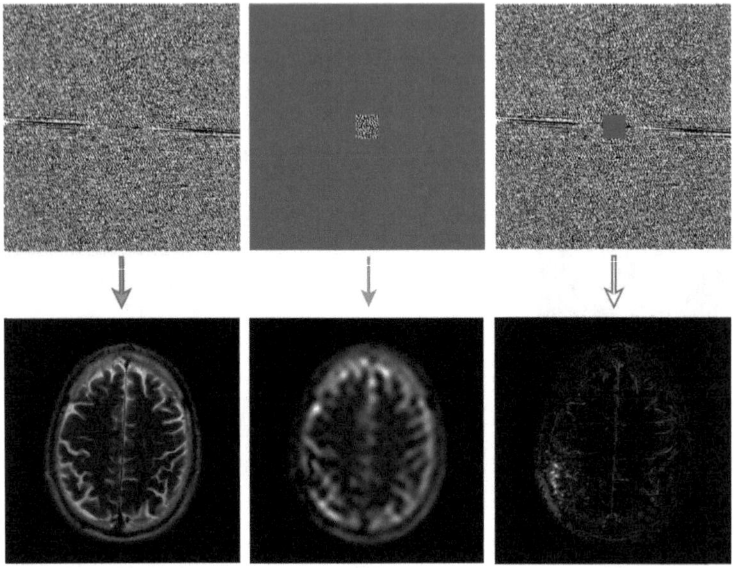

Fig. 2.15 Different parts of the k-space and their relationships with the image-space: reconstruction of the k-space data. (source: fastMRI [293]) from **(a)** all the frequencies **(b)**, low spatial frequencies—10% of the centre of the k-space, and **(c)** high spatial frequencies—90% of the outer k-space (i.e. ignoring 10% of the centre of the k-space)

the MR images soon became the norm for MRIs, as PR resulted in excessive image blurring due to magnetic field inhomogeneity and gradient nonlinearities of early MR scanners. With the development of the MR systems, PR was re-invented as radial imaging.

Radial Imaging: In a radial k-space trajectory (shown in Fig. 2.16), 1D projections, commonly known as spokes, are acquired around the object with different angles ϕ, produced by applying frequency encoding on two axes x and y of the slice simultaneously using two gradients G_x and G_y respectively, while the rotational pattern is produced by varying the amplitudes of the gradients. Thus, the pulse sequence consists of only slice selection and frequency encoding, and the concept of phase encoding is absent. The acquired data is then interpolated on a conventional Cartesian grid and then reconstructed using 2D Fourier Transform. This process is known as regridding, shown in Fig. 2.17.

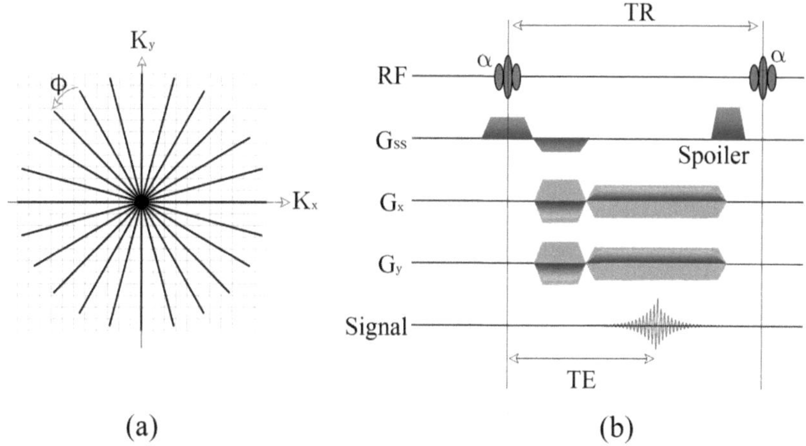

Fig. 2.16 Radial k-space: (**a**) Example of a k-space with radial acquisition with an angle ϕ of 45°. (**b** Pulse sequence for radial acquisition with an angle ϕ of 45° where $G_x = G_{FE} \cos\phi$ and $G_y = G_{FE} \ \sin\phi$

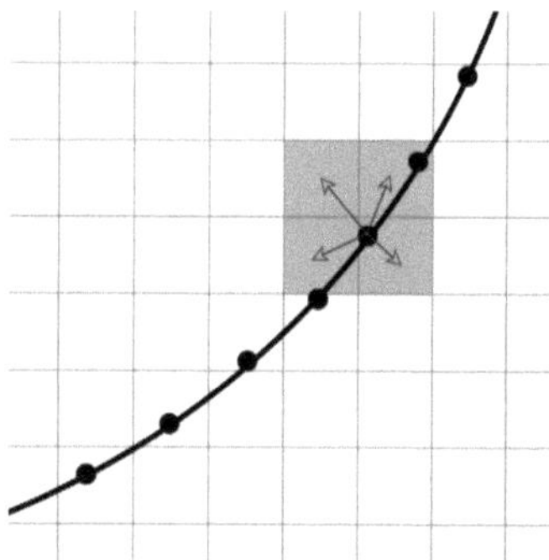

Fig. 2.17 Re-gridding of non-Cartesian trajectories into a Cartesian grid—each point affects four neighbours

It is to be noted that such sampling patterns heavily oversample the k-space centre, requiring the data to be density compensated before reconstruction. The acquired raw data is first multiplied with a density compensation function, which is usually based on a Bessel function, and then reconstructed by using regridding followed by Inverse Fourier Transform, or using Nonuniform Fast Fourier Transform (NUFFT) [46, 94]. One interesting question in radial sampling is the way to optimally choose the angle (ϕ) between the spokes of acquisition. A typical way of performing such acquisition is following the equidistant radial sampling scheme, where the ϕ is calculated as $180°/nSpokes$, where $nSpokes$ is the desired number of spokes to be acquired. One sophisticated solution is to increment each spoke by 111.25°, which is 180° divided by the golden ratio ($\varphi = \frac{1+\sqrt{5}}{2} \approx 1.618$), known as Golden-angle RAdial Sparse Parallel (GRASP) sampling [162, 163], or simply golden-angle sampling. 3D volume can be acquired by stacking multiple GRASP slices acquired with the help of conventional slice-select phase-encoding, known as the Stack-of-Stars [162] radial sampling. Data acquired with such a sampling technique can be reconstructed by applying 1D Inverse Fourier Transform on the slice direction, followed by reconstructing each slice separately using techniques discussed earlier. However, the radial acquisition can also be extended completely to 3D using spherical k-space trajectory using Vastly-undersampled Isotropic Projection Reconstruction (VIPR) [54], commonly known as the Koosh ball trajectory.

The radial sampling technique can also be extended by combining it with Cartesian sampling, forming the Periodically Rotated Overlapping ParallEL Lines with Enhanced Reconstruction (PROPELLER) trajectory. It acquires a number of parallel spokes, known as blades, rotating about the centre of the k-space with an angle ϕ, shown in Fig. 2.18(a).

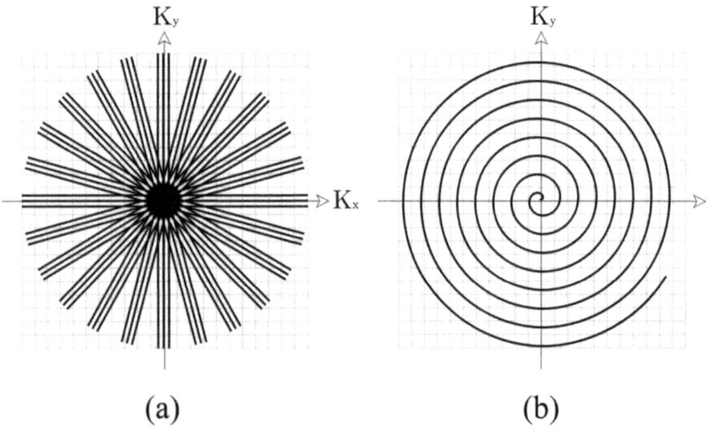

Fig. 2.18 (**a**) Periodically Rotated Overlapping ParallEL Lines with Enhanced Reconstruction (PROPELLER) trajectory and (**b**) Spiral trajectory

Spiral Imaging: Another non-Cartesian k-space trajectory is the spiral trajectory, where the data is sampled starting from the centre of the k-space along an Archimedean spiral. Similar to radial, a spiral pulse sequence also comprises only slice selection and frequency encoding. The spiral acquisition can be of two types—single-shot and multi-shot. For the first, the whole k-space is filled following a single RF excitation using in-plane gradient waveforms—commonly known as a shot, as shown in 2.18(b). Multiple interleaves shots can be performed to fill the k-space more densely, where each shot is being rotated with an angle of $\pm 2\pi/nShot$, where $nShot$ is the number of shots to be acquired—this is known as multi-shot spiral trajectory. Similar to the Stack-of-Stars radial sampling, 3D volumes can be acquired using Stack-of-Spirals. The reconstruction of raw data acquired using spiral sampling is also similar to radial sampling.

2.4 Types of MRI

One of the biggest advantages of MRI is the ability to manipulate the contrast of MRIs by simply modifying the parameters of the pulse sequence (Sect. 2.3.2.2), an example is given in Fig. 2.19. This can be especially useful for identifying a certain type of pathology, as shown in 2.20, as depending on the material present in the pathology—the contrast of the pathology will be different in different types of MRIs. This is because the different types of tissues appear differently in those contrasts, as listed in Table 2.1. A few types of MRIs are mentioned here.

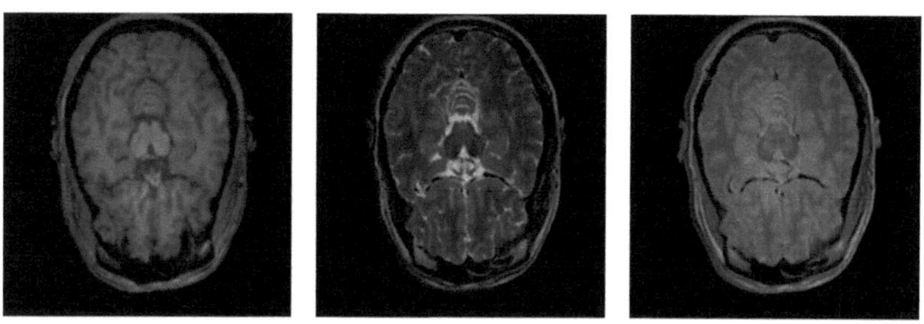

(a) (b) (c)

Fig. 2.19 A few types of MRIs - **(a)** T1w, **(b)** T2w, and **(c)** PDw

Table 2.1 Different tissues in different MRI contrasts

Tissue	T1w	T2w	FLAIR
CSF	Dark	Bright	Dark
White Matter	Light	Dark grey	Dark grey
Cortex	Grey	Light grey	Light grey
Fat	Bright	Light	Light
Inflammation (e.g., infection, demyelination)	Dark	Bright	Bright

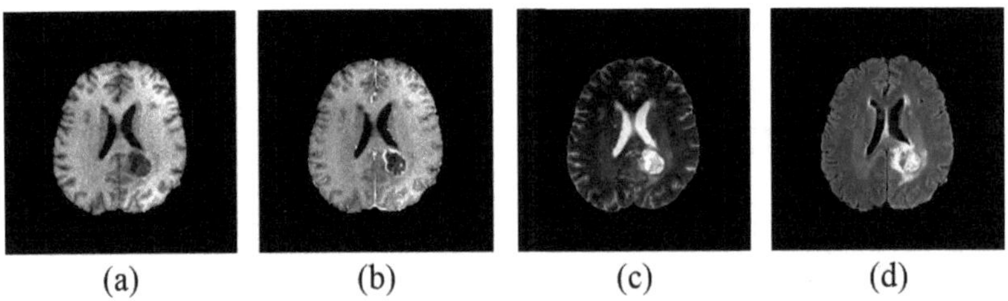

Fig. 2.20 Visualisation of a tumour (data source: BraTS 2020 dataset) in different types of MRIs—**(a)** T1w, **(b)** Contrast-enhanced T1w (after injecting contrast agent), **(c)** T2w, **(d)** FLAIR

T_1-weighted images: T_1 differences between the tissues are enhanced by using a short TR, along with a short TE. In the resultant image, fluids appear dark, water-based tissues appear grey, and fat-based tissues appear white. T_1-weighted images are typically the anatomical scans, as they can delineate the boundaries between the different types of tissues very clearly.

T_2-weighted images: In the pulse sequence, if a long TR is used coupled with a long TE, the resultant image will be a T_2-weighted image, where the fluids appear the brightest and tissues appear grey. These are commonly used for pathological scans, as the accumulated abnormal fluids appear much brighter than the neighbouring tissues.

PD-weighted images: The density of protons present in fluids is higher than in tendon and bone, and this can be useful in some clinical applications, such as the differentiation of articular cartilage from the cortical bone and menisci in knee imaging. This kind of image can be generated by using a TR, but a short TE, and the resultant image would be called a PD-weighted image. During the acquisition of T_2-weighted images, it is possible to create an additional echo with a shorter TE to obtain the corresponding PD-weighted images additionally—this kind of sequence is known as dual-echo.

FLAIR images: FLuid Attenuated Inversion Recovery (FLAIR) sequences use an additional RF pulse called Inversion Recovery (IR) with an interval defined by the Inversion Time (TI). T_2-weighted images with IR can suppress the signal from the CSF, making these sequences frequently used for neurological scans when lesions may be adjacent to the ventricles.

MRA images: To visualise the vessels better in MRIs, Magnetic Resonance Angiography (MRA) can be used. Blood flows continuously during acquisition, and during the TR of a pulse sequence, blood might flow out of the slice, and fresh blood might flow in which has not encountered the same RF pulse. When the RF pulse is applied to this flowed-in blood, it provides a stronger signal. This is known as the in-flow or Time of Flight (ToF) effect. Using various techniques, all the signals apart from this bright blood signal can be suppressed during the acquisition of MRA with the help of a velocity-encoding gradient. The three most commonly used sequences for the same are: ToF-MRA, phase-contrast MRA, and contrast-enhanced MRA.

Diffusion-Weighted images: Diffusion-Weighted Imaging (DWI) is one of the MR techniques, which is employed in diagnosing white matter diseases, cancer or stroke. The proportion of water in the human body is approximately $60 - -70\%$, and it is distributed over intra- and extracellular compartments. Different tissues in the human body exhibit varying diffusion characteristics. DWI is a technique that generates signals based on changes in Brownian motion using Pulsed Gradient Spin Echo (PGSE) method, and it provides information regarding the diffusion properties. Axon membranes present in brain white matter limit the molecular movement perpendicular to the fibre, thus resulting in anisotropic diffusion in white matter. DWI uses this property to provide information on white matter integrity and structural details of white matter tracts [193].

fMRIs: The brain activation can be detected by being sensitive to the changes in regional cerebral blood flow using a specialised MRI technique known as Functional MRI (fMRI), or more technically known as the Blood Oxygenation Level Dependent (BOLD) imaging. Increased blood flow changes the local ratio of deoxyhemoglobin (paramagnetic) to oxyhemoglobin (diamagnetic), thus affecting T_2 and T_2* relaxation times—the property used to generate the BOLD contrast. During fMRI, by comparing the image acquired when the subject was performing a certain task against the image acquired when the subject was resting (RS-fMRI), the brain activation during the task can be visualised.

2.5 Artefacts in MRI

Even though MRI comes with various advantages, it can also be prone to different types of artefacts. These artefacts can be tissue-related, such as chemical shift, bounce point, susceptibility, metal, dielectric, etc.; can be technique-related, for example, undersampling, partial volume, wrap-around, Gibbs (truncation); and can also be due to motion caused by the subject's movements or even by respiration. In this section, two of the technique-related artefacts, undersampling and motion, are discussed in detail.

2.5.1 Undersampling

MRI is an inherently slow process due to consecutive data acquisition [234]. Fast imaging can improve patient compliance, reduce motion artefacts, increase patient throughput, etc. The speed of acquisition can be increased by ignoring parts of the k-space, which is known as undersampling. Taking the inverse Fourier transform from not densely enough sampled k-space frequency data might cause the resultant image to lose resolution and might also have artefacts due to the violation of the Nyquist-Shannon sampling theorem [4, 8].

2.5.1.1 Nyquist Criterion

In signal processing, the process of converting an analogue signal to digital, known as digitisation, is not lossless. The analogue signal is continuous, whereas the digital signal is discrete—that has to be sampled with a certain frequency.

$$T_S = \frac{1}{f_s} \tag{2.46}$$

where f_s is the sampling frequency, or simply the speed of sampling, and T_S is the sample period, the gap between two consecutive measurements. From this, it can be derived that if the sampling frequency f_s is low, then the sample period T_S would be large—which would cause more loss of signal. According to the Nyquist-Shannon sampling theorem [4, 8], the highest frequency signal that can be accurately digitised at a certain f_S, is half of the f_S:

$$f_N = \frac{1}{2} \cdot f_S \quad (2.47)$$

This f_N is known as the Nyquist frequency or folding frequency. If the highest frequency of a signal is higher than the f_N of the sampler, the resultant discretised signal will have aliasing. This has been discussed further in Sect. 3.2.1 (Fig. 2.21 and 2.22)

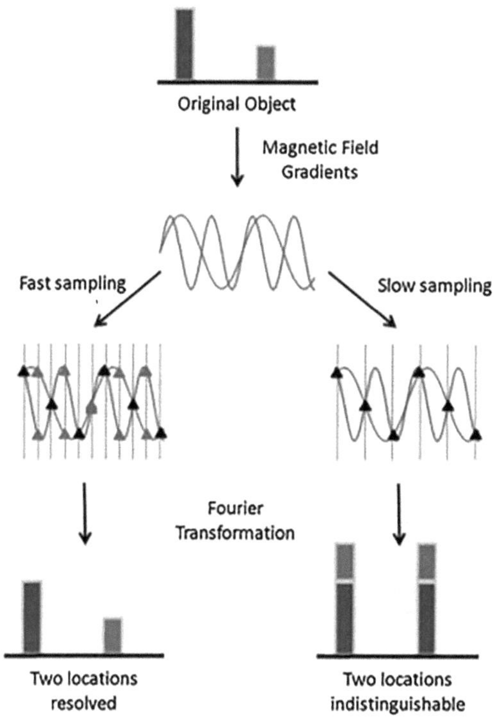

Fig. 2.21 Reason for aliasing artefacts in undersampled MRI. 1st row: Two signal sources from one 1D object—blue and red. 2nd row: They result in two signals oscillating at two different frequencies. 3rd row: The signals are sampled with high (left) and low (right) sampling rates—indicated with the dotted lines. Triangles denote the sampled signal—red and blue denote the distinguishable samples for respective sources, while black denotes samples where both appear the same. If the sampling frequency is high enough (higher than the Nyquist rate), the signals can be distinguished, as they might contain both black as well as their respective triangles. Whereas, if the frequency is too low, the signals are not distinguishable (only black triangles). 4th row: If Fourier transform is applied on these sampled signals, the right one results in two correctly resolved locations, whereas the left one results in two indistinguishable locations. (Fig source: [134])

2.5.1.2 Effect of Undersampling

Undersampling can be performed in different ways—resulting in different types of artefacts/problems. If the k-space coverage is reduced by ignoring phase-encoding lines towards the edges of the k-space, the resulting MRI loses resolution (Fig. 2.23b). However, if the distance between the phase-encoding lines is increased (skipping lines in-between), the resulting image contains artefacts (Fig. 2.23c). Depending upon the acquisition type (e.g., Cartesian, radial, spiral) and undersampling type (e.g. uniform, Variable density), the resulting artefact in the reconstructed image differs—shown in Fig. 2.24. In Cartesian sampling, uniform undersampling results in coherent noise (i.e., aliasing), while Varden undersampling results in more incoherent noise. On the other hand, radial undersampling results in streaking artefacts (similar to computed tomography), and spiral undersampling results in spiral streaks.

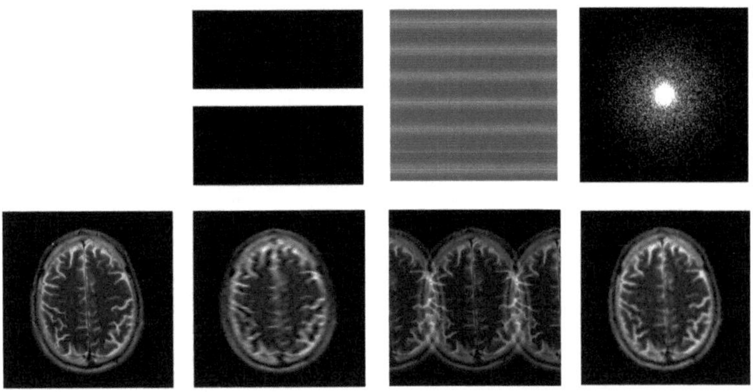

(a) (b) (c) (d)

Fig. 2.22 Effects of Nyquist violation: (a) fully sampled, (b), reduced k_{max}—reduces the image resolution, causing it to blur (c) increased δk—consequently FOV is reduced, and replicas of the original object appear in the image—wrap-around artefacts, and (d) random undersampling—aliasing appears as incoherent noise. Top row shows the sampling patterns, and the bottom row shows the resultant images. (MRI source: fastMRI dataset [293])

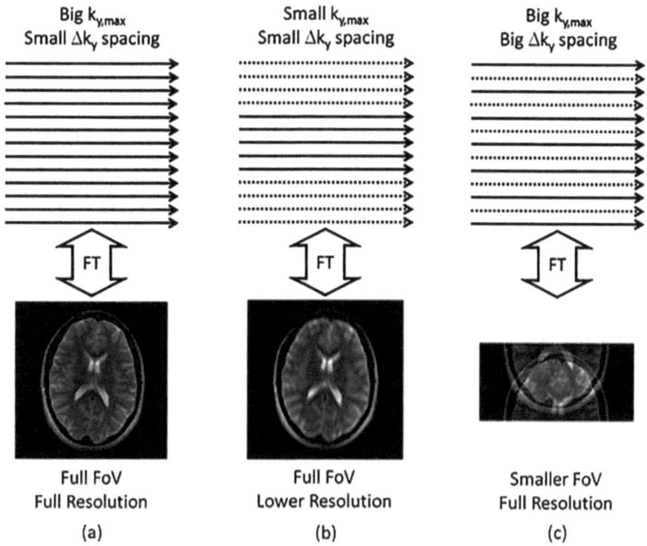

Fig. 2.23 Different ways of performing undersampling and their effects. (a) Fully-sampled MRI, (b) If the k-space coverage is reduced by reducing the number of phase-encoding lines and only acquiring the lines at the centre (reduction of $k_{y,max}$—the resolution is of the image is lowered), (c) If the distance between the phase-encoding lines is increased (increase of Δk_y), the FOV is reduced—causing aliasing artefacts. (Fig source: [134])

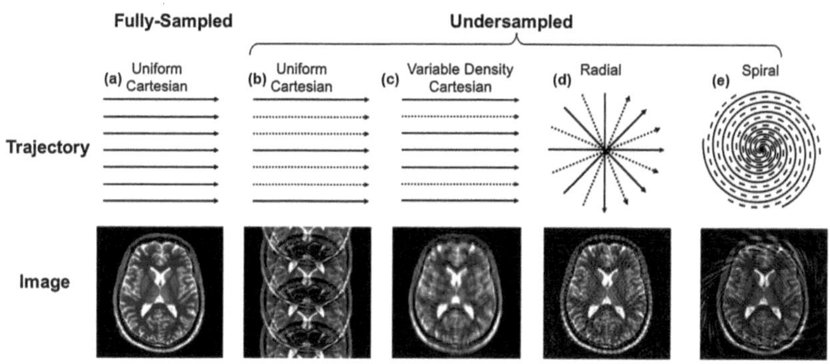

Fig. 2.24 Examples of different k-space trajectories and the corresponding artefacts after undersampling. (a) Fully sampled (uniform Cartesian sampling), (b) Uniform Cartesian undersampling, (c) 1D Variable density Cartesian undersampling, (d) Radial undersampling, and (e) Spiral undersampling. (Fig source: [225])

2.5.2 Motion

Motion is yet another major source of artefacts in MRIs, which can cause image blurring, cause signal drop-outs or even introduce ghosting artefacts [153]. Motion can be physiological (e.g., breathing), flow (e.g., blood flow), or even from the physical movements of the subject. During long acquisitions, the probability of the subject's physical movement also increases due to the discomfort he/she experiences inside the scanner during these longer stays. Motion can be categorised based on type of motion (Fig. 2.25)—rigid and deformable; occurrence—inter-image, inter-scan, and intra-scan, pattern—periodic, quasi-periodic, continuous, and random, direction—in-plane and through-plane.

Several factors can contribute to the motion artefacts in MRI. Three of the major reasons are discussed here.

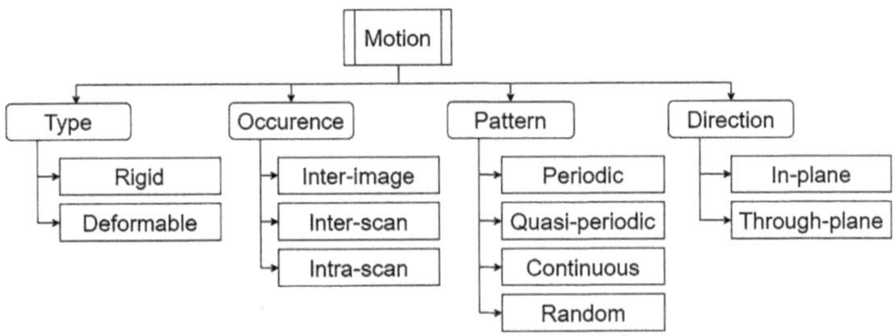

Fig. 2.25 Different categories of motion in MRI. (Content from [197])

Incorrect phase accumulation: During the acquisition, when gradients are switched on, spins along with the gradient direction gain additional phase, shown in Fig. 2.26. Because of the motion, dephasing and refocusing happen at different physical locations—leading to this problem. This leads to variations among the different phase-encoding lines—making the k-space inconsistent [191].

Excitation history: Motion in the slice direction can cause slice-selective RF-pulses to be applied wrongly as showcategorised2.27—resulting in too weak or too strong signal generation, can also alter image contrast or introduce magnitude inconsistencies in the k-space. This is known as the excitation history effect [191].

Rotation during acquisition: Rotation in the image space is equivalent to the rotation in the k-space. If the object rotates during acquisition, the different phase-encoding lines also undergo the same rotation—leading to inhomogeneous k-space

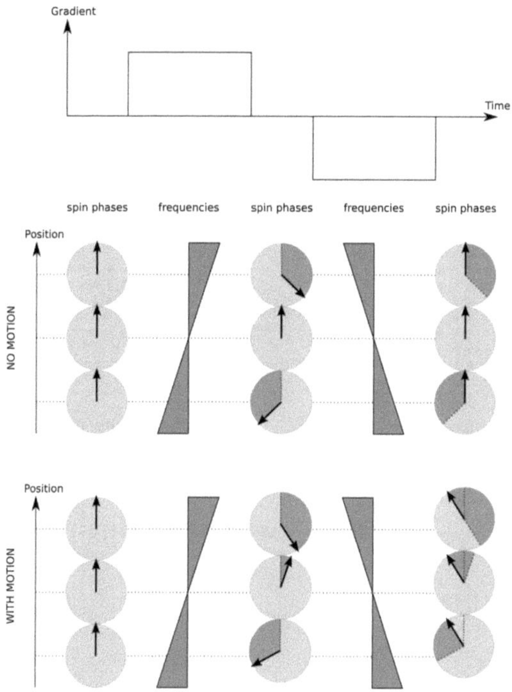

Fig. 2.26 Incorrect phase accumulation—effect of velocity on the phase of the MR signal. (Fig source: [191])

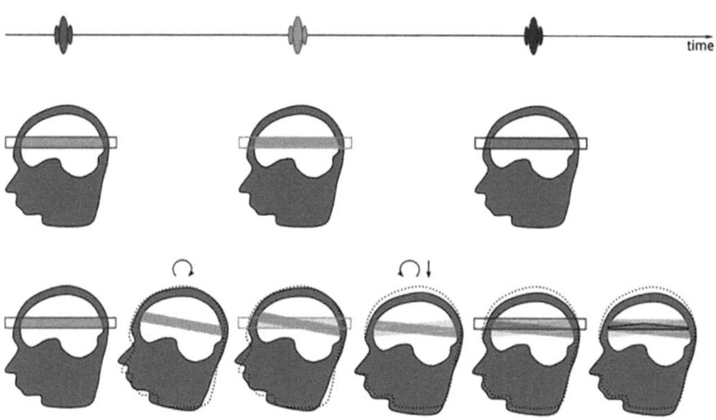

Fig. 2.27 An example of excitation history problem for a 3-pulse MR sequence. (Fig source: [191])

sampling. This can lead to inconsistent sampling density—leading to local Nyquist violations, shown in Fig. 2.28. Consequently, ghosting or streaking artefacts can be seen in the resulting images [191].

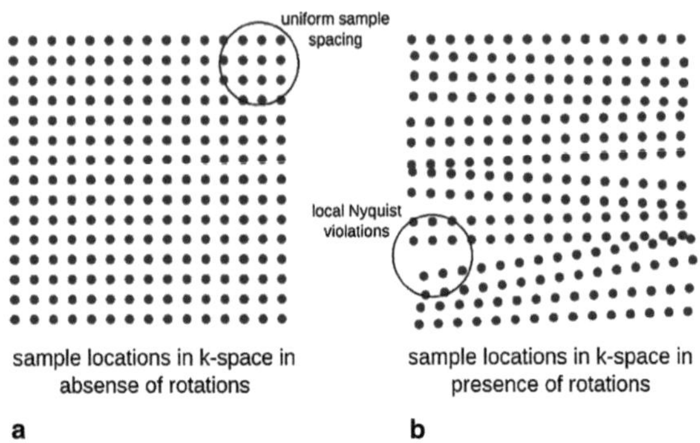

Fig. 2.28 Different categories of motion in MRI. (Fig source: [191])

Image Processing 3

An **image** can be defined as a two-dimensional (2D) light-intensity function $f(x, y)$, where x and y represent the spatial coordinates, and the amplitude of f, A, at a given coordinate x, y is a positive scalar quantity—called the intensity level of the image at that point. $f(x, y)$ can be characterised by two components: illumination $i(x, y)$—the amount of source light incident on the scene and reflectance $r(x, y)$—the amount of light reflected by the object, and can be expressed as a product of these two components: $f(x, y) = i(x, y)r(x, y)$. An image can be called as a **digital image** only if x, y, and A are all finite, the discretisation (or digitisation) of spatial coordinates x, y is called image sampling and discretisation of the amplitude A is called grey-level quantisation. The elements of a such discretised image, or digital image, are called as the image or picture elements—commonly known as the **pixels** [35, 124]. It is noteworthy that this idea of an image can be extended to the 3rd dimension for representing a 3D light-intensity function $f(x, y, z)$, commonly known as a **volume**. Each element of a volume is known as a **voxel**.

3.1 Convolution

The convolution operation is the integral of the product of two functions—one of them is a reversed and shifted version of the other one.

$$s(t) = \int f(\tau)g(t - \tau)d\tau \tag{3.1}$$

which is the area under the function $f(\tau)$, weighted by the function $g(-t\tau)$ shifted by t. Different parts of the input $f(\tau)$ is emphasised by the weighting function $g(t - \tau)$ with the change in t. This operation is typically denoted by an asterisk:

S. Chatterjee, *Deep Learning Innovations in MRI Reconstruction and Analysis*,
https://doi.org/10.1007/978-3-658-50741-1_3

$s(t) = (f \star g)(t)$. If t only contains integer values, the discrete convolution can be defined as:

$$s(t) = (f \star g)(t) = \sum_{a=-\infty}^{\infty} f(a)g(t-a) \tag{3.2}$$

Because of the commutative property of the convolution operation, this can also be written as:

$$s(t) = (g \star f)(t) = \sum_{a=-\infty}^{\infty} f(t-a)g(a) \tag{3.3}$$

Now, for a 2D image $f(x, y)$, convolving with a function $g(x, y)$ can be given by extending Eq. 3.2:

$$S(x, y) = (f \star g)(x, y) = \sum_{m}\sum_{n} f(m, n)g(x-m, y-n) \tag{3.4}$$

An example is shown in Fig. 3.1, where an Input $f(x, y)$ is convolved with a kernel W (i.e. $g(x, y)$).

The convolution operation has a wide array of applications in the field of image processing, from edge detection to image quality improvement. Moreover, convolution has applications in other fields as well. In the field of statistics, a weighted moving average is a convolution operation. In probability theory, the probability distribution of the sum of two independent random variables is the convolution of their individual distributions.

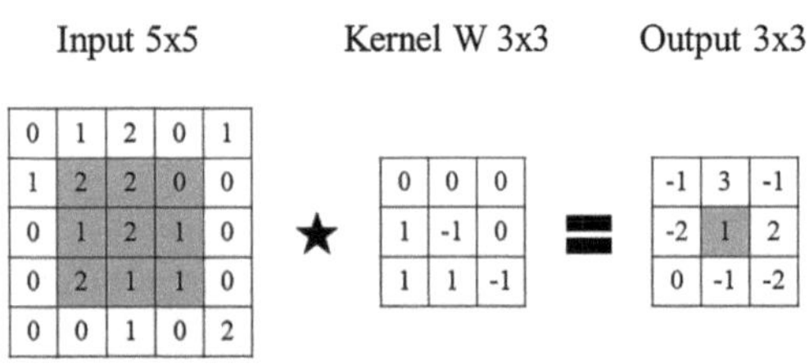

Fig. 3.1 An example convolution operation numerically

3.2 Fourier Transform

Image transformations, such as Discrete Fourier Transform (DFT), Discrete Cosine Transform (DCT), Walsh transform, etc., play a key role in the field of image processing—from image enhancement, restoration, representation (encoding), undersampling etc. Fourier Transform is one of the most widely used image transformation operations in the field of image processing, which is also inherently related to MRI, as discussed in Sect. 2.3.2.3. For a continuous function $f(x)$, where x is a real-valued continuous variable, the Fourier transform of $f(x)$ is $\mathcal{F}\{f(x)\}$, given by:

$$\mathcal{F}\{f(x)\} = F(\mu) = \int_{-\infty}^{\infty} f(x)e^{-i2\pi\mu x}dx \tag{3.5}$$

where $i = \sqrt{-1}$ is the imaginary unit and μ is another continuous variable. As the integration has been performed with respect to t, $\mathcal{F}\{f(x)\}$ is a function of μ. Given $F(\mu)$, the original function $f(x)$ can be obtained with the help of inverse Fourier transform $\mathcal{F}^{-1}\{F(\mu)\}$ as:

$$\mathcal{F}^{-1}\{F(\mu)\} = f(x) = \int_{-\infty}^{\infty} F(\mu)e^{i2\pi\mu x}d\mu \tag{3.6}$$

As the Fourier transform of a real function is complex, it can be denoted using complex notations: $F(\mu) = R(\mu) + iI(\mu)$, where $R(\mu)$ and $I(\mu)$ are the real and imaginary parts of $F(u)$. With the help of this notation, the Eq. 3.5 can be written as:

$$F(\mu) = |F(\mu)|e^{i\phi(\mu)} \tag{3.7}$$

where $|F(\mu)$ and $\phi(\mu)$ are the magnitude and phase of $F(u)$, respectively, given by:

$$|F(\mu)| = \sqrt{R^2(\mu) + I^2(\mu)} \tag{3.8}$$

and

$$\phi(\mu) = \tan^{-1}\left[\frac{I(\mu)}{R(\mu)}\right] \tag{3.9}$$

Now, as the images are discretised, a discrete form of the Fourier transform is also required to work with them. If $f(x)$ is discretised by taking N samples with a uniform distance of ΔX, the Fourier transform pair Eq. 3.5 and 3.6 is given by:

$$\mathcal{F}\{f(x)\} = F(\mu) = \frac{1}{N}\sum_{x=0}^{N-1} f(x)e^{-i2\pi\mu x/N} \tag{3.10}$$

and

$$\mathcal{F}^{-1}\{F(\mu)\} = f(x) = \sum_{\mu=0}^{N-1} F(\mu)e^{i2\pi\mu x/N} \tag{3.11}$$

These equations so far were for a 1D Fourier transform, i.e., a one-variable scenario for x. They can be further extended to a two-variable scenario x, y to 2D discrete Fourier transform and inverse Fourier transform as follows:

$$\mathcal{F}\{f(x, y)\} = F(u, v) = \frac{1}{MN} \sum_{x=0}^{M-1} \sum_{y=0}^{N-1} f(x, y) e^{-i2\pi(ux/M+vy/N)} \tag{3.12}$$

and

$$\mathcal{F}^{-1}\{F(u, v)\} = f(x, y) = \sum_{u=0}^{M-1} \sum_{v=0}^{N-1} F(u, v) e^{i2\pi(ux/M+vy/N)} \tag{3.13}$$

These 2D versions are generally used while working with 2D images $f(x, y)$. Moreover, they also show the relationship with an MR-acquisition, as explained in Sect. 2.3.2.3. Equations 2.41 and 2.42 can be seen as the continuous versions of the equations 3.12 and 3.13, respectively.

The complexity of DFT (Eq. 3.10) algorithm is $O(N^2)$, because for each N values of μ, N complex multiplications of $f(x)$ by $e^{-i2\pi\mu x/N}$ will be performed. This complexity can be reduced to $O(N \log_2 N)$ by proper decomposition of Eq. 3.10. This is known as the **Fast Fourier Transform (FFT)**, and the inverse of this is known as the **Inverse Fast Fourier Transform (iFFT)**. Just like the 2D versions of DFT, there are also the corresponding 2D versions of FFT and iFFT, as FFT2 and iFFT2 [35] (Fig. 3.2).

3.2.1 Sampling

To discretise a continuous function, it has to be sampled with a certain sampling frequency. For a continuous variable x, its unit impulse—a spike of infinity amplitude at the location $x = 0$ is given by:

$$\delta(x) = \begin{cases} \infty & \text{if } x = 0 \\ 0 & \text{if } x \neq 0 \end{cases} \tag{3.14}$$

which also satisfies the identity:

$$\int_{-\infty}^{\infty} \delta(x) dx = 1 \tag{3.15}$$

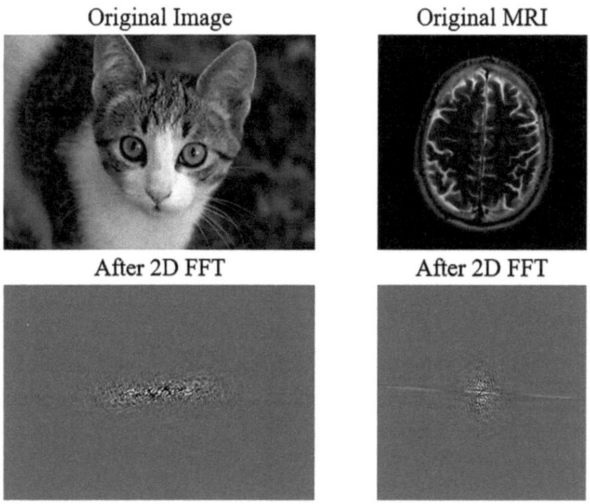

Fig. 3.2 Applying 2D Fast Fourier Transform (FFT) on a digital photo and an MRI. (Source of MRI: fastMRI dataset [293])

For the initial location, $x = 0$, the value of the continuous function $f(x)$ is:

$$\int_{-\infty}^{\infty} f(x)\delta(x)dx = f(0) \tag{3.16}$$

With the help of the shifting property of the impulse, the impulse of an arbitrary point x_0 can be obtained as:

$$\int_{-\infty}^{\infty} f(x)\delta\left(x - x_0\right) dx = f\left(x_0\right) \tag{3.17}$$

A discrete equivalent of the same can be denoted as:

$$\sum_{x=-\infty}^{\infty} f(x)\delta\left(x - x_0\right) = f\left(x_0\right) \tag{3.18}$$

If a 1D continuous function $f(x)$ extending from $-\inf$ to $\inf$ is sampled with a uniform interval of ΔX, the sampled function $\tilde{f}(x)$ is:

$$\tilde{f}(x) = f(x)s_{\Delta X}(x) = \sum_{n=-\infty}^{\infty} f(x)\delta(x - n\Delta X) \tag{3.19}$$

where $s_{\Delta X}(x)$ is known as the impulse train and each component of the summation is an weighted impulse of that location, the value of which is given by:

$$\begin{aligned} f_k &= \int_{-\infty}^{\infty} f(x)\delta(x - k\Delta X)dx \\ &= f(k\Delta X) \end{aligned} \tag{3.20}$$

The convolution theorem states that the multiplication of two functions in the spatial domain is the convolution of them in the frequency domain. Hence, the Fourier transform of this sampled function is:

$$\begin{aligned} \tilde{F}(\mu) &= \mathcal{F}\{\tilde{f}(x)\} \\ &= \mathcal{F}\{f(x)s_{\Delta X}(x)\} \\ &= F(\mu)\star S(\mu) \\ &= \int_{-\infty}^{\infty} F(\tau)S(\mu - \tau)d\tau \\ &= \frac{1}{\Delta X}\sum_{n=-\infty}^{\infty} F\left(\mu - \frac{n}{\Delta X}\right) \end{aligned} \tag{3.21}$$

Eq. 3.21 shows that the Fourier transform $\tilde{F}(\mu)$ of the sampled function $\tilde{f}(x)$ is an infinite periodic sequence of copies of $F(\mu)$, separated by $1/\Delta X$—a lower value of which would make the periods to merge and a higher value would provide a clean separation between them. For a band-limited function within a finite interval $(-\mu_{max}, \mu_{max})$, sufficient separation of the periods can be achieved following the Nyquist-Shannon sampling theorem [4, 8]:

$$\begin{aligned} \frac{1}{\Delta X} &> 2\mu_{\max} \\ => \frac{1}{2\Delta X} &> \mu_{\max} \end{aligned} \tag{3.22}$$

This can be extended to a 2D scenario, where $\tilde{f}(x, y)$ is a 2D sampled function sampled with an uniform distance of ΔX and ΔY, and $\tilde{F}(\mu, \nu)$ is its corresponding Fourier transform. If $\tilde{f}(x, y)$ is band-limited within a finite interval of

$(-\mu_{max}, \mu_{max})$ and $(-\nu_{max}, \nu_{max})$, the Nyquist criterion is as follows:

$$\begin{aligned}\frac{1}{\Delta X} &> 2\mu_{\max}\\ \frac{1}{\Delta Y} &> 2\nu_{\max}\end{aligned} \tag{3.23}$$

This implies that if the sampling rate is more than twice the highest frequency, the band-limited function can be received without any overlapping periods. This is also known as the Nyquist criterion, or the sampling rate is known as the Nyquist rate. If the sampling rate is lower than the Nyquist rate, commonly known as undersampling, the periods overlap—causing the resultant image to have aliasing artefacts. The term aliasing comes from the word alias, which means "false identify"—as the overlap of the periods causes the false introduction of the high-frequency components of a continuous function as low-frequency components in the sampled function, creating false copies of the object in the resultant image [124]. This aliasing is a type of artefact also encountered in MR imaging, as discussed in Sect. 2.5.1.

3.3 Image Enhancement

The main objective of image enhancement techniques is to improve the quality of the image to make it more suitable for a specific application. One method of enhancing an image is with the help of filters, in the spatial domain or in the frequency domain.

3.3.1 Sharpening

Highpass Filter: The edges of the image in terms of the grey level values in the spatial domain correspond to the high-frequency components in the frequency domain. Hence, the image sharpening can be performed by enhancing the edges in the frequency domain with the help of a highpass filter which attenuates the low-frequency components. The transfer function $H(x, y)$ of a 2D Ideal Highpass Filter (IHPF) satisfies the relationship:

$$H(x, y) = \begin{cases} 0 & \text{if } D(x, y) \leqslant D_0 \\ 1 & \text{if } D(x, y) > D_0 \end{cases} \tag{3.24}$$

where $D_0 >= 0$ is a specified cutoff distance from the origin of the frequency plane and $D(x, y)$ is the distance between the origin of the frequency plane and a specific point (x, y), calculated as:

$$D(x, y) = \sqrt{x^2 + y^2} \tag{3.25}$$

Another highpass filter is the Butterworth Highpass Filter (BHPF). The transfer function of a BHPF of order n is defined by:

$$H(x, y) = \frac{1}{1 + [D_0/D(x, y)]^{2n}} \tag{3.26}$$

The filtered image obtained after filtering in the frequency domain with the help of the transfer function $H(x, y)$ can be denoted as:

$$G(x, y) = H(x, y)F(x, y) \tag{3.27}$$

where $F(x, y)$ is the Fourier transform of the image $f(x, y)$ and $G(x, y)$ is the final filtered image. Since the multiplication of two functions in the frequency domain is the convolution operation in the spatial domain, this filtering can be performed in the spatial domain with the help of a convolution operation:

$$g(x, y) = h(x, y) \star f(x, y) \tag{3.28}$$

where $h(x, y)$ is the inverse Fourier transform of the filter transfer function $H(x, y)$ (Figs. 3.3, 3.4, 3.5, 3.6).

Laplacian Filter: The derivative of an image enhances the edges and other discontinuities and de-emphasises areas with slowly varying grey-level values. The Laplacian operator performs the second-order derivative of an image $f(x, y)$, by performing partial derivatives along the two spatial axes:

$$\nabla^2 f = \frac{\partial^2 f(x, y)}{\partial x^2} + \frac{\partial^2 f(x, y)}{\partial y^2} \tag{3.29}$$

The discrete form of the Laplacian filter is:

$$\nabla^2 f = [f(x+1, y) + f(x-1, y) + f(x, y+1) + f(x, y-1) - 4f(x, y)] \tag{3.30}$$

The filtered image with this operation $g(x, y)$ can be obtained as:

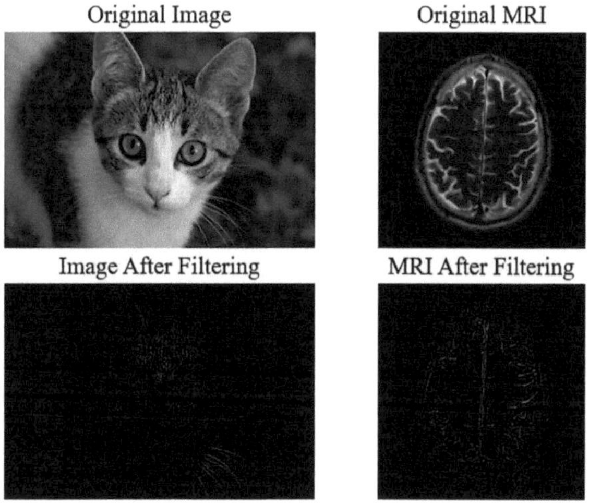

Fig. 3.3 Applying an Ideal Highpass Filter (IHPF) on a digital photo and an MRI. (Source of MRI: fastMRI dataset [293])

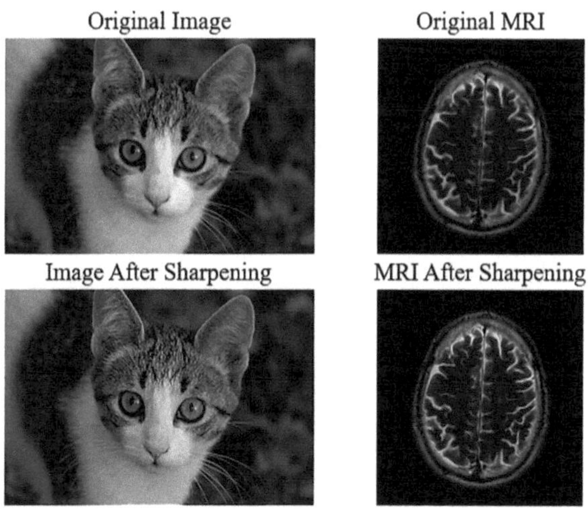

Fig. 3.4 Sharpening a digital photo and an MRI. (Source of MRI: fastMRI dataset [293]) using an Ideal Highpass Filter (IHPF)

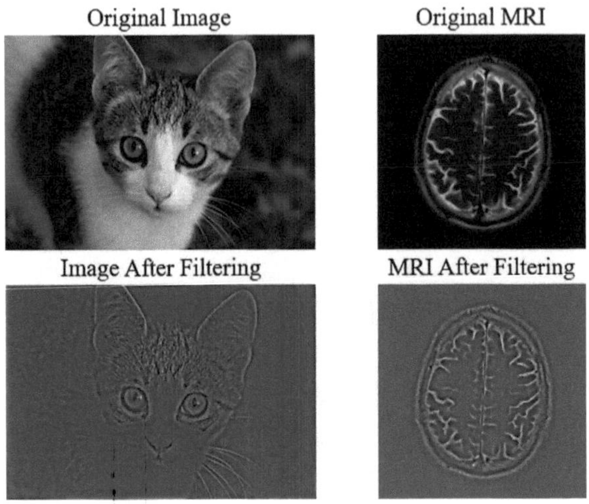

Fig. 3.5 Applying a Butterworth Highpass Filter (BHPF) on a digital photo and an MRI. (Source of MRI: fastMRI dataset [293])

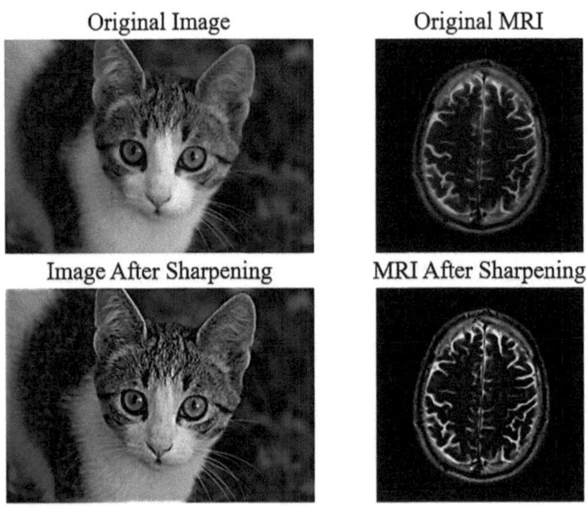

Fig. 3.6 Sharpening a digital photo and an MRI. (Source of MRI: fastMRI dataset [293]) using a Butterworth Highpass Filter (BHPF)

$$g(x, y) = f(x, y) - \nabla^2 f(x, y) \tag{3.31}$$

High-boost Filter: Unsharp masking is a technique of image sharpening where the sharpened image is obtained by subtracting a blurred version of the image from the original image:

$$f_s(x, y) = f(x, y) - \bar{f}(x, y) \tag{3.32}$$

where $\bar{f}(x, y)$ is a blurred version of the image $f(x, y)$. High-boost filter is a generalised form of unsharp masking:

$$f_{hb}(x, y) = Af(x, y) - \bar{f}(x, y) \tag{3.33}$$

where A is a scaling factor with the value ≥ 1 (Figs. 3.7, 3.8, 3.9, 3.10).

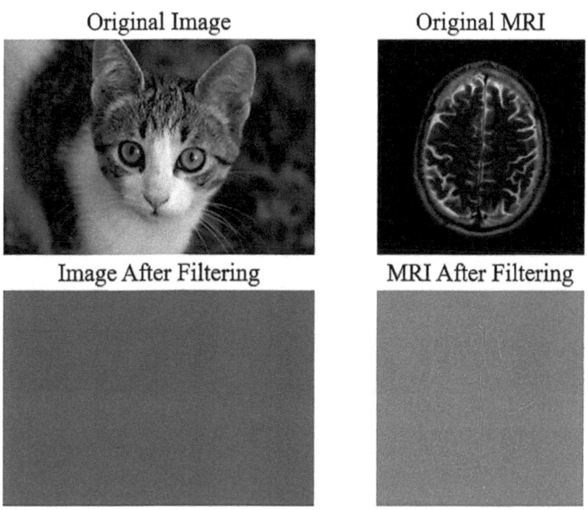

Fig. 3.7 Applying a Laplacian Filter on a digital photo and an MRI. (Source of MRI: fastMRI dataset [293])

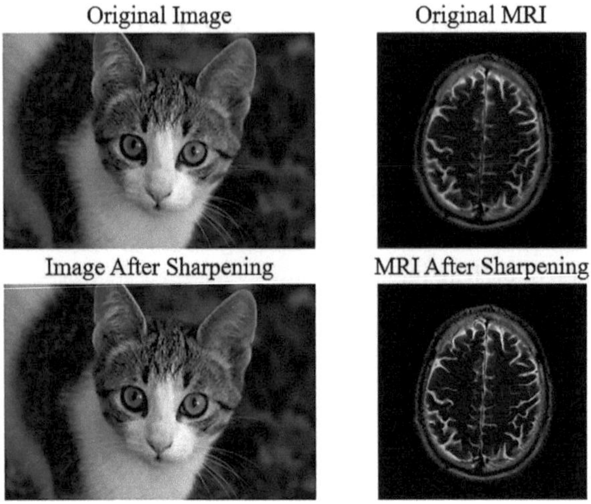

Fig. 3.8 Sharpening a digital photo and an MRI. (Source of MRI: fastMRI dataset [293]) using a Laplacian Filter

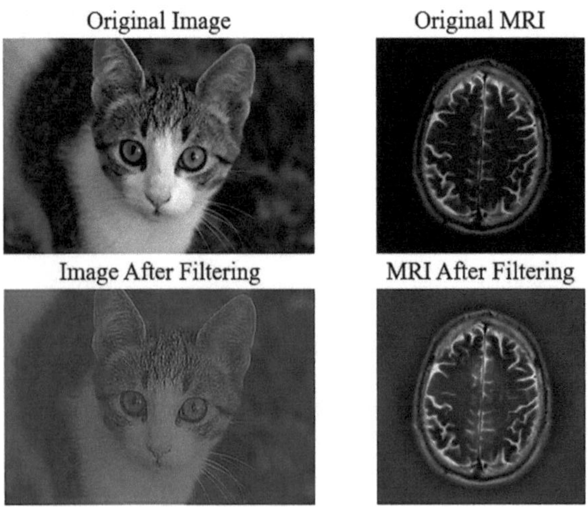

Fig. 3.9 Applying a High-boost Filter on a digital photo and an MRI. (Source of MRI: fastMRI dataset [293])

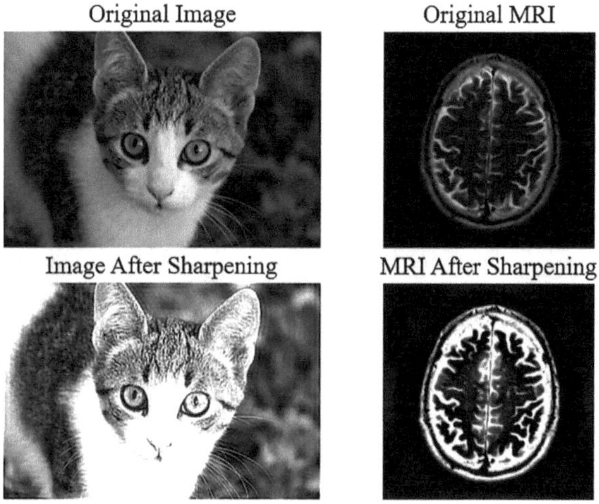

Fig. 3.10 Sharpening a digital photo and an MRI. (Source of MRI: fastMRI dataset [293]) using a High-boost Filter

3.3.2 Denoising

Lowpass Filter: Edges and other sharp transitions (such as noise) in the grey levels in the spatial domain of the image correspond to the high-frequency components in the frequency domain. Hence, attenuating high-frequency components in the frequency domain can smooth the image - this is known as the lowpass filter. The transfer function $H(x, y)$ of a 2D Ideal Lowpass Filter (ILPF) satisfies the relationship:

$$H(x, y) = \begin{cases} 1 & \text{if } D(x, y) \leqslant D_0 \\ 0 & \text{if } D(x, y) > D_0 \end{cases} \tag{3.34}$$

where D_0 is a non-negative cutoff distance from the origin of the frequency plane and $D(x, y)$ is given by Eq. 3.25. Another lowpass filter is the Butterworth Lowpass Filter (BLPF). The transfer function of BLPF of order n is given by:

$$H(x, y) = \frac{1}{1 + [D(x, y)/D_0]^{2n}} \tag{3.35}$$

These filterings are performed using their respective transfer functions (Eq. 3.34 or Eq. 3.35) in the frequency domain following Eq. 3.27 or in the spatial domain using Eq. 3.28 (Figs. 3.11, 3.12, 3.13).

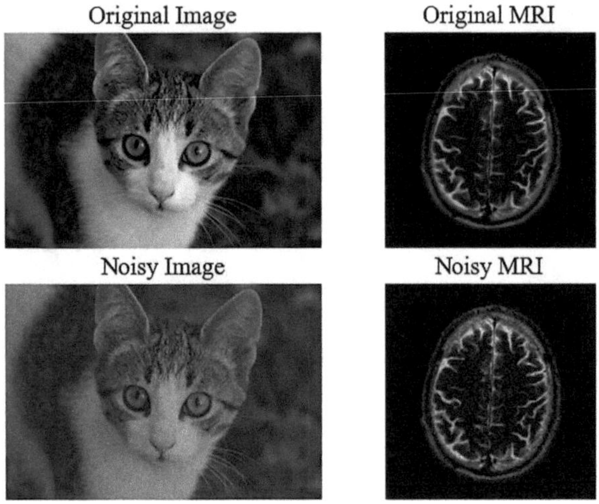

Fig. 3.11 Adding Gaussian noise to a digital photo and an MRI. (Source of MRI: fastMRI dataset [293])—these noisy images have been used for all the subsequent figures related to denoising

Anisotropic diffusion: The Perona-Malik equation [30] has been widely used in image processing for various tasks, such as image segmentation and edge detection, as well as for image enhancement and denoising [53]. Perona-Malik diffusion, commonly known as the anisotropic diffusion, is given for the input image $I(\cdot, 0)$ and the filtered image $I(\cdot, t)$ by:

$$\frac{\partial I}{\partial t} = \operatorname{div}\left(c(x, y, t)\nabla I\right) = \nabla c \cdot \nabla I + c(x, y, t)\Delta I \tag{3.36}$$

where div) is the divergence operator, $c(x, y, t)$ is the diffusion coefficient, ∇ denotes the gradient, and Δ denotes the Laplacian operator (Eq. 3.29). Smoothed image can be obtained with $t > 1$, while t is determining the amount of blur (Fig. 3.14).

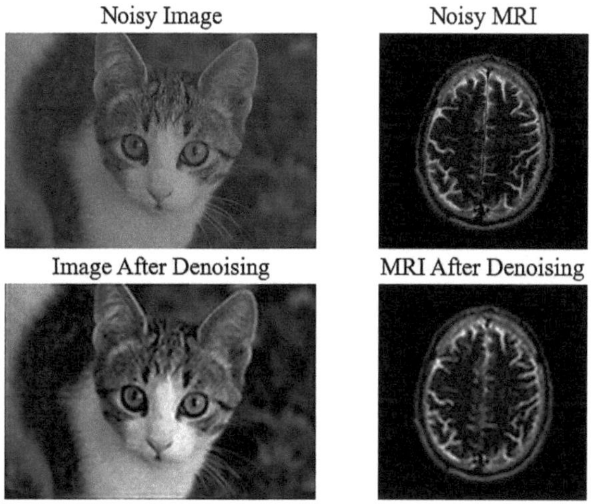

Fig. 3.12 Denoising a digital photo and an MRI. (Source of MRI: fastMRI dataset [293]) using an Ideal Lowpass Filter (ILPF)

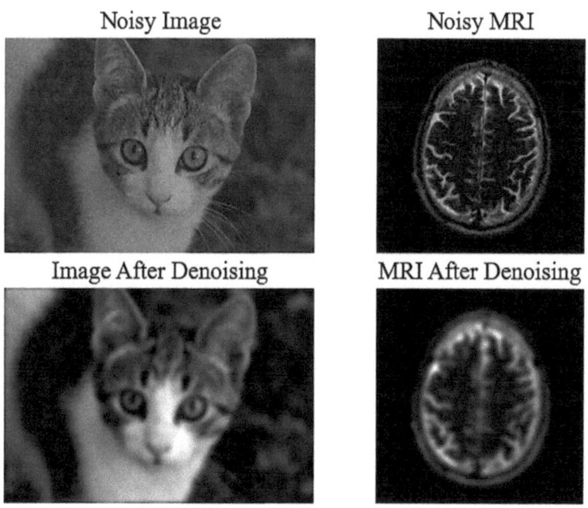

Fig. 3.13 Denoising a digital photo and an MRI. (Source of MRI: fastMRI dataset [293]) using a Butterworth Lowpass Filter (BLPF)

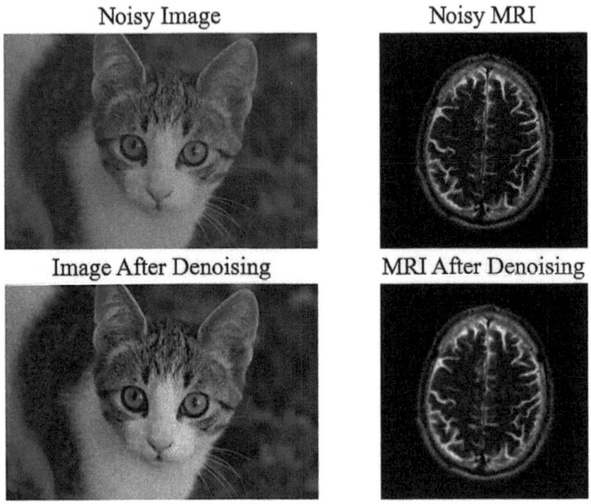

Fig. 3.14 Denoising a digital photo and an MRI. (Source of MRI: fastMRI dataset [293]) using Anisotropic diffusion

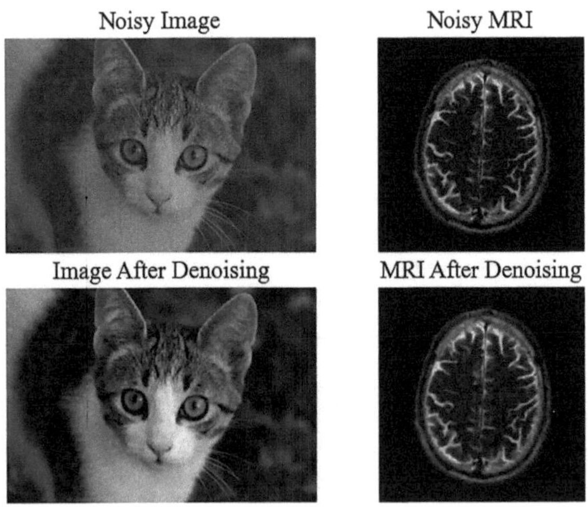

Fig. 3.15 Denoising a digital photo and an MRI. (Source of MRI: fastMRI dataset [293]) using Wavelet filtering

Wavelet filtering: In the wavelet domain, the noise is usually evenly spread throughout the wavelet coefficients, whereas the object is represented by some large coefficients. Hence, by filtering in the wavelet domain, the noise can be removed while preserving the features. This can be achieved by thresholding the coefficients [50]. This technique has widely been used in the field of MRI (e.g. L1 Wavelet regularised reconstruction [95]) (Fig. 3.15).

Filtering using Convolutional Neural Network: Eq. 3.28 has shown the relationship of the convolution operator with image filtering operations (denoising and sharpening). With the help of machine learning techniques, to be precise, with the help of Convolutional Neural Network (CNN), the optimal filter $h(x, y)$ can be learnt for a certain given task [288, 302]. Moreover, the wavelet thresholding might also be improved with the help of CNNs [275]. CNN has been further discussed in Sect. 4.6.1.1.

Neural Networks

4

From the day a child is born, they start learning from experiences and examples—either self-reliantly or taught by its parents. Learning is what makes it smart, what makes it a human being. A human being cannot simply function based on a set of rules—life is too complicated for that, but functions with the help of whatever it has learnt throughout its life. Researchers took a clue from this to start developing intelligent machines/algorithms: by teaching them with examples. For solving complicated tasks, it is not always possible to create algorithms with a fixed set of rules; instead, it is easier to provide examples of the task, including the solution, and then ask the algorithm to learn from it—to figure out the path from problem to solution. This is known as Machine Learning (ML), and the described scenario here is precisely called **supervised learning**. To solve a task using supervised learning, pairs of input data and their corresponding correct answers, known as the ground-truths, are supplied to a certain algorithm, and the algorithm tries to figure out the relationship between them— "learns" from examples. Mathematicians or statisticians might not agree with the term "machine learning"; rather, they might prefer to call it fitting a statistical model to a dataset or simply data-fitting. Whichever term is used, these algorithms have demonstrated their abilities to solve complex problems over the years. A rudimentary example of a supervised learning task can be to classify a given image—whether it is of a dog or of a cat. For the same, many images with the correct labels (dog or cat) can be supplied as pairs to an algorithm, and once it has learnt properly, it should be able to classify a supplied unseen image.

Providing input-groundtruth pairs might not be possible for every task. So, the algorithm might be just given some input and is then driven to reach a certain goal. For example, it divides a collection of images into five groups based on their similarities. In this case, there will not be any specific label provided for each image—instead, the algorithm is asked to find out which images are related. This is known as **unsupervised learning**. It is noteworthy that there are other types of learn-

S. Chatterjee, *Deep Learning Innovations in MRI Reconstruction and Analysis*, https://doi.org/10.1007/978-3-658-50741-1_4

ing techniques as well, such as **semi-supervised learning** or **weakly-supervised learning**—a combination of supervised and unsupervised learning—where one set of examples are provided to the algorithm in a supervised manner and another set of examples in an unsupervised manner. There are also **few-shot learning** and **one-shot learning**, which are techniques to learn from only a small set of examples or even with a single example, respectively.

4.1 Machine Learning to Deep Learning: A Brief History

The journey of Machine Learning started with the discovery of various statistical methods. The commencement of the journey can be attributed to the work of Thomas Bayes: "An Essay towards solving a Problem in the Doctrine of Chances" [1], published in 1763, which was then later extended by Pierre-Simon Laplace to be the Bayes' Theorem in his 1812 publication "Théorie Analytique des Probabilités" (Analytical Probability Theory) [2]. In 1805, Adrien-Marie Legendre came up with the "méthode des moindres carrés" or the least squares method. Moving forward more than a century, in 1913 Andrey Markov invented a technique to analyse poems—which later became widely known as the Markov Chain.

The concept of artificial intelligence started in 1950 by Alan Turing when he proposed a "learning machine" that could "learn" and become artificially intelligent. He also proposed a method to judge a machine's abilities to exhibit intelligence similar to a human being—when a machine can fool a human interrogator into thinking that it is a human and not a machine—commonly known as the Turing Test, proposed initially as the imitation game. In 1951, Marvin Lee Minsky designed the Stochastic Neural Analogue Reinforcement Calculator or SNARC—the first neural network machine. Arthur Samuel, in 1952, developed a computer program that could play checkers, and in 1963 Donald Michie created a machine that could play Tic-Tac-Toe using reinforcement learning.

Frank Rosenblatt proposed the perceptron in 1958 [11], which took two inputs and provided an output. By connecting and stacking multiple perceptrons together, in 1989, Hornik et al. proposed a more complex model capable of handling more complex tasks—this is known as a Multilayer perceptron (MLP) [27]. When the perceptrons of an MLP are connected to other perceptrons of the same model through some non-linear function, known as an activation function, this model can be a universal function approximator [27]. This whole idea was inspired by the powerhouse of human intelligence—the human brain. The most fundamental component of a human brain is a neurone, which is then connected to other neurones through axons and dendrites— forming a neural network. Because of this analogy, an MLP is more commonly known as an artificial neural network (ANN). Hence, all the individual perceptrons of an MLP are known as artificial neurones—or simply neurones.

4.2 Multilayer Perceptron—A Biologically Inspired Mathematical Model

The journey of the neural network, and, ipso facto, deep learning, started with Frank Rosenblatt proposing the first idea of a "Perceptron" [11, 13]. The motivation behind the creation of the perceptron, commonly referred to as an artificial neurone, was to mimic some parts of the real biological neurone—the building block of the human brain, like dendrites, cell bodies, and axons, using a simple mathematical model. Fig. 4.1 shows a comparison of a biological neurone and a perceptron. Unlike now, the original perceptron was intended to be a custom-built hardware rather than a software programme. The initial implementation, however, was a software one, for IBM 704, and followed by that, custom hardware was built for the task of image recognition—called the Mark 1 perceptron.

A perceptron, also known as an artificial neurone, consists of a summing junction—which performs a weighted sum of the different inputs, followed by a step function—to generate the final output.

The perceptron (Fig. 4.2) can be explained with the following equation:

$$\hat{y}\left(x_1, \ldots, x_n\right) = \begin{cases} 1 & \text{if } w_0 + w_1 x_1 + \cdots + w_n x_n > 0 \\ -1 & \text{otherwise} \end{cases} \tag{4.1}$$

where x_1 to x_n are the inputs, w_0 to w_n are the weights—where w_0 is the bias term, and $\hat{y}$ is the output of the perceptron. Due to the presence of the step function at the end, the output of the perceptron will be 1 if the weighted sum is more than 0; otherwise, -1. This can also be represented using vector notation as:

$$\hat{y}(\mathbf{x}) = \begin{cases} 1 & \text{if } \mathbf{w} \cdot \mathbf{x} > 0 \\ -1 & \text{otherwise} \end{cases} \tag{4.2}$$

where $\mathbf{x}$ is the input vector, and $\mathbf{w}$ is the weight vector—collection of all the neurones.

The equation of the neurone can also be written as:

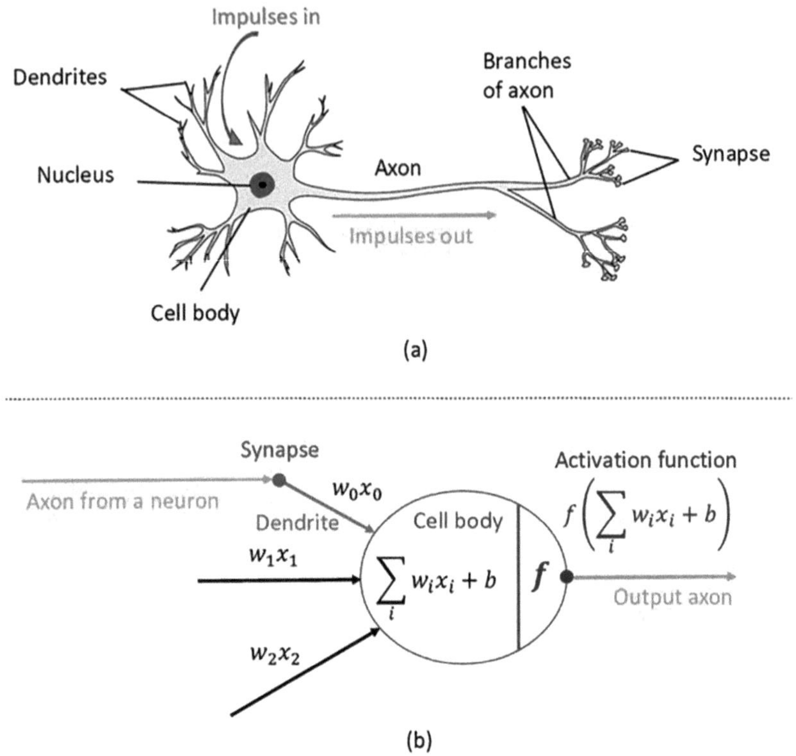

Fig. 4.1 Comparison of a biological neurone and an artificial neurone. (Illustration by Giorgio Roffo [241])

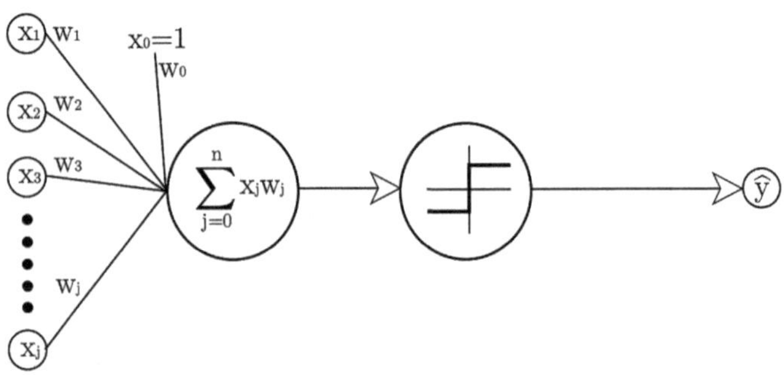

Fig. 4.2 A Perceptron

$$\hat{y} = \phi\left(\sum_{j=1}^{n} w_j x_j + b\right) \quad (4.3)$$

where x_j are the inputs, w_j are the weights, b is the bias term (w_0 in Eq. 4.1), ϕ is the activation function—a step function in case of a simple perceptron like here, and finally, $\hat{y}$ is the output. This can be simplified using vector notation as:

$$\hat{y} = \phi\left(\mathbf{Wx} + \mathbf{b}\right) \quad (4.4)$$

One big limitation of perceptrons is that they can only learn linearly separable functions. The paper [27] has shown that a network of perceptrons—multilayer feedforward networks or also known as Multilayer perceptron (MLP) are universal function approximators—which means, given a sufficient number of neurones, any Borel measurable function can be represented with them. An MLP tries to mimic the biological neural network—the interconnectivity of neurones—the building block of intelligence.

Fig. 4.3 shows a simple 3-layer MLP, with input-hidden-output layers. By increasing the number of hidden layers and the number of neurones in each layer, the complexity of the network can be increased—making it possible to learn complex relationships between input and output. This 3-layer MLP can be represented as:

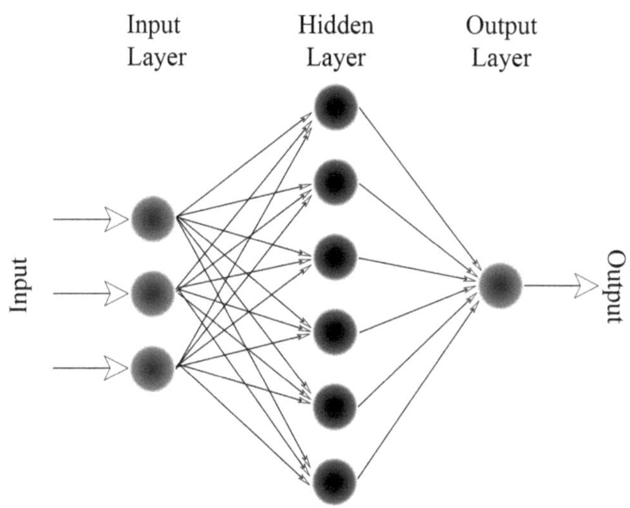

Fig. 4.3 Multilayer perceptron (MLP)

$$\begin{aligned} \mathbf{i} &= \phi_i\left(\mathbf{W_i x} + \mathbf{b_i}\right) \\ \mathbf{h} &= \phi_h\left(\mathbf{W}_h \mathbf{i} + \mathbf{b_h}\right) \\ \hat{y} &= \phi_o\left(\mathbf{W}_o \mathbf{h} + \mathbf{b_o}\right) \end{aligned} \tag{4.5}$$

where **i** and **h** are the output vectors of input and hidden layers respectively, $\hat{y}$ is the final output, $\mathbf{W_i}$, $\mathbf{W_h}$, $\mathbf{W_o}$ represent the weight vectors and $\mathbf{b_i}$, $\mathbf{b_h}$, $\mathbf{b_o}$ represent the bias vectors, ϕ_i, ϕ_h, ϕ_o represent the activation functions—for the input, hidden, and output layers, respectively. To increase the complexity of the model, the number of hidden layers can be increased—this is known as going **deep**. For the same, the **h** in Eq. 4.5 has to be replaced with:

$$\begin{aligned} \mathbf{h_0} &= \phi_{h0}\left(\mathbf{W_{h_0} i} + \mathbf{b_{h_0}}\right) \\ \mathbf{h_n} &= \phi_{h_n}\left(\mathbf{W_{h_n} h_{n-1}} + \mathbf{b_{h_0}}\right) \end{aligned} \tag{4.6}$$

where n is the number of hidden layers. Then this $\mathbf{h_n}$ is supplied to the equation of the output layer in Eq. 4.5.

It is noteworthy that the network's complexity can also be increased by increasing the number of neurones in each layer of the network— this is known as going **wide**.

4.2.1 Activation Functions

In a typical perceptron, a step function is used as the activation function ϕ. But that is not it. There are several activation functions in existence to introduce non-linearity in the network. The universal approximation theorem states that if the activation function is non-linear, a 2-layer neural network can be a universal function approximator [27, 36].

Sigmoid and Hyperbolic Tangent (TanH) are two of the oldest activation functions used in neural networks, which provide output in the range of (0, 1) and (−1, 1), respectively. Given x as input, Eq. 4.7 and 4.8 represent sigmoid and TanH, respectively.

$$\phi(x) = \frac{1}{1 + e^{-x}} \tag{4.7}$$

$$\phi(x) = \frac{e^x - e^{-x}}{e^x + e^{-x}} \tag{4.8}$$

Nair and Hinton (2010) [117] introduced a new activation function known as the Rectified Linear Unit or ReLU, which later became the most widely used activation function in the field of neural networks—because of sparsity and the reduced likelihood of the vanishing gradient, along with a lower computational load than sigmoid and hyperbolic tangent. This is defined as:

$$\phi(x) = \begin{cases} 0 & \text{if } x \leq 0 \\ x & \text{if } x > 0 \end{cases} \tag{4.9}$$

As can be observed from the equation, the output range of a ReLU is $(0, \infty)$.

Even though it is the most popular activation function, it can suffer from the dying ReLU problem. ReLU introduces sparsity, but it also is a problem. Neurones with large negative values might never recover during training because the ReLU's output for negative is 0, and the gradient (discussed in the next section) of 0 is 0. There are variations of ReLU, like Leaky ReLU [152] and Exponential Linear Unit (ELU) [178], are given by:

$$\phi(x) = \begin{cases} \alpha x & \text{if } x < 0 \\ x & \text{if } x \geq 0 \end{cases} \tag{4.10}$$

$$\phi(x) = \begin{cases} \alpha \left(e^x - 1\right) & \text{if } x \leq 0 \\ x & \text{if } x > 0 \end{cases} \tag{4.11}$$

In both equations, α is a pre-supplied parameter. This parameter can also be learnt during training, rather than pre-supplying a fixed one—which can improve the performance of the network. Such activations are known as Parameterised Rectified Linear Unit (PReLU) [179] and Parametric Exponential Linear Unit (PELU) [244].

There are also many other activation functions, such as Gaussian Error Linear Unit (GELU) [201], Scaled Exponential Linear Unit (SELU) [229], and Sigmoid-Weighted Linear Unit (SiLU) [261]. Fig. 4.4 illustrates the graphical profiles of several widely utilised non-linear activation functions.

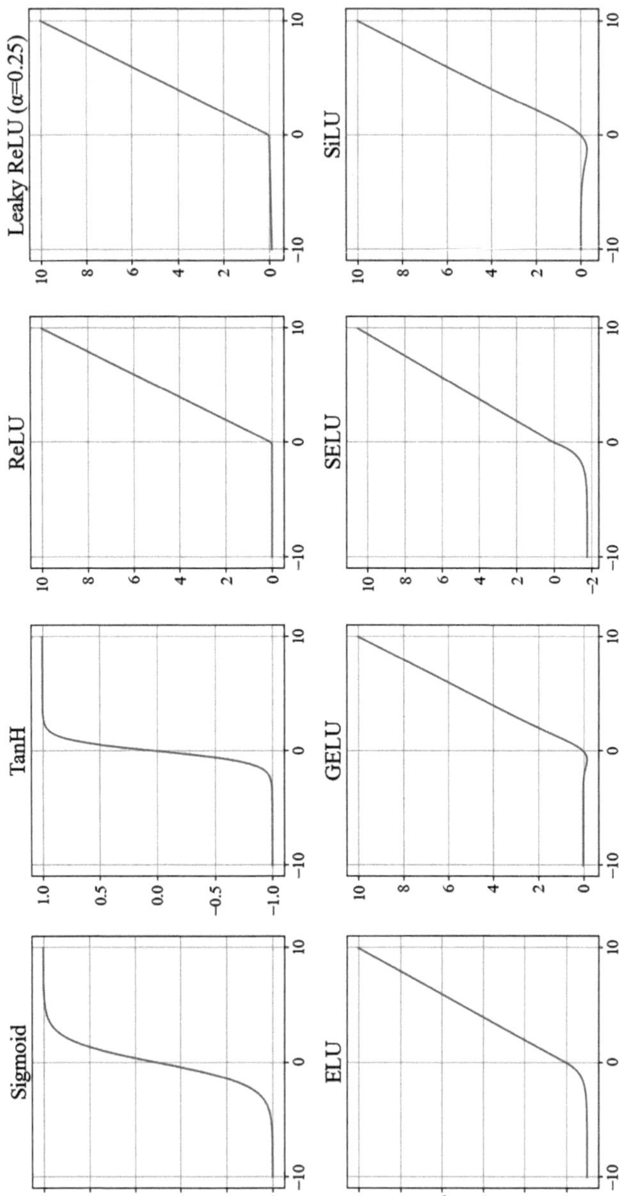

Fig. 4.4 Few examples of activation functions

4.3 Let's Learn: Backpropagation

When input x is supplied to an MLP, it flows forward in the network (from the input layer, through the hidden layers, to the output layer) and finally produces the output $\hat{y}$. This is known as **forward propagation**, and the network is known as a feedforward network. Once the network produces an output $\hat{y}$, this is used to calculate the error $J(\boldsymbol{\theta})$ using a given error function J—also known as the loss function, where $\boldsymbol{\theta}$ is the set of parameters (in this case, $\mathbf{W}, \mathbf{b}$). One example of an error function in a supervised learning setup can be the mean-squared error to calculate the difference between the output $\hat{y}$ and the ground-truth y:

$$J(\boldsymbol{\theta}) = \frac{1}{n} \sum_{i=1}^{n} \left(Y_i - \hat{Y}_i\right)^2 \tag{4.12}$$

where n is the number of examples in the training dataset. This error function signifies the distance of the network's output from the desired output. In order to reduce this error, the network's parameters $(\mathbf{W}, \mathbf{b})$ have to be tuned—this is commonly known as "learning". This learning algorithm can be divided into two steps: computation of gradients and learning (optimising $\mathbf{W}, \mathbf{b}$) using those gradients.

To compute the gradients, information is flowed backwards in the network—from the error to the initial input layer. This is known as **back-propagation** [23]. This can be defined as $\nabla_{\mathbf{x}} f(\mathbf{x}, \mathbf{y})$, where f is an arbitrary function, x is a set of parameters for which the gradient has to be calculated, and y is a set of variables which will affect the result—but their corresponding gradients are not required to be computed. During neural network training, the gradient of the error function J is calculated with respect to the network parameters $\boldsymbol{\theta}$ as $\nabla_{\boldsymbol{\theta}} J(\boldsymbol{\theta})$. An MLP is composed of multiple functions as shown in Eq. 4.5. Derivatives have to be computed for each of these functions, starting from the error function. With the help of the chain rule of calculus, the derivative of a composite function can be calculated if the derivatives of its "child functions" are known. Hence, the backpropagation is started by computing the gradient of the error function, and then it is continued backwards in the network. In simple terms, if they are two related functions $y = g(x)$ and $z = f(g(x)) = f(y)$, the gradient of z with respect to x can be computed as:

$$\frac{dz}{dx} = \frac{dz}{dy}\frac{dy}{dx} \tag{4.13}$$

Now, if x, y, z are not scalar values, rather they are vectors where $\boldsymbol{x} \in \mathbb{R}^m$ and $\boldsymbol{y} \in \mathbb{R}^n$, this equation has to be extended. If g is a function that maps from $\mathbb{R}^m$ to $\mathbb{R}^n$ and f

maps from $\mathbb{R}^n$ to $\mathbb{R}$, then this equation can be expressed for the gradient computation of z with respect to each element of x, with the help of partial derivatives, as:

$$\frac{\partial z}{\partial x_i} = \sum_j \frac{\partial z}{\partial y_j} \frac{\partial y_j}{\partial x_i} \tag{4.14}$$

where x_i denotes each element of x, whereas y_j denotes each element of y. This can be expressed in vector notation as:

$$\nabla_{\boldsymbol{x}} z = \left(\frac{\partial \boldsymbol{y}}{\partial \boldsymbol{x}} \right)^{\top} \nabla_{\boldsymbol{y}} z \tag{4.15}$$

where $\frac{\partial \boldsymbol{y}}{\partial \boldsymbol{x}}$ is the Jacobian matrix of g. This is applied recursively to each of the functions present in the neural network [198].

Once the gradients are computed, they are utilised to optimise the weights and biases of the network—so that the error is minimised. Since the error function and the network are very complex functions —it is not possible to find the minima analytically. Hence, gradient descent is employed, which can minimise differentiable functions. Once the gradient of a function f with respect to parameters $\boldsymbol{\theta}$ has been computed, then the $\boldsymbol{\theta}$ is updated by moving in the direction of the negative of the gradient. The updated $\boldsymbol{\theta}$ can be obtained by:

$$\boldsymbol{\theta}_{n+1} = \boldsymbol{\theta}_n - \gamma \nabla f(\boldsymbol{\theta}_n) \tag{4.16}$$

where γ is the step size at each iteration while moving towards the minima—commonly known as the learning rate, for which $f(\boldsymbol{\theta}_{n+1}) < f(\boldsymbol{\theta}_n)$. This process is repeated until a local minima is reached or the gradients have converged—became smaller than a predefined threshold. Fig. 4.5 portrays a simple error minimisation process for a convex function. A is the initial state of $\boldsymbol{\theta}$ in the contour of the error function, where as B is the local minima. x, y, z axes of the plot signify the weights, biases, and error, respectively.

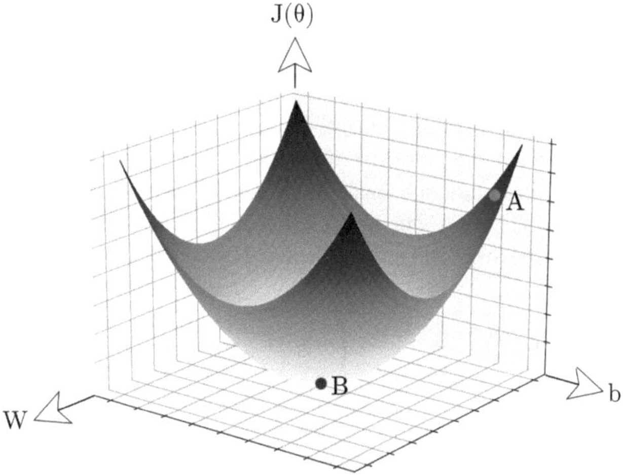

Fig. 4.5 Contour of a loss function J, where $J(\boldsymbol{\theta})$ is the loss with respect to the model parameters $\boldsymbol{\theta}$ comprising of the weights W and the biases b, A is the initial state and B is the local minima

4.4 Loss Function

The error function, or as it is commonly known, the loss function, is used to calculate how far the model's prediction is from the desired output, and then the backpropagation tries to reduce this during training to learn on the given training dataset—as explained in Sect. 4.3, denoted by J. The selection of this loss function depends on the task at hand. For supervised learning, Mean-Squared Error (MSE) is one of the most commonly used and versatile loss functions that can be used for classification tasks, as well as for image reconstruction tasks. The MSE loss between the model's prediction $\hat{y}$ and the ground-truth y with for the parameter set $\boldsymbol{\theta}$ can be calculated using Eq. 4.12.

4.4.1 Classification

In supervised learning for the task of classification, the Cross Entropy (CE) loss is a popular choice that can be computed as:

$$J(\boldsymbol{\theta}) = \frac{1}{N}\sum_{i=1}^{N} - \log \frac{\exp\left(\hat{y}_{i,y_i}\right)}{\sum_{c=1}^{C} \exp\left(\hat{y}_{i,c}\right)} \tag{4.17}$$

where N is the total number of elements in the training dataset (or, commonly, in the given batch of data—i.e., a subset of the training set) and C is the total number of classes. If there is a class imbalance in the training set, the performance of this loss can be improved by introducing a class weight term—forming the Weighted Cross Entropy (wCE) loss that can be calculated using the following formula:

$$J(\boldsymbol{\theta}) = \sum_{i=1}^{N} \frac{1}{\sum_{i=1}^{N} w_{y_i}} \left(-w_{y_i} \log \frac{\exp\left(\hat{y}_{i,y_i}\right)}{\sum_{c=1}^{C} \exp\left(\hat{y}_{i,c}\right)} \right) \tag{4.18}$$

where w represents the weight and w_{y_i} denotes the weight given to the class of the particular true label y_i. This weight can be calculated in several ways depending upon the problem at hand. One straightforward way is to calculate the weight based on the number of available elements in each class following the equation:

$$W_c = \left[1 - \left(\frac{samples_c}{\Sigma samples_t} \right) \right] \tag{4.19}$$

where W_c is the weight for class c, $samples_c$ is the number of samples of class c and $samples_t$ is the total number of samples in the training set (all classes combined).

For binary classification problems, the Binary Cross Entropy (BCE) can be used that can be calculated as:

$$J(\boldsymbol{\theta}) = \frac{1}{N}\sum_{i=1}^{N} - \left[y_i \cdot \log \hat{y}_i + (1 - y_i) \cdot \log\left(1 - \hat{y}_i\right) \right] \tag{4.20}$$

A more numerically stable version of this loss function is the BCE with Logits loss, which combines a Sigmoid layer σ with the BCE loss to take advantage of the log-sum-exp trick for numerical stability [url:4]. This is equivalent to applying a Sigmoid layer on the model's final output and then supplying that to the BCE loss function.

$$J(\boldsymbol{\theta}) = \frac{1}{N}\sum_{i=1}^{N} - \left[y_n \cdot \log \sigma\left(\hat{y}_n\right) + (1 - y_n) \cdot \log\left(1 - \sigma\left(\hat{y}_n\right)\right) \right] \tag{4.21}$$

4.4.2 Image Segmentation

For the image segmentation task also, CE loss can be used. But the extreme scarcity of foreground examples, especially during medical image segmentation, will make the model biased towards the background [377]. Dice loss can address this problem by balancing the background and the foreground and hence is one of the most popular loss functions for image segmentation. The Dice loss between a predicted image probability P and the ground-truth segmentation G for the model parameters $\boldsymbol{\theta}$ can be computed as:

$$J(\boldsymbol{\theta})_{P,G} = 1 - \frac{2\sum_{i=1}^{N} p_i g_i}{\sum_{i=1}^{N} p_i^2 + \sum_{i=1}^{N} g_i^2} \tag{4.22}$$

where $p_i \in P$ is the predicted probability of the i-th pixel or voxel and $g_i \in G$ is the ground-truth value of the same. This formulation of the Dice loss computes the loss during a binary segmentation task. For multi-class segmentation, Dice loss is computed for each class, and their mean is taken as the final loss value, as shown here:

$$J(\boldsymbol{\theta})_{P,G} = \frac{1}{C}\sum_{c=1}^{C} Dice_c \tag{4.23}$$

where $Dice_c$ is the Dice loss computed using Eq. 4.22 for class c and C is the total number of classes.

One limitation of the Dice loss is that it gives equal importance to the false positives and the false negatives, resulting in high precision but low sensitivity. To improve sensitivity in the case of highly imbalanced datasets with small Region Of Interest (ROI), more importance (i.e., weight) needs to be given to the false negatives than to the false positives [300]. The Tversky similarity index can provide this required flexibility and can be computed as:

$$TI_c = \frac{\sum_{i=1}^{N} p_{ic} g_{ic} + \epsilon}{\sum_{i=1}^{N} p_{ic} g_{ic} + \alpha \sum_{i=1}^{N} p_{i\bar{c}} g_{ic} + \beta \sum_{i=1}^{N} p_{ic} g_{i\bar{c}} + \epsilon} \tag{4.24}$$

where p_{ic} is the predicted probability of a pixel or voxel i belonging to class c and $p_{i\bar{c}}$ is the predicted probability of that pixel/voxel not belonging to class c, g_{ic} $g_{i\bar{c}}$ are similarly for the ground-truth labels—1 if i belongs to c or else 0 and vice versa, respectively. α and β are hyperparameters which can be adjusted to tune the network's sensitivity—especially when there is a class imbalance in the dataset, and finally, ϵ is to provide numerical stability—to avoid division by zero. This can then be used to formulate the Tversky loss function:

$$J(\boldsymbol{\theta})_{P,G} = \frac{1}{C}\sum_{c=1}^{C}(1 - TI_c) \tag{4.25}$$

Both these loss functions can have a significant negative impact if the ROIs are small as they will not be contributing significantly to the final loss value. To address this issue, [300] introduced an additional hyperparameter γ in their proposed Focal Tversky Loss (FTL) that controls the network's focus on the misclassifications. FTL is calculated using the following equation:

$$J(\boldsymbol{\theta})_{P,G} = \frac{1}{C}\sum_{c=1}^{C}(1 - TI_c)^{\frac{1}{\gamma}} \tag{4.26}$$

4.4.3 Image Reconstruction

When it comes to the task of image reconstruction, any of the image quality metrics discussed in Sect. 4.5.1 can be employed as the loss function—because the final loss of the model is the amount of image degradation in the prediction when compared against the ground-truth image. Mean-Squared Error (MSE), also known as the L2 loss, is also one of the most frequently used loss functions in this case, including for undersampled MR reconstruction [269]. For each image, pixel-wise or voxel-wise loss is calculated using the squared error, and then finally, the mean is taken to obtain the MSE loss for that particular image output. For a 2D image, this can be computed by adapting Eq. 4.12 as:

$$J(\boldsymbol{\theta})_{I,\hat{I}} = \frac{1}{mn}\sum_{i=0}^{m-1}\sum_{j=0}^{n-1}[y_{ij} - \hat{y}_{ij}]^2 \tag{4.27}$$

where I is the ground-truth image, $\hat{I}$ is the predicted image, m and n are the number of pixels in that image in height and width directions.

Even though it is widely used, MSE loss can result in blurry images as applying a blur filter will reduce the MSE and learning a blur filter might be easier for the network during gradient descent [url:10]. Mean Absolute Error (MAE), also known as the L1 loss, can produce better results from the perspective of a human observer. This can be computed using the following equation:

$$J(\boldsymbol{\theta})_{I,\hat{I}} = \frac{1}{mn} \sum_{i=0}^{m-1} \sum_{j=0}^{n-1} |y_{ij} - \hat{y}_{ij}| \tag{4.28}$$

Since medical images are mainly used for diagnosis, evaluation using perception-based metrics is more suitable than pixel-wise metrics. Structural Similarity Index (SSIM) [74, 99] is such a perception-based method that can be used to judge image quality in terms of the amount of degradation in visible structures. SSIM is based on the computations of luminance, contrast and structure terms between a given pair of images. The value of SSIM ranges between 0 and 1, where higher SSIM implies more image similarity. The negative of this SSIM value can be treated as the SSIM loss for training, given by:

$$J(\boldsymbol{\theta})_{I,\hat{I}} = -\frac{(2\mu_x\mu_y + C_1)(2\sigma_x y + C_2)}{(\mu_x^2 + \mu_y^2 + C_1)(\sigma_x^2 + \sigma_y^2 + C_2)} \tag{4.29}$$

The different components of this formulation is discussed in detailed later in Sect. 4.5.1.

4.4.3.1 Perceptual Loss

Even though SSIM is a perception-based metric, it still relies upon pixel-based comparisons of different terms. Perceptual loss [203] on the other hand, employs a pre-trained network as the Perceptual Loss Network (PLN) and uses that model to extract features from the given pair of images and has also demonstrated the ability to improve image quality perceptually, yielding superior results and reducing blurriness than classical pixel-based metrics such as L1 or L2 [196, 313]. A recent study from [296] presented that deep feature extractions, which were obtained from the trained network, could be utilised to deal with excessive blurry images and showed that perceptual similarity is the rising property that has been shared among deep visual representations. In case of perceptual loss, the predicted image $\hat{I}$ and the ground-truth image I are supplied as input to a pre-trained network known as the PLN to obtain features from the different layers of the network. Then these features obtained from $\hat{I}$ and I are compared using any of the earlier mentioned loss functions and then summed or averaged to obtain the final loss value.

4.5 Model Evaluation

Once a model is trained by optimising the loss function on a training set, the performance of the model is evaluated on a validation set—to check if the generalisation capabilities of the model (e.g., the model is overfitting or underfitting) and, most importantly—to check the accuracy of the model. When the model neither can learn properly on the training set nor can generalise well on new unseen data—the validation set, it is known as underfitting. In contrast, overfitting refers to the phenomenon when the model learns "too well" on the training set and fails to perform well on the unseen validation set (i.e., fails to generalise). The available dataset is typically split into training and validation sets randomly to avoid any human bias. However, there can still be unknown biases present in those splits, which were not recognised during the split. To avoid any such bias, multiple random splits of the same dataset are created, and the model is trained and tested multiple times using those splits. This is performed to judge how a model will generalise to an independent data set. This is known as the k-fold cross-validation, where k is the number of folds (i.e., the number of random splits). This improves the trust on an ML model as it shows how might a model might perform on unseen data later in production. Nevertheless, it might be challenging to perform such cross-validations when the training time of the model is very long. For the same, an alternate method for model evaluation can be employed—the hold-out method can be employed. In this case, the dataset is split into three subsets— training, validation, and testing sets. The training and validation sets are used for model selection, while the testing set is used for model evaluation. In this manner, the model is initially tested and selected on the validation set, then tested again for generalisability on the test set. For exhaustive evaluation, a combination of both k-fold cross-validation and hold-out testing can be performed, where the dataset is first randomly split into two subsets: training-validation and testing sets and then the training-validation set is randomly split into k folds of the training set and validation set. Another important aspect of model evaluation is the statistical significance test. When two models are compared and the quantitative evaluation show that one model is performing better than the other one, it is important to know whether or not that difference in performance is statistically significant or not. To do so, several statistical tests are available, one of which can be chosen based on the data and the problem at hand. Independent 2-sample t-test, paired t-test, Wilcoxon signed-rank test [6], and Mann-Whitney U-test [7] are to name of few of such tests.

Evaluations of the models are usually performed quantitatively and qualitatively. During the model evaluation following the strategies discussed earlier, evaluation is done employing quantitative metrics to compare the model's prediction against the available ground-truth. However, when it comes to models for the medical domain, the ultimate evaluation should always be done using clinicians. A few of the commonly used quantitative metrics are discussed here. The choice of the metrics depends on the task at hand. As The principal focus of this thesis is on image reconstruction models—undersampled image reconstruction and motion correction, the first set of metrics presented here are for this task. Then the metrics for classification and image segmentation tasks are furnished, as this thesis also presents automatic image processing pipelines dealing with image classification and segmentation tasks.

4.5.1 Image Reconstruction

Similar to the loss functions, one of the most common methods to evaluate the reconstructed images, is using the Mean-Squared Error (MSE):

$$MSE = \frac{1}{n}\sum_{i=1}^{n}(Y_i - \hat{Y}_i)^2 \tag{4.30}$$

where Y_i denotes the pixels of the ground-truth fully-sampled image, $\hat{Y}_i$ denotes the pixels of the undersampled image or the reconstruction (depending upon the comparison being performed), and n denotes the number of pixels in the image. A modified version of MSE, Root Mean-Squared Error (RMSE) can also be used for evaluating the image quality, which is simply the square root of MSE:

$$RMSE = \sqrt{MSE} \tag{4.31}$$

But, as the MSE does not have a fixed range (other than being limited by the intensity values), it might be difficult to comprehend the performance simply looking at a MSE value. For the same, Normalised Root Mean-Squared Error (NRMSE) can be employed following:

$$NRMSE = \frac{\sqrt{MSE} * \sqrt{N}}{||y||} \tag{4.32}$$

where $|| \cdot ||$ denotes the Frobenius norm, N is the number of elements in the data, and y is the ground-truth.

Another metric which can be used to evaluate the image quality, especially for super-resolution problems, is the Peak Signal to Noise Ratio (PSNR). This is the ratio between the maximum possible power of a signal and the power of corrupting noise that affects the fidelity of its representation. It is calculated using the MSE as:

$$PSNR = 10 \log_{10} \left(\frac{R^2}{MSE} \right) \tag{4.33}$$

where R is the maximum fluctuation in the input image.

Universal Quality Index (UQI) [64] attempts to evaluate the perceptual image quality by modelling the distortion as a combination of loss of correlation, luminance distortion, and contrast distortion using the following equation:

$$UQI = \frac{\sigma_{xy}}{\sigma_x \sigma_y} \cdot \frac{2\bar{x}\bar{y}}{(\bar{x})^2 + (\bar{y})^2} \cdot \frac{2\sigma_x \sigma_y}{\sigma_x^2 + \sigma_y^2} \tag{4.34}$$

where the first term is the correlation coefficient between the images x and y, the second term measures the closeness of the mean luminance between the images, and finally, the third term measures the contrast similarity between the images.

As discussed earlier in the Sect. 4.4, Structural Similarity Index (SSIM) [74, 99] is a perception-based method that is widely used to evaluate image quality, including for radiological images [240]. This considers image degradation as a perceived change in structural information and includes important perceptual parameters using luminance masking and contrast masking terms. SSIM is based on the computations of luminance, contrast and structure terms between image x and y, is computed using the following formula:

$$SSIM(x, y) = \frac{(2\mu_x \mu_y + C_1)(2\sigma_x y + C_2)}{(\mu_x^2 + \mu_y^2 + C_1)(\sigma_x^2 + \sigma_y^2 + C_2)} \tag{4.35}$$

where x and y are the two images between which the structural similarity is to be calculated, $\mu_x, \mu_y, \sigma_x, \sigma_y$ and σ_{xy} are the local means, standard deviations, and cross-covariance for images x and y, respectively. $c_1 = (k_1 L)^2$ and $c_2 = (k_2 L)^2$, where L is the dynamic range of the pixel-values, $k_1 = 0.01$ and $k_2 = 0.03$.

4.5.2 Classification

The most straightforward metric that is used for evaluating the classification performance of a model, is the accuracy—which is computed between the predicted values $\hat{y}$ and the ground-truth values y using:

$$\text{Accuracy}(y, \hat{y}) = \frac{1}{n_{\text{samples}}} \sum_{i=0}^{n_{\text{samples}}-1} 1 \left(\hat{y}_i = y_i \right) \tag{4.36}$$

where $1(x)$ is the indicator function. Accuracy can be computed in this manner for binary classification, as well as for multi-class or multi-label classification problems. For the latter cases, class-wise accuracy can also be computed.

Accuracy, however, does not provide an in-depth analysis of the result and can also be misleading, especially in the case of data imbalance. Hence, several other metrics are employed to get more insights regarding the model's performance—how good a model is in recognising the real negatives or how good a model is in recognising the real positives. These metrics are calculated using four different counts—the number of true positives TP, the number of true negatives TN, the number of false positives FP, and the number of false negatives FN.

The first metric is the precision ($\frac{TP}{TP+FP}$), which gives information regarding how many of the positively classified pixels were actually relevant. The next metric is sensitivity ($\frac{TP}{TP+FN}$), also known as hit rate or recall or true positive rate, it is a common metric used in classification to judge how good a classifier is at detecting the true positives. Specificity ($\frac{TN}{FP+TN}$), also known as true negative rate, provides information regarding the model's capability to avoid false positives.

One of the most commonly used metrics in classification problems is the F1-score, also known as the balanced F-score or F-measure, which is the harmonic mean of the precision and recall. This is calculated for problems as:

$$\text{F1} = 2 * (\text{precision} * \text{recall})/(\text{precision} + \text{recall}) \tag{4.37}$$

These metrics presented here are computed for binary classification problems. For multi-class or multi-label problems, averaging is performed. This averaging can be done in three ways: micro-computing the metric by globally counting TP TN FP FN, macro—by computing separate metrics for each class and then considering their unweighted mean as the final metric value, and weighted—by computing separate metric for each class and then considering their average weighted by support. The weighted average accounts for the class imbalance in the data and is usually preferred when there is a large class imbalance.

4.5.3 Image Segmentation

The most common metric which is used to quantitatively evaluate the segmentation performance of a model by comparing the results against the available ground-truth (typically manually segmented), is the Sßrensen-Dice coefficient or Dice Similarity Coefficient (DSC)—commonly known as the Dice score, given by the following equation for Boolean data:

$$\mathrm{DSC} = \frac{2TP}{2TP + FP + FN} \tag{4.38}$$

Another metric frequently applied for image segmentation tasks is the Jaccard index, also known as the Intersection over Union (IoU)—which provides less importance to the true positives (counts them once in the numerator and denominator), which is given by the following equation for binary segmentation tasks :

$$\mathrm{IoU} = \frac{TP}{TP + FP + FN} \tag{4.39}$$

Furthermore, as an image segmentation task can also be viewed as a pixel-level classification task, the classification metrics discussed earlier can also be used to evaluate the models' performance in terms of pixel-level classification.

4.6 Deep Learning

The term **Deep Learning (DL)** was introduced to the Machine Learning community by Rina Dechter in 1986 [21]. However, this term later in the 2010s became popular and analogous with the Deep Learning pioneering trio: Yoshua Bengio, Geoffrey Hinton, and Yann LeCun—who were awarded the Turing Award for the same. Even though it had a long history, initially, complex neural networks did not become popular because of the computational requirements. In 2004, it was shown that the neural networks could be considerably accelerated with the help of Graphical Processing Unit (GPU) [71] and in 2005, the importance of General-Purpose computing on Graphical Processing Unit (GPGPU) in the field of Machine Learning was shown [80]. With the development of fast and affordable GPUs, training deep complex models became easier, and they became more widely used Machine Learning techniques in various domains [188].

Deep learning is a family of ML methods based on representation learning. Deep-learning architectures are built by stacking multiple non-linear modules (layers), each of which transforms the input to the module into a more abstract representation [184]. Deep neural networks [101, 188], deep belief networks [106], deep reinforcement learning [185, 209], recurrent neural networks [188] and convolutional neural networks [49] architectures are part of the Deep Learning family.

4.6.1 Convolutional Neural Network

Just like a perceptron tries to mimic a biological neurone, and an MLP tries to imitate a biological neural network, a Convolutional Neural Network (CNN) tries to mimic the visual cortex of a brain. In the visual cortex, the receptive field size decreases from the IT to secondary visual cortices (V2, V4) to the primary visual cortex (V1). Consequently, V1 focuses on more abstract features like edges and lines, whereas IT focuses on the whole brain (illustrated in Fig. 4.6). A CNN tries to mimic this working mechanism of the visual cortex mathematically using pairs of convolutional layers and pooling layers, as shown in Fig. 4.7 [241, 390].

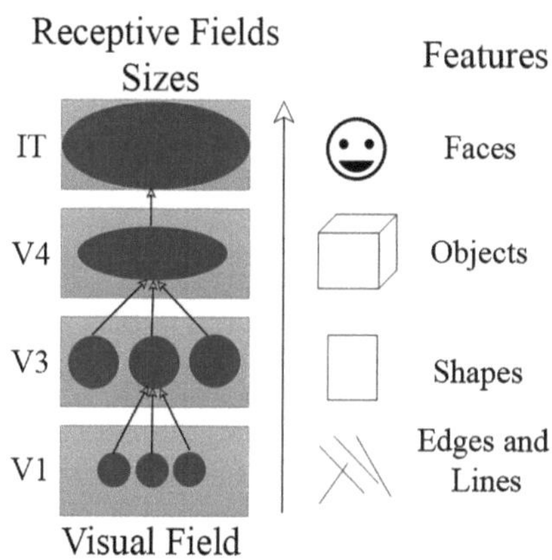

Fig. 4.6 Four Brodmann areas of the visual cortex of a brain and their respective focus on the different features of an image

CNNs were originally introduced by Yann LeCun in 1989, when he used the backpropagation algorithm to learn convolutional kernels for recognising handwritten numbers [28, 29]— called the **LeNet**—which became a foundation of modern computer vision. However, it is noteworthy that the origin of the CNNs can be traced back to 1979, but with a different name—Neocognitron—proposed by Kunihiko Fukushima [16, 18].

4.6.1.1 Convolution Layer

In simple words, CNN is a special type of neural network which employs convolution operation (in at least one layer) instead of matrix multiplication operation as the basic building block of the network. As explained in Sect. 3.1, convolution operation is

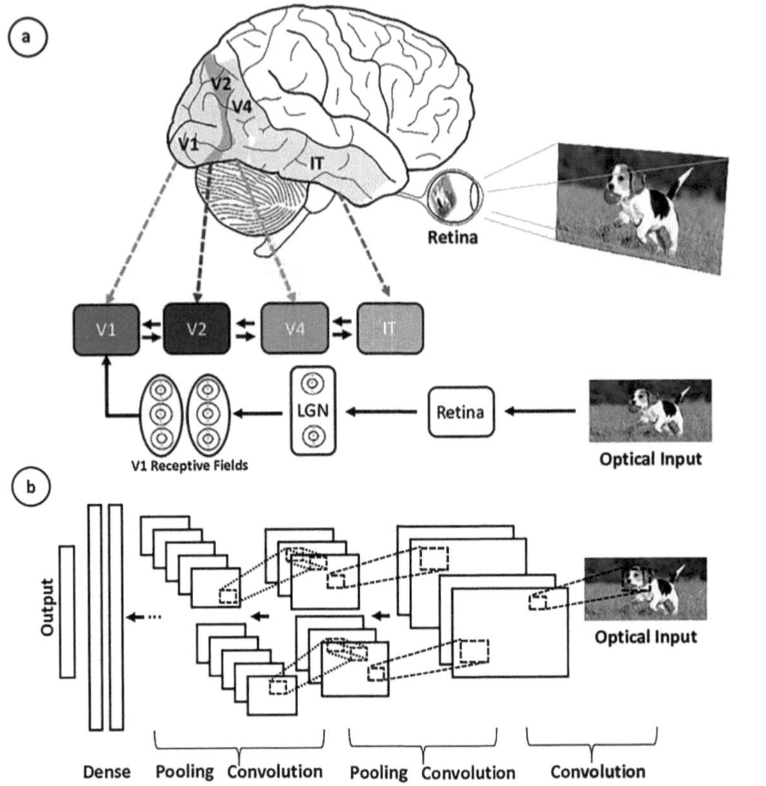

Fig. 4.7 Comparison of the visual cortex with a Convolutional Neural Network. (a) Four Brodmann areas. (Illustration by Giorgio Roffo [241])

given by the integral of the product of two functions, given by Eq. 3.1. In the regime of CNNs, f and g of Eq. 3.1 are referred to as the input (x) and kernel (w), respectively, and the output is referred to as the feature map or activation map, and the equation can be re-written as:

$$s(t) = (x \star w)(t) \tag{4.40}$$

This kernel w, which is called the weights in the general terminology of neural networks, is learnt during the training process. The standard form of implementation uses the flipped version of the discrete convolution operation, given by Eq. 3.3, as the range of valid values of a has less variation [198]. Moreover, it is also noteworthy that the standard version of convolution, which is implemented in most Machine Learning libraries, is without flipping the kernel ($t + a$ instead of $t - a$), which is mathematically called the cross-correlation operation [198, 327][url:5] and is given by:

$$s(t) = (w \star x)(t) = \sum_a x(t + a)w(a) \tag{4.41}$$

In the field of computer vision, the convolution operation is often used over more than one axis at a time—when the input I is a 2D image or 3D volume. In that case, typically, a 2D or 3D kernel K is employed, respectively (however, it is possible to use even a 1D kernel on 2D, 3D input or a 2D kernel on 3D input). The Eq. 4.41 is for 1D convolution and can be extended to 2D and 3D as:

$$\begin{gathered} S(i, j) = (K \star I)(i, j) = \sum_m \sum_n I(i + m, j + n)K(m, n) \\ S(i, j, k) = (K \star I)(i, j, k) = \sum_m \sum_n \sum_o I(i + m, j + n, k + o)K(m, n, o) \end{gathered} \tag{4.42}$$

Fig. 4.8 demonstrates the functionality of the convolution operation using a numerical example.

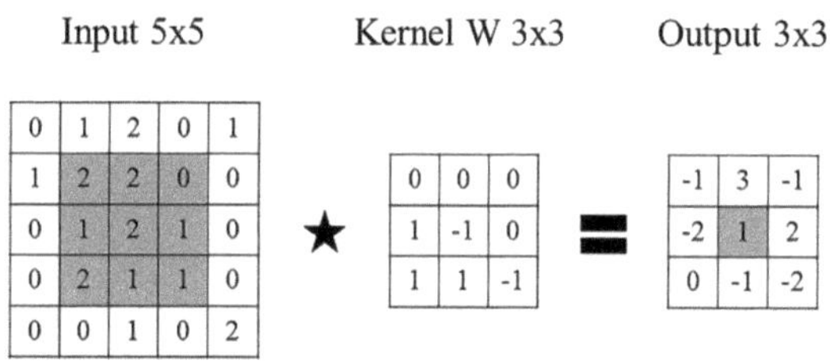

Fig. 4.8 An example convolution operation numerically

4.6.1.2 Pooling Layer

Pooling or downsampling layers help to obtain a more abstract representation of the feature map (output of a convolution layer) by reducing the spatial dimension. Pooling also makes the representation approximately invariant to small translations of the input—which is useful when the presence of a certain feature of the input has to be judged, rather than the exact location of the feature [198]. The first and most common pooling layer is the max pooling layer [25], which returns the maximum value within a rectangular neighbourhood— the size of which is defined by a kernel size parameter. There are several other types of pooling operations that can be used in a CNN architecture: average pooling—taking the average within a rectangular neighbourhood, weighted-average pooling—where the weight is calculated based on the distance from the centre, or even by taking the L^2 norm of a rectangular neighbourhood. Fig. 4.9 provides an example of the max pooling and the average pooling operations with the help of a numerical example.

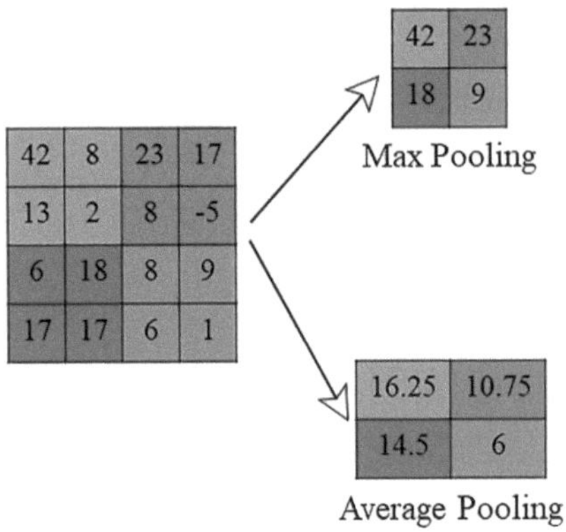

Fig. 4.9 Examples max-pool and average-pool operations

4.6.1.3 Architectures

Since the inception of the LeNet, numerous CNN architectures have been proposed over the years. With the publication of a publicly available large-scale benchmark dataset for image recognition task: ImageNet [104], benchmarking of the models became more accessible, and most of the early architectures were initially proposed by comparing their results on this dataset.

AlexNet: Dominance of CNNs for the task of image recognition started with AlexNet [140], shown in Fig. 4.10. This network won the ILSVRC (ImageNet Large-Scale Visual Recognition Challenge) 2012, achieving a top-5 test error rate of 15.3%. This network consists of eight layers in total: five convolution+pooling layers, followed by three fully-connected layers (MLPs). Data augmentation techniques and dropout [137] were employed to prevent the network from overfitting.

ReLU was used as the activation function in the inner layers, and softmax activation was used on the final layer to produce a distribution over the required 1000 class labels.

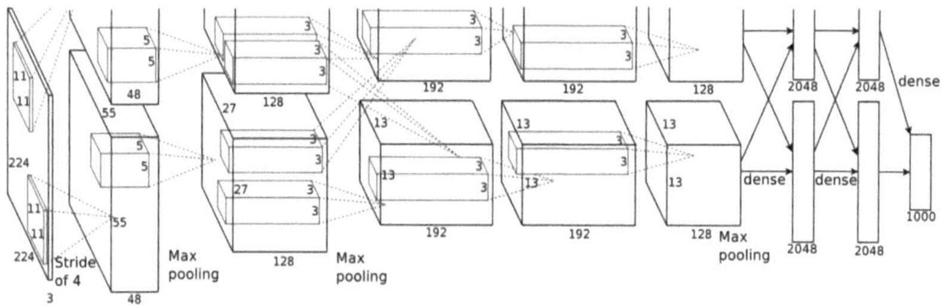

Fig. 4.10 AlexNet architecture [140]. (Illustration by Giorgio Roffo [241])

ZF Net and Deconvnet: A modified version of the AlexNet was used to win the ILSVRC 2013, known as the ZF Net [174]. The contribution of this paper was more than just winning the challenge with their network—employment of the deconvolutional network (Deconvnet) [132] as a visualisation technique to important insights into the working mechanism of the network—a step towards interpretability and explainability (discussed in Sect. 4.7).

VGG Net: Over the years, deeper and more complex models have been proposed. 2014 introduced VGG-19 in a paper which had "very deep convolutional networks" in the title [172]: 19 convolution layers (filters: 3x3, stride: 1, padding: 1) with 2x2 max-pooling layers with a stride of 2. Even though this model was not the winner of ILSVRC 2014, this model became one of the most popular models in existence and also became the inspiration for many other models later on.

ResNet: As the networks were becoming deeper—one problem became really challenging to handle: the vanishing gradient problem—where the amount of gradient received by the outermost layers is significantly small to none, which can prevent the network from learning. Scientists from Microsoft Research proposed a new idea in 2015—called the residual block, where they added an identity mapping x by adding the input to the block with the output of the block, shown in Fig. 4.11. With the help of this block, they created a network with 152 layers. They called this technique "Deep Residual Learning" and the network: ResNet [200]. Fig. 4.12 shows a comparison of a plain VGG network against a ResNet architecture. They won the ILSVRC 2015 by securing a 3.57% top-5 test error rate.

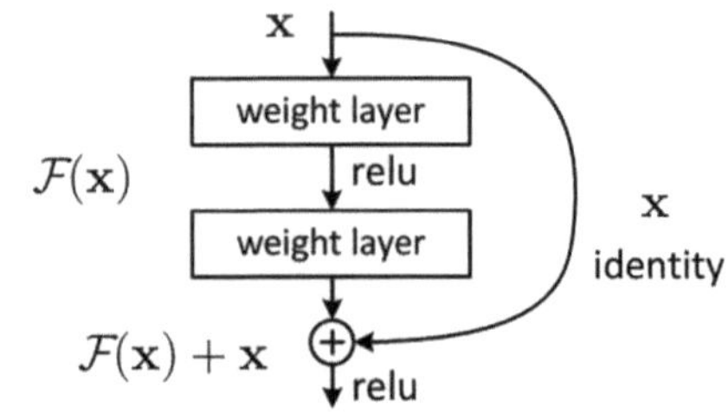

Fig. 4.11 Residual block. (Fig source: [200] © 2016 IEEE)

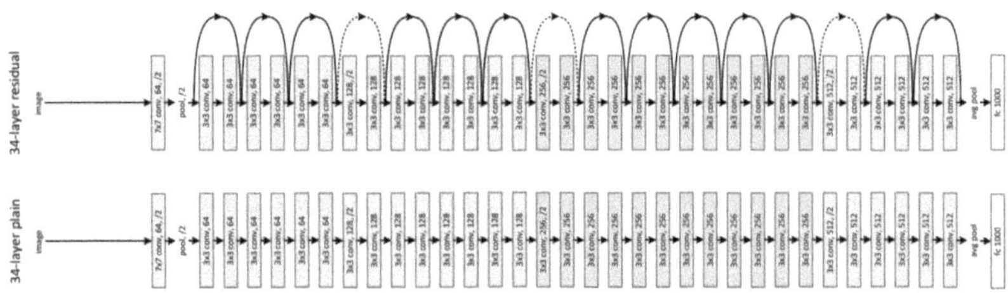

Fig. 4.12 VGG vs ResNet. (Fig source: [200] © 2016 IEEE)

UNet: Another network architecture proposed in 2015 that revolutionised the field of computer vision is UNet [186], shown in Fig. 4.13. Originally proposed for the task of biomedical image segmentation—UNet (and later, its variations) has been used for several image segmentation tasks, as well as for solving inverse problems like image reconstruction and super-resolution.

4.6.1.4 Complex-valued Convolution

The convolution layer discussed earlier (Sect. 4.6.1.1) is meant for real-valued data. A special type of convolution was proposed by Trabelsi et al. [286] in 2018 for complex-valued data. A complex number can be expressed as $a + ib$, where a and b are independent real numbers and i is the imaginary unit, which is: $i = \sqrt{-1}$. a and b are the real and imaginary parts of that complex number $a + ib$. In a complex-

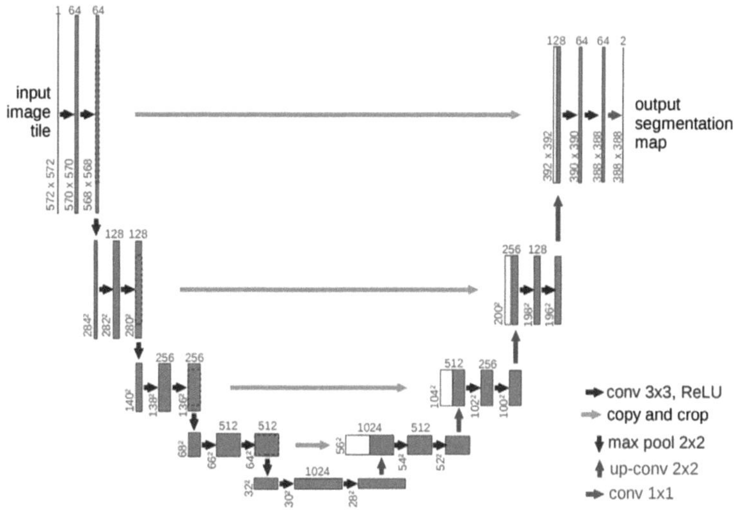

Fig. 4.13 UNet architecture. (Fig Source: [186])

valued convolution operation, both the kernel and the input are made of complex numbers. Hence, the kernel $K = K_r + iK_i$, the input $I = I_r + iI_i$, and the output $S = S_r + iS_i$. The real and imaginary parts of the output are obtained by applying the real and imaginary parts of the kernel on the real and imaginary parts of the input by modifying Eq. 4.41 (while also modifying the variables as: $K = w$ and $I = x$) as:

$$\begin{aligned} S_r(t) &= (K_r \star I_r)(t) - (K_i \star I_i)(t) \\ S_i(t) &= (K_i \star I_r)(t) + (K_r \star I_i)(t) \end{aligned} \tag{4.43}$$

This can also be written using matrix notation as:

$$\begin{bmatrix} \Re(\mathbf{K} \star \mathbf{I}) \\ \Im(\mathbf{K} \star \mathbf{I}) \end{bmatrix} = \begin{bmatrix} \mathbf{K_r} & -\mathbf{K_i} \\ \mathbf{K_i} & \mathbf{K_r} \end{bmatrix} \star \begin{bmatrix} \mathbf{I_r} \\ \mathbf{I_i} \end{bmatrix} \tag{4.44}$$

Similarly, Eq. 4.42 for 2D and 3D convolutions can also be extended as complex-valued convolution.

Activation Functions: A Complex-valued Convolutional Neural Network (CV-CNN) would also require specialised actuation functions. Several activation functions have been proposed to work in complex-valued scenarios. An ordinary ReLU activation function (Eq. 4.9) can be applied separately on the real and imaginary parts of the data, known as the Complex Rectified Linear Unit or $\mathbb{C}$ReLU, which can be expressed as:

$$\mathbb{C}\text{ReLU}(z) = \text{ReLU}(\Re(z)) + \text{ReLU}(\Im(z))i \tag{4.45}$$

where z is the complex-valued input to the activation function. A modified version of the ReLU activation function for complex-valued data, modReLU, was proposed by Arjovsky et al.[192], which can be given by:

$$\text{modReLU}(z) = \text{ReLU}(|z| + b)e^{i\phi(z)} = \begin{cases} (|z| + b)\frac{z}{|z|} & \text{if } |z| + b \geq 0 \\ 0 & \text{otherwise} \end{cases} \tag{4.46}$$

where $|z|$ and $\phi(z)$ are the magnitude and phase of z, respectively. Another form of complex ReLU was introduced by Guberman [199], known as the zReLU:

$$z\,\text{ReLU}(z) = \begin{cases} z & \text{if } \phi(z) \in [0, \pi/2] \\ 0 & \text{otherwise} \end{cases} \tag{4.47}$$

4.7 Blackbox Problem

With the increasing complexity of the Deep Learning models, one issue became more problematic—understanding the inner working mechanism and reasoning done by such model—known as the black-box problem of Deep Learning. Due to this, it might be difficult to trust the predictions of such black-box models in high-impact domains like medicine.

Interpretability vs Explainability: Interpretability and explainability methods aim to unravel the black-box nature of decision-making in Machine Learning and Deep Learning algorithms, commonly known as a model. Explainability focuses on explaining the internal working mechanisms of the model. In contrast, interpretability focuses on observing the effects of changes in model parameters and inputs on the model prediction, hence attributing properties to the input, and it also requires a context, considering the audience who is to interpret the properties and characteristics for the model outcomes [358]. It is to be noted that earlier work failed to draw a clear line of distinction between interpretability and explainability, as these terms are subjective and pertain to the stakeholders, considering the audience who must understand the model and its outcomes. Interpretability and explainability nurture a sense of human-machine trust as they help the users of the machine/deep-learning models understand how certain decisions are made by the model and are not limited to statistical metric-based approaches such as accuracy or precision. This, in turn, allows employing such models in mission-critical problems such as medicine, autonomous driving, legal systems, banking etc. The basis of measuring model transparency is said to comprise simulatability, decomposability, and algorithmic transparency [221]; and model functionality which consists of textual descriptions of the model's output and visualisations of the model's parameters. The goals to be attained through interpretability of models [358] are trust, reliability, robustness, fairness, privacy, and causality. Explainability can be formulated as the explanation of the decisions made internally by the model that in turn generates the observable, external conclusions arrived at by the model. This promotes human understanding of the model's internal mechanisms and rationale, which helps build user trust in the model [349]. The evaluation criteria for model explainability include comprehensibility by humans, fidelity and scalability of the extracted representations, and scalability and generality of the method [344].

Some models are transparent and hence inherently understandable, such as decision trees, K-nearest neighbours, rule-based and Bayesian models, but some are opaque and require post-hoc explanations. A post-hoc model explanation can be through model-agnostic and model-specific techniques that aim to explain a black-box model in human-understandable terms. Through interpretability and explainability, to some degree, the following goals can be achieved: trustworthiness, causality, confidence, fairness, informativeness, transferability and interactivity, which in turn can help improve the model [339].

Data scientists might generate interpretability-explainability results for their opaque models and present the model's reasoning or the model's inner mechanism to the domain experts (e.g., clinicians). If the reasoning is the same as a human domain expert would argue, then the experts can have more faith in that model, which implies that the model can be incorporated into real-life workflows. Apart from building trust in the models, interpretability and explainability techniques can be used by the data scientists to improve their models by discussing the outcomes with the domain experts and improving the model based on expert feedback. Moreover, accurate (verified by experts) interpretability-explainability results can be used as part of automatic or semi-automatic teaching programmes for trainees.

uncertainty: AI models are typically evaluated by comparing their output with the corresponding ground-truth. As most of the models suffer from the "black-box problem" of Deep Learning, evaluations on large test datasets with as many use-case scenarios as possible are performed to check the model's stability, generalisability, and applicability in clinical scenarios. Nevertheless, when used in production, there is no available ground-truth to compare and evaluate the predictions. Uncertainty quantification is a possible solution to this problem: evaluating the quality of prediction without comparing it against any ground-truth. For example, one of the systematic methods for evaluating DL-based image reconstruction models is creating a difference image by subtracting the predicted image from the ground-truth, followed by visual inspection to understand which regions of the image have been mispredicted by the network. It is one goal of this project to generate ground-truth-free images similar to the difference image with the help of uncertainty quantification.

Part II
Current Techniques

Summary
This part introduces the different State Of The Art (SOTA) technologies currently available for dealing with the artefacts in MRI (undersampling and motion) and presents the open questions. Moreover, this part also introduces different SOTA automatic MR image processing pipelines. The initial chapter presents the techniques currently employed for undersampled MRI reconstruction, which is divided into two main sections—non-deep learning based and deep learning based techniques. The section on non-DL based techniques presents parallel imaging and iteration reconstruction techniques. DL-based methods are then grouped into three different groups: methods working in the image space, methods working in the k-space, and finally, the methods working in a mix of image and k-space. The methods which work in the image space are further sub-divided based on their problem statements: artefact reduction and super-resolution. Finally this chapter presents some open questions regarding DL-based undersampled MRI reconstruction. The following chapter provides a concise view of the current motion correction techniques. Initially, a brief explanation of different techniques is provided that can be employed to track motion during MR acquisitions. Then the next two sections present the two main types of motion correction technologies—Prospective Motion Correction (PMC) and Retrospective Motion Correction (RMC). RMC techniques are further divided into non-DL based techniques and DL-based techniques. The final section of this chapter presents a few of the open questions related to motion correction with deep learning. The final chapter introduces four different automatic image processing tasks—brain tumour classification and segmentation, anomaly detection in brain MRIs, vessel segmentation, and image registration. Each of these sections introduces the task, furnishes several State Of The Art, and presents some of the open questions. Finally, a general set of open questions applicable to all these automatic image processing pipelines is provided.

Undersampled MRI Reconstruction 5

5.1 Non-Deep Learning Methods

5.1.1 Parallel Imaging

Parallel Imaging (PI) is one of the most popular techniques for speeding-up MRI acquisitions. In PI, an array of receiver coils is utilised to acquire signals from different regions of the object that is being scanned while also ignoring phase-encoding lines (for 3D imaging, partition-encoding steps can also be skipped) of the acquired k-space. The undersampling factor AF is the ratio between the amount of data acquired and the total amount of data in the corresponding fully-sampled k-space—the maximum of which is limited by the number of available receiver coils [134]. Each receiver coil has a different sensitivity to the different regions of the object—known as the sensitivity profiles or sensitivity maps, demonstrated in Fig. and 5.1 5.3. After acquiring these undersampled images, the images contain artefacts due to the violation of the Nyquist criterion 2.5.1. Special PI algorithms are then applied to the aliased images to obtain the final image. These algorithms can be divided into two groups based on their working mechanism—working with the aliased images (e.g. SENSE [52]) and synthesising missing k-space points (e.g. GRAPPA) [134, 225].

SENSE: **SENSE**tivity encoding (SENSE) [52] tries to "unfold" the undersampled images acquired using an array of coils with the help of the sensitivity maps. An example scenario of image reconstruction using SENSE for $AF = 2$ using four coils has been demonstrated here. Fig. 5.2 presents the basic idea of a coil profile and a coil image, while Fig. 5.3 shows the coil-image and the respective coil-profiles acquired around the object. For $AF = 2$, SENSE obtains the actual image pixel values I_A and I_B by solving Eq. 5.1 using linear algebra and matrix inversion. In

S. Chatterjee, *Deep Learning Innovations in MRI Reconstruction and Analysis*, https://doi.org/10.1007/978-3-658-50741-1_5

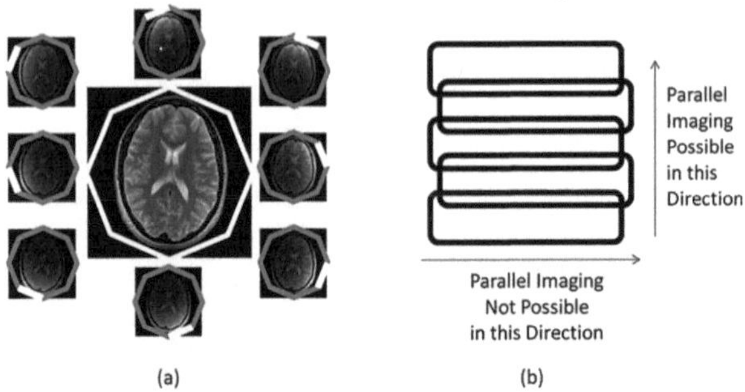

Fig. 5.1 Parallel imaging. (Fig source: [134])

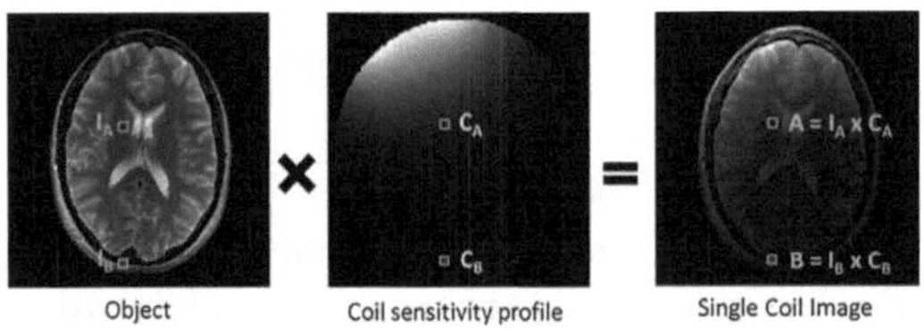

Fig. 5.2 Any given pixel in the final image (object), multiplied by the coil-profile, results in the coil-image. In parallel imaging, such coil images are acquired for the different regions of the object, and then the final pixel values of the object are obtained by computing the weighted sum of the coil-images by the coil-profiles. (Fig source: [134])

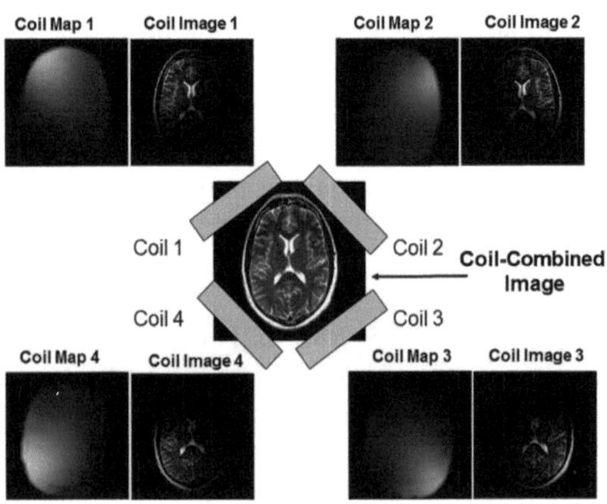

Fig. 5.3 Coil-images acquired by the different coils are combined to obtain the final image. Each coil has different sensitivities to different regions of the object. (Fig source: [225])

Eq. 5.1, F_n denotes the individual aliased pixel with respect to coil n—comprising of the coil-pixels for the location A and B, A_n and B_n, respectively, overlapped due to the reduced FOV, each of which are a multiplication of the real pixel value I_A and I_B with the coil-profiles at the location A and B—C_{An} and C_{Bn}.

$$F_n = A_n + B_n = I_{An} \times C_{An} + I_{Bn} \times C_{Bn} \tag{5.1}$$

PI techniques like SENSE are widely used in clinical applications, including for real-time and dynamic imaging [134, 225], but suffers from one major drawback of amplification of noise in the image. The drop in the SNR can be calculated as:

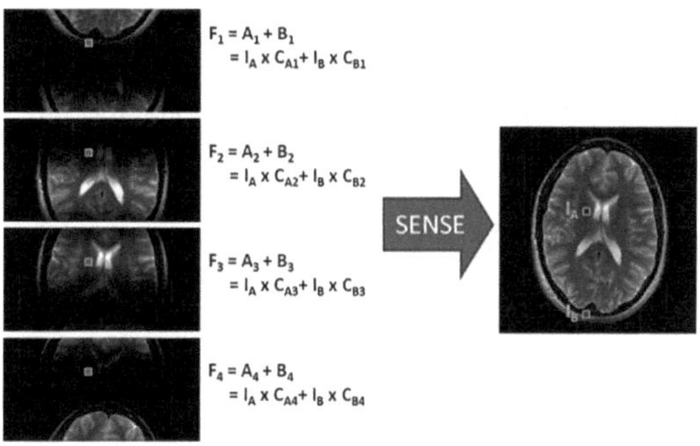

Fig. 5.4 A representative example of a SENSE reconstruction using 4-channel linear array for $AF = 2$, where the pixels I_A and I_B are aliased. F_n are the images from the individual coils—each of which are the weighted summation (A_n, B_n) of both the pixels, weighted by the sensitivity profiles ($I_A \times C_{An}$ and $I_B \times C_{Bn}$). (Fig source: [134])

$$SNR_{PI} = \frac{SNR_{full}}{g\sqrt{AF}} \tag{5.2}$$

where the SNR_{full} denotes the SNR of the fully-sampled image, g signifies the coil geometry factor, and AF denotes the acceleration factor (Fig. 5.4).

GRAPPA: Gene**R**alized **A**utocalibrating **P**artially **P**arallel **A**cquisition (GRAPPA) [62] is another PI technique which attempts to synthesise the missing phase-encoding lines of the undersampled k-space. The k-space points which are not acquired are synthesised (known as the target points) with the help of the acquired neighbouring points (known as the source points). Each source point from every coil is multiplied with a coefficient, known as the GRAPPA weight, and the summation of them is used to estimate the corresponding coil-combined target point—can be represented as follows:-

$$s_{\text{targ},i}\left(k_x, k_y + \Delta k_y\right) = \sum_{j=1}^{n_c} \sum_{\tau_x} \sum_{\tau_y} w\left(i, j, \tau_x, \tau_y\right) \cdot s_{src,j}\left(k_x + \tau_x, k_y + \tau_y\right) \tag{5.3}$$

where i represents the coil index for the target point, j represents the coil index for the source point, τ_x and τ_y correspond to the position of the source point within the GRAPPA kernel—which is the spatial arrangements of the source and target points, w is the GRAPPA weight for a given source point, k_x and k_y are the frequency and phase encoding respectively. The same GRAPPA weights can be applied to the whole k-space in case of the Cartesian acquisitions as they are shift-invariant to the first approximation. Additional lines are acquired at the centre of the k-space, known as the autocalibration signal (ACS) region, and utilised to calculate the GRAPPA weights. Afterwards, they are employed to estimate the missing k-space points (source points) [225]. Fig. 5.5 shows a schematic diagram of the GRAPPA reconstruction process.

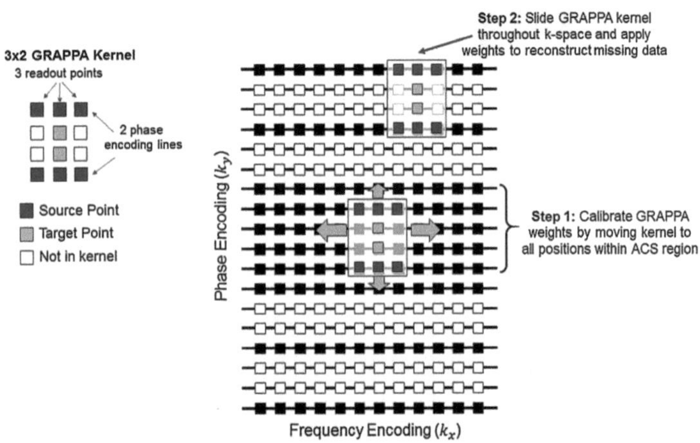

Fig. 5.5 Example of a GRAPPA reconstruction for $AF = 3$, using a 3×2 GRAPPA kernel constitutes of three readout points and two phase-encoding lines. Then the GRAPPA weights are estimated in Step 1 and applied to the k-space to obtain the missing points in Step 2. The source points are collected from all the coils to obtain the corresponding one coil-combined target point. (Fig source: [225])

GRAPPA is one of the most widely used PI techniques used in clinical routines [134, 225]. However, similar to SENSE, GRAPPA also suffers from a drop of SNR depending upon the GRAPPA g-factor and the AF [103]. Apart from SENSE and GRAPPA, there are several other parallel imaging techniques in existence for Cartesian trajectories, for example, SPIRiT [116] and its extension ESPIRiT [173] which build a bridge between GRAPPA and SENSE. Parallel imaging is not limited to Cartesian trajectories only—several approaches have been proposed for non-Cartesian trajectories as well, for example, Conjugate Gradient SENSE (CG-SENSE) [59], radial GRAPPA [129], and spiral GRAPPA [130].

5.1.2 Iterative Reconstruction

Iterative reconstruction techniques try to improve the reconstruction quality by iteratively optimising the reconstruction process by employing strong constraints that use prior information and assumptions about the acquired k-space. Compressed Sensing (CS) has been one of the most important developments for the task of accelerating MRIs [182, 242, 336], which exploits the redundancy present in the k-space. The three key components employed in CS-MRI are incoherent sampling, sparsifying transform, and non-linear reconstruction [95]. Incoherent sampling ensures that the resulting image does not contain a superposition of shifted copies of the original signal, like in the case of equidistant sampling. Random undersampling causes incoherent (noise-like) artefacts, similar to the additive noise in the image [95]—making it possible to be recovered using CS techniques. Then that sampled data needs to be converted to a domain where it is sparse—known as the sparsifying transform (Ψ). Finally, a non-linear construction technique is employed to find the final image. Mathematically, the final image can be obtained by minimising the following equation:

$$\min_{x} (\|S - Fx\|_2 + \lambda\|\Psi x\|_1) \tag{5.4}$$

where S is the acquired signal, F is the Fourier transformation operator, x is the reconstructed image, λ is the regularisation parameter—determines the trade-off between the data consistency and sparsity terms, Ψ is the sparsifying transform, and finally, $|...|_1$ and $|...|_2$ represent the absolute values and the squares of the absolute values, respectively [182]. The final solution is obtained using a constrained version of the reconstruction problem, can be denoted as:

$$\begin{aligned} min\ &\|\Psi(x)\|_1 \\ s.t.\ &\|Fx - S\|_2 < \epsilon \end{aligned} \tag{5.5}$$

where ϵ is the constrain or threshold parameter that controls the fidelity of reconstruction to the measured data. The l_1 norm is defined as $\|x\|_1 = \sum_i |x_i|$ Minimising the l_1 norm of $\|\Psi(x)\|_1$ increases sparsity, while the constraint $\|Fx - S\|_2 < \epsilon$ enforces data consistency [95].

One of the most common sparsifying transforms (Ψ) employed in CS-MRI is the wavelet transform, and it constitutes the **L1 Wavelet regularised reconstruction** [95]. Total Variation (TV) penalty is yet another method that can be used as the sparsifying transform, where the finite-differences transform - which is the summation of the gradients of intensity between adjacent pixels [79, 92, 95]. This reconstruction can be denoted as:

$$min_x \|Fx - S\|_2 + \lambda \|TV(x)\|_1 \tag{5.6}$$

where the TV penalty $TV(x)$ is defined as:

$$TV(x) = \|\nabla x\|_1 = \sum \|\nabla x\|_1 \tag{5.7}$$

where ∇ represents the finite differences operator. TV penalty can also be employed along with another sparsifying transform [95] in Eq. 5.5, can be defined as:

$$\begin{aligned} min\ \|\Psi(x)\|_1 + \alpha TV(x) \\ s.t.\ \|Fx - S\|_2 < \epsilon \end{aligned} \tag{5.8}$$

where α is the trade-off parameter between the sparsity achieved from the sparsifying transform Ψ and the sparsity from the finite differences, even though iterative algorithms have proven to be a powerful tool for the reconstruction of undersampled MRIs, they come with two inherent problems. The first and biggest problem is in their inherent nature - as they are iterative algorithms, the optimisation process is time-consuming—making them unsuitable for real-time applications. The second limitation of these algorithms is the requirement of incoherent sampling—making them limited in terms of the possible sampling patterns that can be reconstructed using such an algorithm [95].

5.2 Deep Learning Methods

Deep learning has proven to be a powerful tool for a wide array of applications, including for undersampled MRI reconstruction. [212] was one of the first to employ deep learning for undersampled MRI reconstruction. Since then, several approaches have been proposed which are based on deep learning. Depending upon the working mechanism of those models, they can be grouped into three classes: image-space models, k-space models, and mixed-space models. Many of the models try to take inspiration from the classical non-deep learning approaches, and some even combine with them.

5.2.1 Working in the Image Space

Models working in the image-space can be sub-grouped based on the task they accomplish: the denoising or artefact reducing networks—networks which are supplied with undersampled MRIs containing artefacts predominantly due to the violation of the Nyquist criterion, and they try to remove those artefacts in the given image, and the super-resolution networks—which try to synthesise high-resolution counter-parts from the given low-resolution images acquired with lower k-space coverage.

5.2.1.1 Artefact reduction

UNet [186] model, originally proposed for the task of image segmentation, and its variants have become one of the most popular classes of network architectures in the field of computer vision since the inception of the UNet. It has also been employed for the task of undersampled MRI reconstruction [269]. In this paper, the authors work with both uniform undersampling with a dense k-space centre for an acceleration factor $AF = 4$—causing the resultant images to have aliasing artefacts. The undersampled zero-filled k-space data is first reconstructed using iFFT and supplied to the UNet model, which attempts to remove the artefacts. The output of the network is then brought back to the k-space using FFT, and the real acquired k-space points present in the undersampled k-space are then replaced in the k-space obtained from the UNet—as the data consistency step, inspired by Compressed Sensing MRI (CS-MRI). This final k-space data is then reconstructed using iFFT to obtain the final output image. The workflow of this framework is shown in Fig. 5.6. The paper [269] reported SSIM values of 0.8782 ± 0.0411 and 0.9039 ± 0.0431, for without and with the data consistency step while reconstructing brain MRIs with an acceleration factor of 4.

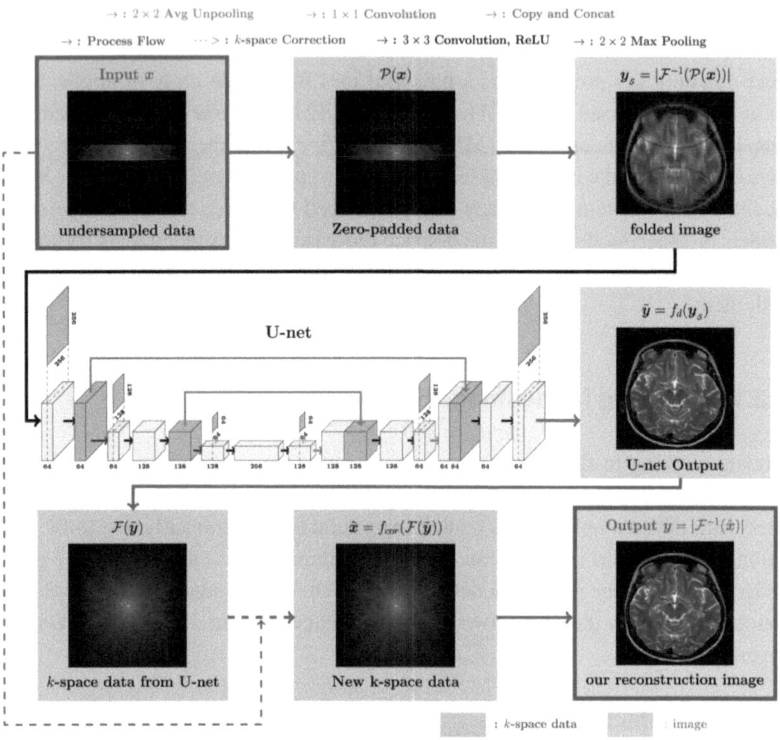

Fig. 5.6 UNet-based undersampled MRI reconstruction framework. (Fig source: [269])

[246] proposed a GAN-based solution - de-aliasing GAN (DA-GAN)—for removing artefacts in the image-space from the undersampled MRIs. [355] employed Denoising Autoencoder (DAE) to remove artefacts from undersampled MRIs, and the paper reported an average SSIM of 0.8319±0.0254 while working with pseudo-radial undersampling for an acceleration factor of 10.

On the other hand, [279, 357] employed a convolutional recurrent neural network to remove artefacts from undersampled dynamic MRIs.

5.2.1.2 Super resolution

Another direction of undersampled reconstruction of MRIs is to treat them as a Super-Resolution (SR) problem. These so-called SR algorithms aim to restore images with high spatial resolution from the corresponding low-resolution images.

Applying SR approaches to undersampled MRIs began before the deep learning era by applying image processing techniques [143, 146, 180]. Since then, several DL-based SR techniques have been applied to super-resolve low-resolution MRIs. [294] utilised CNNs to perform single-contrast and multi-contrast SR-MRI. [238] performed SR of brain MRIs using 3D CNN, which was further extended as a multi-scale approach [330] by incorporating a prior high-resolution image using image registration. In [330], the authors exploited the available high-resolution prior image of different contrast but of the same subject. The network is trained to predict the difference image between the low-resolution input and the high-resolution ground-truth, which is then added to the low-resolution input to obtain the final result. [353] integrated a deeply-supervised attention model with a GAN-based framework to perform SR-MRI. SR-MRI approaches have also been used for dynamic MRIs. [253] used a UNet-like model coupled with convolutional LSTM layers. [360] uses a 3D UNet-like model to perform SR-MRI of 2D dynamic MRIs by using the temporal domain as the third dimension.

5.2.2 Working in the k-space

Robust **a**rtificialneuralnetworks for **k**space **i**nterpolation (RAKI) [301] is a scan-specific deep learning approach inspired by GRAPPA, which learns from the Auto-calibration Signal (ACS) region of the multi-coil k-space using a 3-layer CNN and attempts to predict the missing k-space points. rRAKI [337] extends that approach using residual connections. LORAKI [319] is a similar approach that utilises RNNs. These approaches come with an inherent limitation: the network has to be trained repeatedly for each scan—making it difficult to implement in real working scenarios and also making it impossible for real-time applications.

[316] proposed a UNet-based approach that tries to predict the missing k-space points from the given undersampled k-space. The network is real-valued, which works with the complex-valued k-space data by splitting it into real and imaginary parts. The output of the network is also a k-space split into real and imaginary parts, which are then stacked to obtain the complex-valued k-space, which is then converted to the image space by applying iFFT. The loss is then calculated in the image space by comparing the output against the reconstructed image from the fully-sampled k-space. [384] also proposes a similar approach but tries to learn not only from the slice that is to be reconstructed but also using the adjacent slices. The model used in this paper is a UNet model with self-attention layers.

5.2.3 Mix k-space and Image Space Models

One of the first approaches which proposed working in a mixed-space setup was AUTOMAP [297], which works by taking the k-space as input and providing the reconstructed image as output. But the network design of AUTOMAP creates a hindrance to scalability—which dAUTOMAP [331] tries to address. This paper [331] has shown experiments up to an acceleration factor of 7, for which the reported SSIM value is 0.82±0.02.

[267] proposed the compressed sensing reconstruction as a variational model embedded in an unrolled gradient descent scheme. This Variational Network (VN) utilises not only the zero-filled reconstructed undersampled image but also utilises the undersampled k-space data and the coil sensitivity maps to provide the final reconstruction. The authors demonstrated that the proposed approach performs significantly better than the classical approaches while working with undersampled knee MRI using uniform undersampling with an acceleration factor of 4. An extension of this work, End-to-End Variational Network (E2E-VN) [368] attempts to perform filtering in the image space, estimate the coil sensitivities, and maintain data consistency. The workflow of E2E-VN is shown in Fig. 5.7.

E2E-VN starts working with the undersampled k-space, which is supplied to the Sensitivity Map Estimation (SME) UNet and the refinement UNet. For SME, only the ACS lines are kept in the undersampled k-space, and iFFT is applied to that before supplying it to the SME-UNet, the output of which is divided by the Root Sum of Squares (RSS) to obtain the coil-sensitivity estimations. These estimated coil-sensitivity maps are used to coil-combine the undersampled coil images (iFFT applied to the undersampled k-space), and that undersampled coil-combined image is forwarded to the refinement UNet to remove artefacts. The coil-combined output of the refinement UNet is then multiplied with the estimated coil-sensitivity maps to obtain the coil-images, on which an FFT is applied to obtain the reconstructed k-space. The real acquired data points present in the undersampled k-space are then replaced in this reconstructed k-space to obtain the data-compensated k-space as the data consistency step. This k-space is the supplied again to the SME and refinement UNets to refine it further. Several of these iterations are performed before finally applying iFFT to the final data-compensated k-space to obtain the final coil-images, which are then combined using RSS to obtain the final reconstructed coil-combined image. The paper reported an SSIM value of 0.943 while reconstructing brain MRIs from the fastMRI [293] dataset with an acceleration factor of 8.

GrappaNet [369] attempts to combine GRAPPA-based parallel imaging with deep learning. They apply supplies undersampled coil k-space to a CNN model, the output of which is reconstructed by applying the estimated GRAPPA kernel, which

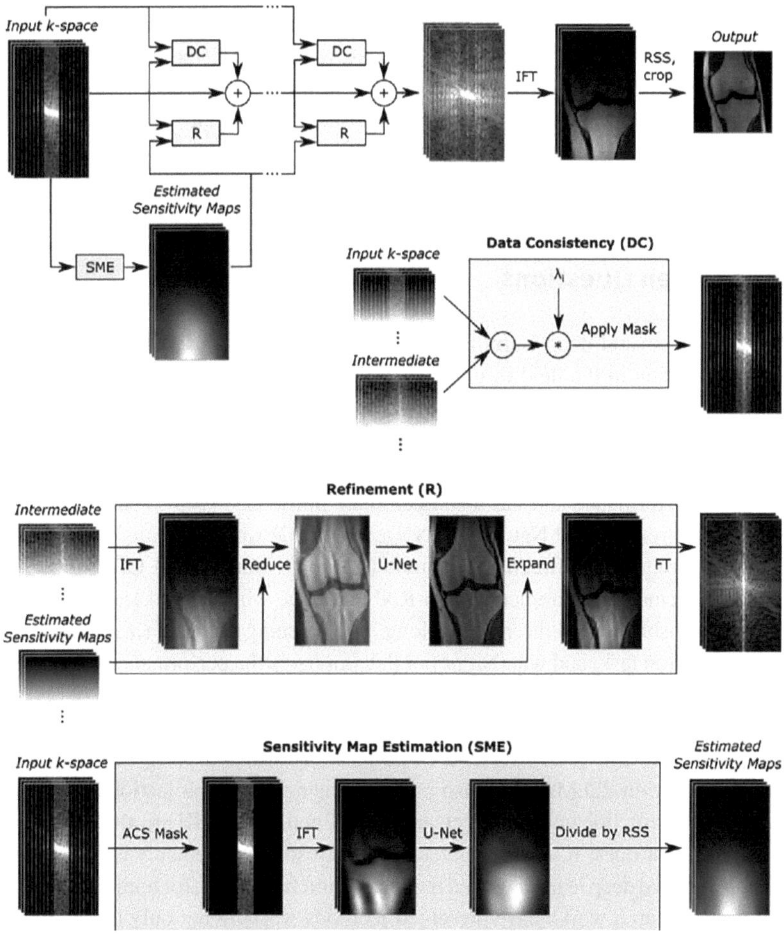

Fig. 5.7 End-to-end variational network for undersampled MRI reconstruction. (Fig source: [368])

is again supplied to another CNN model, the output of which is finally reconstructed using iFFT and RSS. Each of the CNN models contains a UNet working in the k-space, followed by data consistency and iFFT, followed by a UNet working in the image space, followed by FFT and data consistency steps. The paper reported an SSIM value of 0.922 while working with PDw Knee MRIs with an acceleration factor of 8.

5.3 Open Questions

Even though several deep learning methods have been proposed, there is always an open question in the field of deep learning—is it possible to design a network architecture that can outperform the existing methods? (***[RQ:2]***, ***[RQ:3]***, ***[RQ:4.2]***, ***[RQ:5]***). When it comes to dynamic MRIs, incorporating the temporal information in the network model can be useful to improve the artefact reduction or super-resolution performance and has also been done in the past with the help of some type of Recurrent Neural Network (RNN). But, while working with 3D volumetric MRIs, it might be difficult to accommodate RNN trainings on GPUs—due to their high memory requirements. Non-RNN models, which try to learn the temporal relationship within the model, along with learning the artefact reduction or super-resolution task, and whether or not this improves the performance—have not been explored yet (***[RQ:3.1]***). When it comes to the task of reconstruction of undersampled radial MRI, CT-inspired techniques might be applicable by performing spoke-wise 1-dimensional inverse fast Fourier transform as a pre-processing step. Even the first-ever 2D MRI was also produced using a back-projection algorithm. But, the reconstruction tasks of undersampled CT and radial MRI are always treated as two different ones. It will be interesting to investigate whether it is possible to develop a unified deep learning based reconstruction framework for both (***[RQ:5.1]***).

Recent research works have presented methods working not only in the image space, but also in the k-space or in a hybrid space of both of them. However, most of these methods employ real-valued CNNs even though the k-space is inherently complex-valued. Employing Complex-valued Convolutional Neural Network (CV-CNN) might be more suitable for this task (***[RQ:4.1]***) and it will be interesting to investigate whether this improves the reconstruction performance compared to a purely image space model (***[RQ:4]***).

With the developments of several DL-based methods for undersampled MRI reconstruction, one interesting question always pops up—what are the limits in terms of the acceleration factors, what amount of undersampling such a method can handle (*[**RQ:6**]*). On the other hand, non-DL techniques, such as Compressed Sensing MRI (CS-MRI), can benefit from using prior knowledge to regularise them better [pub:43]. An interesting research direction is to explore what kind of prior knowledge can be used and how effectively that can be incorporated into DL-based methods (*[**RQ:1**]*), and whether they benefit from the same (*[**RQ:1.1**]*).

6 Motion Correction

High-resolution acquisitions typically have long acquisition times—even the most compliant subjects move unintentionally—known as the physiological movements [166, 190]. Such movements lead to motion artefacts in the reconstructed images. There are two different classes of techniques commonly used to combat the problem of motion artefacts: Prospective Motion Correction (PMC) and Retrospective Motion Correction (RMC) [40].

PMC performs real-time adaptive updates of the data acquisition to avoid motion artefacts in the final reconstructed image, with the help of various tracking modalities [197, pub:15], while RMC techniques can be considered as post-processing techniques where the MR image data is modified after the k-space data acquisition has been performed—during or after the reconstruction process [197, 372].

6.1 Motion Tracking

The tracking systems for motion correction can be grouped as MR-based, RF-based, and optical. Navigators are commonly-used techniques for motion tracking, which is an MR-based tracking system. They can be broadly divided into three classes based on their working principles: self-navigation, navigator echoes, and coil sensitivities [197]. In self-navigation, the motion is estimated directly from the acquired image data with the help of k-space trajectories. PROPELLER (shown in Fig. 2.18(a)) [51], SNAILS [70], and MOJITO [102] are some of the trajectories for self-navigation. Ehman and Felmlee [26] demonstrated that the motion could also be estimated by using additional echoes, typically at the beginning or the end of the repetition loops—known as navigation echoes. Since then, several methods have utilised this principle [37, 65, 114]. However, this requires additional acquisition time, consequently increasing the minimum TR—making it less-efficient [197].

S. Chatterjee, *Deep Learning Innovations in MRI Reconstruction and Analysis*,
https://doi.org/10.1007/978-3-658-50741-1_6

Bydder et al. [61] proposed the third idea of navigators—estimation of in-plane motion from the coil sensitivities (SENSE). Later, other papers also introduced several coil-sensitivity based navigators [91, 126]. Motion tracking can also be performed by comparing consecutively-acquired volumes from a single study—this is known as image-based tracking, which is another MR-based motion tracking system [197].

The second type of motion tracking system performs RF-based tracking by utilising additional pickup coils with local sensitivities whose signal is recorded using additional receivers. This kind of system was first introduced by Ackerman et al. [20] and was first applied for motion tracking by Derbyshire et al. [42]. Since then, several innovations have taken place which improve the quality of tracking. Techniques, such as active markers [111] estimate the motion by comparing the updates of the marker position—a technology which has been further developed by several researchers [154, 155, 171].

Optical tracking systems [87, 90] have also been employed for tracking motion and have become one of the most popular techniques. Ever since Tremblay et al. [81] introduced an optical tracking system for PMC utilising two infrared cameras along with reflective markers, several techniques have been proposed [112, 123, 144].

6.2 Prospective Motion Correction

Prospective Motion Correction (PMC) techniques employ different tracking technologies to track the movements and correct for them in the 3D space by updating the imaging volume in real-time [22]. The objective of the PMC techniques is to keep the acquisition FOV constant with respect to the object—by adapting it based on the motion [197]. The first PMC approach was proposed by Haacke et al. [22], where they used navigators to monitor breathing motion and then adapted the phase-encoding gradients accordingly to reduce motion artefacts. Since then, several navigator based PMC systems have been proposed [65, 89, 145]. Apart from navigators, several other techniques [153], such as, pickup-coils or active markers [111] and optical tracking systems [87, 90, 121, 189], have also been employed for PMC. One of the optical tracking systems is the Moiré Phase Tracking (MPT) proposed by Maclaren et al. [141], which utilises an external in-bore single-camera optical tracking system, is considered to be the current gold-standard [309, 351] and several studies have proven the effectiveness of this technique [190, 325, pub:15].

6.3 Retrospective Motion Correction

Retrospective Motion Correction (RMC) are post-processing techniques that aim to remove or reduce motion artefacts after the acquisition of the k-space data has already taken place without involving any consideration during the acquisition regarding the possibility of motion [372]. Non-DL based methods aim to modify the acquired k-space data or image data during the reconstruction process [197], while DL-based techniques attempt to remove artefacts from already-reconstructed images [372].

6.3.1 Non-Deep Learning Methods

In RMC, the motion can be estimated either from the acquired data or using additional tracking systems. Tracking systems can be similar to the ones for PMC, which include—MR-based tracking, e.g. navigators, RF-based tracking, and optical tracking systems. By knowing the motion in terms of translation and rotation, some of its effects can be removed from the image by modifying the k-space data, exploiting the k-space properties [197]. Bookwalter et al. [102] has shown (Eq. 6.1) that the points of the motion-corrected k-space $S\left(k_x, k_y, k_z\right)_{\text{cor}}$ from the tracking data regarding translation in the x, y, z directions—Δx, Δy, and Δz, and phase correction $\Delta\phi_{\text{cor}}$, by exploiting the property that translation of the object causes linearly-increasing phase-shift in the k-space.

$$\begin{aligned} S\left(k_x, k_y, k_z\right)_{\text{cor}} &= S\left(k_x, k_y, k_z\right)_{\text{orig}} \cdot \mathrm{e}^{-\mathrm{i}\Delta\phi_{\text{cor}}} \\ &= S\left(k_x, k_y, k_z\right)_{\text{orig}} \cdot \mathrm{e}^{-\mathrm{i}2\pi\left(k_x\Delta x + k_y\Delta y + k_z\Delta z\right)} \end{aligned} \tag{6.1}$$

On the other hand, another property of the k-space (Image-Fourier relationship) is that any rotation α in the image space causes the same α rotation in the k-space. If the rotation happens around the z-axis in the image space, this relationship can be defined as:

$$\begin{aligned} &I(x\cos\alpha + y\sin\alpha, -x\sin\alpha + y\cos\alpha, z) \supset \\ &S\left(k_x\cos\alpha + k_y\sin\alpha, -k_x\sin\alpha + k_y\cos\alpha, k_z\right) \end{aligned} \tag{6.2}$$

This type of motion is more difficult to handle retrospectively—several factors need to be considered to correct for motion caused by rotation—phase-correction, Nyquist violation caused by the correction, re-gridding as the corrected k-space might be on non-rectangular grid [197]. Over the years, several retrospective motion correction techniques have been proposed based on these properties [114, 157, 197].

Autofocusing is another technique used in RMC, which does not rely upon any motion tracking information. Instead, motion models are employed to improve the image quality by optimising a certain image quality metric [41, 197] in an iterative manner. Using this technique, several RMC methods have been proposed over the years—for both rigid motion as well as deformable motion [135, 150, 197]. Moreover, several approaches have been proposed combining autofocusing techniques with motion trackings, such as using navigator echoes [57] or self-navigation sequences [133].

6.3.2 Deep Learning Methods

With the developments in the field of deep learning, several approaches have been proposed based on deep learning techniques to perform motion correction retrospectively. Pawar et al. [278] employed a UNet model to remove artefacts from the reconstructed images. Duffy et al. [260] proposed HR3DNet and HR3DNetGAN to perform RMC using 3D CNNs. Jiang et al. [317] presented a UNet-based generator, coupled with a discriminator and VGG19-based perceptual loss—to remove motion artefacts from MRIs. Küstner et al. [320] proposed a VAE-GAN based approach, MedGAN, which utilises a cascade of six UNets. Another GAN-based approach has been proposed by Usman et al. [372], where the authors applied CG-SENSE based reconstruction on the motion-corrupted k-space data to reconstruct the images before applying a GAN-based model for artefact removal. In another work, Pawar et al.[328] took a different path—without using GAN-based approaches, employed an Inception-ResNet model for RMC. Lee et al. proposed MC^2-Net[388]—which combines a registration network to align MRIs of different contrasts, followed by a motion correction network. This framework has been shown to work when one of the three supplied contrasts is corrupted by motion, and the others are not—it uses the available knowledge from these non-corrupted contrasts to correct the corrupted one. Zhao et al. [397] proposed an idea of a motion patch localisation network using Single Shot MultiBox Detector (SSD), followed by a Convolutional Autoencoder (CAE) to correct the corrupted patches.

6.4 Open Questions

Several non-DL, as well as DL-based methods have been proposed to perform Retrospective Motion Correction. One interesting research direction that has not been explored much so far is the incorporation of prior knowledge. It is interesting to investigate what kind of prior knowledge can be used and how effectively that can be incorporated into DL-based methods (***[RQ:1]***), and most importantly, whether these methods would benefit from the same (***[RQ:1.1]***). Given both undersampling and motion result in image artefacts, it might be possible to utilise similar methods for the same. A thorough evaluation of multiple such methods is yet to be performed (***[RQ:7]***). Similar to the task of undersampled reconstruction, it will be interesting to investigate whether k-space or hybrid space models can improve the quality of motion correction compared to a pure image space model—another research direction that has not been explored much (***[RQ:4]***).

7 Automatic MR Image Processing Pipelines

MRIs are not only used by radiologists to visually inspect as a part of clinical routines or used by surgeons during MR-guided interventions, but they are also used with many automatic image processing pipelines. Such a pipeline can be tasked to analyse MRIs to classify diseases, or to segment some region of interest for a better understanding of the disease, or even used to standardise large MRI datasets for large-scale analysis. With the advancements in the field of machine learning, several machine learning based MR processing pipelines have been proposed. The possibility of achieving human-level accuracy using such a pipeline by employing CNNs has been demonstrated in several research works [216, 251, 387]. Four different machine learning tasks were explored and discussed in detail as a part of this thesis—brain tumour classification and segmentation, vessel segmentation, image registration, and anomaly detection.

7.1 Brain Tumour Classification and Segmentation

A brain tumour is the growth of abnormal cells in the brain which can be classified based on their speed of growth and the likeness of them growing back after treatment. They are mainly divided into two overall categories: malignant and benign. Benign tumours are not cancerous; they grow slowly and are less likely to return after treatment. Malignant tumours, on the other hand, are essentially made up of cancer cells, they have the ability to invade the tissues locally, or they can spread to different parts of the body, a process called metastasise [58]. Each of these tumours behaves differently, and they are graded accordingly. Grades I and II gliomas are considered Low-Grade Glioma (LGG), while grades III and IV are known as High-Grade Glioma (HGG). The LGGs are benign tumours, and they can be excised using surgical resection. In contrast, HGGs are malignant tumours that are hard to excise

S. Chatterjee, *Deep Learning Innovations in MRI Reconstruction and Analysis*,
https://doi.org/10.1007/978-3-658-50741-1_7

by surgical methods because of their extent of nearby tissue invasion. Fig. 7.1 shows an example MRI of LGG and HGG, while Fig. 7.2 shows an example of the different tumour tissues in HGG as visible in three different contrasts.

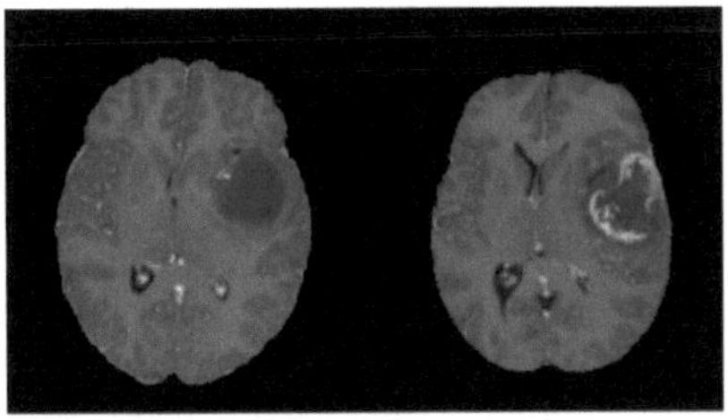

Fig. 7.1 An example MRI of Low-grade glioma (LGG, on the left) and High-grade glioma (HGG, on the right). (Source: BraTS 2019 [pub:10])

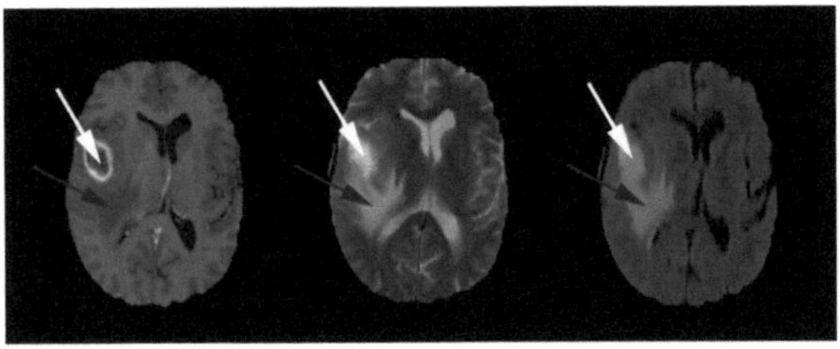

Fig. 7.2 High-grade glioma structure on T1ce, T2 and FLAIR contrast images (from left to right), (·) Necrotic core, (·) Perifocal oedema. (Source: BraTS 2019 [pub:10])

The prognosis of a brain tumour depends on many factors, such as the tumour's location, the histological subtype of the tumour, and the tumour margins. In many cases, the tumour reoccurs and progresses to grade IV even after treatment [177]. Modern imaging methods such as MRI can be used for multiple diagnostic purposes; they can be used to identify the tumour location—which is used for investigating tumour progression and surgical pre-planning. Detection of cancer, specifically an earlier detection, holds the potential to make a difference in treatment. Deep learning methods can help automate the process of detecting and classifying brain lesions—they can also reduce the radiologists' burden of reading many images by prioritising only malignant lesions. This will eventually improve the overall efficiency, and it can reduce diagnostic errors [169]. Therefore an automated deep-learning method might offer a better solution and improve the overall outcome of the treatment. Several deep learning models have been proposed for the task of brain tumour classification and segmentation, while some models address both problems in a unified fashion. This has been discussed earlier in [pub:10, pub:1].

7.1.1 Classification Models

Without any prior region-based segmentation, [299] applied the most straightforward possible architecture of CNN on the brain tumour dataset with a validation accuracy of 84.19% at best. [332] proposed a CNN architecture for classifying brain cancers into the three classes of tumours in the brain tumour dataset, as well as distinguishing between the three glioma categories (Grade II, Grade III, and Grade IV). [362] proposed an approach using volumetric CNNs to classify high-grade glioma and low-grade glioma using T1 contrast-enhanced images. Most of the literature on glioma tumour classification and grading used one single MR contrast image at a time, but [264] used a fusion framework that uses T1 contrast-enhanced, T2, and FLAIR images simultaneously for classifying the tumour. [393] used a novel fusion method for the inclusion of multiple MRI contrasts, first, the T1 images are transformed by non-subsampled shearlet transform (NSST) into low frequency (LF) and high frequency (HF) sub-images, essentially separating principle information in the source image from edge information, then the images are fused by predefined rules to include the coefficients, resulting in fusion of T1 and T2 or FLAIR images. [340] created a 22-layered CNN architecture for brain tumour type classification. [380] proposed a CNN-based Computer-assisted Diagnosis (CAD) system using an 18-weighted layered CNN model for brain classification on two distinct datasets. As for models based on transfer learning, [228] developed a modified version of the AlexNet CNN model to classify brain MRI images into healthy, low-grade glioma and high-grade glioma. [307] employed a pre-trained GoogleNet CNN model to distinguish between the three types of tumours. [364] proposed using three pre-trained CNN models known as AlexNet, GoogleNet, and VGG16 to classify tumours in the brain tumour dataset.

7.1.2 Segmentation Models

[208] developed a 3x3 kernels-based CNN and yielded 88%, 83%, 77% dice coefficients in the BraTS 2013 challenge for complete, core, and enhancing tumour,

respectively. They won second place in the BraTs 2015 challenge with a Dice coefficient of 78%, 65%, and 75%. Razzak et al. (2018) presented a new model two-pathway-group CNN architecture for brain tumour segmentation that concurrently uses local and global contextual data. They include the cascade design into a two-pathway-group CNN, in which the output of a basic CNN is considered a second source and concatenated at the final layer. Cascade CNN was Superior to the CNN alone and yielded 88.9%, 81.1%, 73.7% Dice coefficients on the BraTS 2013 dataset for complete, core, and enhancing tumours, respectively. They also achieved 89.2%, 79.18%, 75.1% dice coefficients on the BraTS 2015 dataset [280]. Li et al. (2019) proposed a novel CNN based on UNet, which improves information flow in the network by using a unique structure called an up-skip connection between the encoding and decoding paths. Furthermore, each block includes an inception module to aid the network in learning richer representations in addition to effective cascade training techniques for successively segmenting brain tumour sub-regions. The model scores a dice value of 89% for the complete tumour of the BraTS 2017 dataset [321].

7.1.3 Unified Models

[329] first performed segmentation of the tumours, then they classified the segmented images into tumour grades HGG and LGG. [383] developed a tumour classification and segmentation model based on pixel-wise classification into four classes, including healthy, meningioma, glioma, and pituitary tumour. This method was trained and tested on the brain tumour dataset achieved 5-fold cross-validation average values of 94% for classification sensitivity and 0.82 for segmentation dice coefficient. Additionally, they scored a 97.3% classification accuracy.

7.1.4 Open Questions

For the task of image classification, several 2D and 3D models have been employed over the years. For 3D volumetric MRIs, either pure 3D models are used—treating each of the dimensions equally, or 2D models are used—splitting the 3D volumes as 2D slices and working with each of those slices. The spatiotemporal models introduced by Tran et al. [287] are somewhat in between. They are effectively 3D models proposed for video classification, but they treat one of the dimensions differently than the other two. In the case of video, this differently-treated dimension is the time dimension. These models achieved significantly better classification performance on videos. It is worth exploring whether for 3D volumetric MRIs, such a spatiotemporal model would work well—treating the slice dimension differently than the in-plane dimensions.

On the other hand, some critical issues must be addressed before deep learning methods can be widely accepted and adopted in the medical field. Because of the complexity of the CNNs reasoning process, they tend to be obscure and are frequently referred to as black boxes, as discussed in Sect. 4.7. While some researchers came up with post-hoc methods to explain and interpret such models [pub:9], others try to create inherently explainable-interpretable models. Additionally, most deep learning based segmentation models are based on supervised learning and require manually annotated data, which is time-consuming and imposes a big challenge. Nevertheless, labelling the data with the type of tumour or whether it is present or not is much easier for radiologists and, hence, less time-consuming. [224] has shown the potential of training a segmentation model in a weakly-supervised manner using regression labels for lesion counts. This approach might be possible to be extended for supervised tumour classification and weakly-supervised tumour segmentation—yet to be widely explored.

As a final note, most of the proposed methods focus on classifying the type of the tumour—knowing that the subject already has a tumour present. But, if an additional class—"tumour-free" can be added in the process, the pipeline can be used as an automatic screening system for patients with possibility of a brain tumour based on their symptoms, but the clinicians are not yet aware whether or not that patient has a tumour or the symptoms are from something else.

7.2 Anomaly Detection

Acquiring and analysing MRIs is an integral part of the work of neuroradiologists. MRI is free from harmful ionising radiation and its ability to show excellent soft-tissue contrast, making it the preferred imaging modality for brain imaging in clinical practice [315]. MR images are used to examine numerous brain pathologies, and manually analysing or annotating such images is part of the radiologists' regular duties. Not only such duties are time-consuming and laborious, but they are also error-prone [220]. Noting from [176], diagnostics of brain pathology remain undiscovered in up to 5–10% of cases. The medical image analysis community has made numerous contributions toward automating specific steps, focusing on achieving human-level accuracy at economically adequate computational complexity. With advancements in artificial intelligence, outstanding achievements have been made in the detection and segmentation of lesions. Furthermore, these scientific results display abilities on a par with experts [169]. Such automated methods try to assist the radiologists in the decision-making process—reducing the workload and observer dependency while improving the workflow and diagnostic accuracy [265]. This has been discussed earlier in [pub:12].

Most of these automatic methods are based on supervised learning techniques, relying on the quality and quantity of labelled data. They utilise a vast amount of annotated training data—ranging from thousands to millions of samples. These trained models are usually specialists in dealing with domain-specific tasks or the tasks they have been optimised for—which can also be considered as the bottleneck in such a method as they typically follow a 'focus point solution' design, which means they are often very focused on one type of pathology or lesion. Considering an example of cerebral small vessel disease, supervised solutions are challenging to develop since the damage is complex in terms of lesion size, contrast or morphology [158]. Considering the complex field of medicine, procuring annotated data is a tedious manual process—while it might be challenging to annotate concerning all the possible pathologies.

An alternative solution to the conventional approach of the supervised machine learning method to detect brain pathologies is an Unsupervised Anomaly Detection (UAD) system. Anomaly detection is an approach that distinguishes anomalies completely based on characteristics that describe regular data. Since the feature of possible irregularities is not learned, they stand apart from the normal distribution and can be consequently detected. Anomaly detection solutions are useful in cases where anomalies are to be detected, but their manifestation is not known, and they even might occur very rarely—making it difficult to create a large dataset for supervised training [170]. Anomaly detection techniques have been applied for multiple tasks, including credit card fraud detection [72], aerospace monitoring during flight [93], illegal object detection at airports [250] or detecting faulty microprocessors [139].

Anomaly detection in medical imaging is defined as identifying relevant indicators of diseases and differentiating them from those of typical healthy tissue characteristics. UAD systems can directly learn the data distribution from a large cohort of un-annotated healthy subjects to detect out-of-distribution or anomalous samples to identify cases for further inspection. As anomaly detection does not depend on reference annotations, this approach does not require any human input and can therefore be used to identify any medical condition. Unsupervised segmentation of brain regions for anomaly detection is clinically relevant as it can assist radiologists. It is also of great interest to the medical image analysis community as it eliminates the need for pixel-level annotations, which is an expensive process and is often a prerequisite for supervised training of the networks. However, it is worth mentioning that system, such as UAD, does not differentiate between different types of anomalies, i.e., lesions and artefacts—both will be treated as anomalies by a UAD system. Variational Autoencoders (VAEs) are one of the most commonly used techniques for anomaly detection [385].

Autoencoders (AEs) are trained by minimising its reconstruction error [161]. Different types of reconstruction errors are commonly used in unsupervised anomaly detection, acknowledging that regular distribution models cannot effectively reproduce anomalies [281]. Therefore, regular anomaly-free data is distinguishable from anomalous data with the degree of error differences in terms of reconstruction. The majority of the methods assume reconstruction errors (losses) are optimised for average distribution data, whereas anomalies are impacted with significant losses when their dimensions are reduced.

Generative Adversarial Network (GAN) [165] and Variational Autoencoder (VAE) [149] are two essential methods based on latent space information modelling. However, both these methods have different approaches for modelling the distribution. VAE applies variational inference to approximate the data distribution. The posterior distribution is encoded into latent space using the encoder network of the VAE, whereas the decoder network is responsible for modelling the likelihood. Unlike VAE, the generator network of the GAN transforms the prior distribution of the random samples in the latent space into data samples that an optimised classifier cannot distinguish. The obtained data distribution is inferable, and GAN is mainly a sampler. Recently some researchers [381] have coupled an autoencoder with GAN to derive enhanced latent representations in the domain of anomaly detection. Between GANs and VAEs, VAEs are more intuitively understandable, but the reconstruction quality might be inferior to GANs, while GANs might suffer from stability problem during training, and in several studies VAEs have been preferred over GANs [310, 333]—including for anomaly detection [395]. Hence, this thesis focuses on AE- and VAE-based anomaly detection and a comprehensive collection of some of the techniques are discussed here in detail.

7.2.1 Adversarial Autoencoder

Autoencoders [19, 33] try to learn an underlying function—which is an approximation to the identity function—trying to reproduce the input as its output, learning a Latent Space Representation (LSR) of the input in the process [127]. The model's output is compared against the input with the help of reconstruction loss, which is then backpropagated to train the network. Adversarial autoencoder [206] converts an autoencoder into a generative model, which has also been used for the task of anomaly detection [304, 322]. The objective is two-fold—a regular reconstruction loss and an adversarial training loss. The second loss criterion tends to match the aggregated posterior distribution of the latent space of the autoencoder to an arbitrary prior distribution. An outcome of this architecture bounds the encoder module to learn the data distribution close to the prior distribution, and the decoder maps the imposed prior to the data distribution. The encoding function of the encoder is formally represented as

$$q(z) = \int_x q(z \mid x)p_d(x)dx \tag{7.1}$$

where x is the input and z is its latent representation vector. $p(z)$ be the prior distribution to be imposed on the latent vectors, $q(z|x)$ is an encoding dimension, and p(x| z) is the decoding distribution. Also, $p_d(x)$ is the data distribution and $p(x)$ is the model distribution. The adversarial autoencoder is regularised by matching this aggregated posterior $q(z)$ to an arbitrary prior, $p(z)$. The whole training takes place in two phases—*reconstruction* phase, where the encoder and decoder are updated to minimise the reconstruction error in the input. Later in the *regularisation* phase, the adversarial network updates the discriminator to classify better actual samples (from the prior) from the generated samples (from the encoder). Finally, the generator or the encoder is updated to confuse the discriminator.

7.2.2 Scale-Space Autoencoder

Many variations of autoencoders have been proposed. One of them is Scale-Space Autoencoder (SSAE) [342].This framework is used for unsupervised anomaly detection based on the Laplacian pyramid, allowing efficient compression and reconstruction of brain MRI images with high fidelity and successfully suppressing anomalies. Contrary to other frameworks, the SSAE framework models the distribution of the scale-space representation of a healthy brain MRI. The Laplacian pyramid is a multi-scale representation of the input image; it allows efficient segmentation of anomalies at different resolutions and then aggregates the results. SSAE architecture is an encoder-decoder network that aims to localise anomalies from reconstruction residuals. The frequency band of the input data is split by compressing and reconstructing the Laplacian pyramid of the healthy brain MRI. The input image x is repeatedly smoothed and downsampled to obtain a Laplacian pyramid with K levels. Reconstruction can be recursively determined using

$$\hat{x} = \sum_{k=0}^{K-1} u(I_{K-k}) + H_{K-1-k} \tag{7.2}$$

where I_k is the low-resolution representation of the image x after K downsamplings and high-frequency residuals $H_0, .., H_{K-1}$ which are obtained at each level k using:

$$H_k = I_k - u(I_{k+1}); \forall 0 \leq k < K \tag{7.3}$$

The overall loss can be defined as the weighted sum of losses at all scales:

$$L = \sum_{k=0}^{K} \lambda_k L_k \tag{7.4}$$

where,

$$L_k = l2_{norm}(I_k, \hat{I}_k) = l2(I_k, u(\hat{I}_k + \hat{H}_k) \tag{7.5}$$

7.2.3 Bayesian Skip autoencoder

Bayesian Skip autoencoder [341] is formulated on a concept that has been used and proven advantageous for biomedical image segmentation. This framework makes use of skip connections that allow the models with limited capacity to be trained and utilised for unsupervised anomaly detection. Skip connections in autoencoders facilitate the reconstruction of high-resolution and complex medical data with high fidelity. A drop-out-based identity mitigation technique is employed to prevent the model from learning an identity mapping and allowing it to bypass the lower levels of the autoencoder. The loss function for the model can be defined as:

$$L(x, \hat{x}) = L1(x, \hat{x}) - gdl(x, \hat{x}) \tag{7.6}$$

Here, $L1$ is defined as the L1 Manhattan distance between the input and its reconstruction x and $\hat{x}$ respectively, $gdl(.,.)$ is the gradient difference loss.

7.2.4 Variational Autoencoder

Variational Autoencoder (VAE) [149] belongs to a class of generative models that provides a probabilistic way to represent the observations in the latent space. VAE and its variants are one of the first and most popular models for unsupervised anomaly detection [175, 291] Unlike an autoencoder, the encoder of the VAE describes the probability distribution of each latent attribute. During training, the model computes a Latent Space Representation (LSR) of the non-anomalous input image. The latent space in VAEs is continuous by design and allows easy random sampling and interpolation. This is achieved by designing the encoder to output two vectors: a vector of means μ of data samples and a vector of standard deviations σ. The encoding is generated within these distributions. The decoder learns that not just a single point but all the points around it are referred to as a sample of the same class, thus decoding even slight variations of the encoding. Typically, the network is trained by optimising the Evidence Lower Bound (ELBO) L, which can be defined as:

$$\log p(x) \geq L = -D_{KL}(q(z \mid x) \| p(z)) + E_{q(z|x)}[\log p(x \mid z)] \tag{7.7}$$

where, q(z| x) and p(x| z) are diagonal normal distributions that are parameterised by neural networks f_{μ}, f_{σ}, and g_{μ}, and constant c such that:

$$\mathrm{q}(\mathrm{z} \mid x) = N\left(z;\, f_{\mu,\theta 1}(x),\, f_{\sigma,\theta 2}(x)^2\right) \tag{7.8}$$

$$p(x \mid z) = N\left(x;\, g_{\mu,\gamma}(z),\, I * c\right) \tag{7.9}$$

7.2.5 Vector-Quantised Variational Autoencoder

Vector-Quantised Variational Autoencoder (VQ-VAE) [237, 359] framework encodes the images in categorical latent space. Further, an auto-regressive (AR) model is used to model the prior distribution of latent codes. The prior probability approximated by the AR model is then used for unsupervised anomaly detection and gives the sample and pixel-wise anomaly scores. The encoder of the VQ-VAE framework is based on a dictionary mapping K discrete keys into a D-dimensional embedding space, i.e., observed variables are mapped to the embedding space, $z_e(x) \in R^D$. The posterior in VQ-VAE is defined as :

$$q(z = k \mid x) = \begin{cases} 1, & fork = \mathrm{arg\,min}_j ||z_e(x) - e_j||_2 \\ 0, & \text{otherwise} \end{cases} \tag{7.10}$$

The embedding in the dictionary nearest to the encoder's output is given to the decoder as an input, i.e., $z_q(x) \sim q(z \mid x)$. The decoder then reconstructs the distribution $p(x \mid z_q(x))$.The VQ-VAE Loss function is therefore defined as:

$$L = log(x \mid z_q(x)) + ||\ sg[z_e(x)] - e\ ||_2^2 + ||\ sg[e] - z_e(x)\ ||_2^2 \tag{7.11}$$

where, $sg[.]$ is the stop gradient operator. Further, an AR model is employed to learn the prior distribution of VQ-VAE, and the probability of samples is approximated. The samples with low probability are identified as anomalies. AR models generate multiple restorations as they are generative models and allow sampling of one variable at a time in an iterative manner.

7.2.6 Gaussian Mixture Variational Autoencoder

The Gaussian Mixture Variational Autoencoder (GMVAE) [195, 347] approach combines image restoration techniques with unsupervised learning—used in various anomaly detection approaches [266, 350]. A GMVAE is first trained on an anomaly-free dataset to learn a normative prior distribution. A Maximum A Posteriori (MAP) restoration model then uses this prior to detecting the outliers. In GMVAE, the ELBO is formally expressed as:

$$\begin{aligned}\mathbb{E}q(z \mid X)[logp(X \mid z)] - \mathbb{E}q(\omega \mid X)p(c \mid z, \omega) \\ [KL[q(z \mid X)|p(z \mid \omega, c)]] \\ -[KL[q(\omega \mid X)|p(\omega)]] - \mathbb{E}_{q(z|X)q(\omega|X)} \\ [KL[p(c \mid z, \omega)|p(c)]]\end{aligned} \tag{7.12}$$

where the first part is the reconstruction term, the second part ensures the encoder distribution fits the prior, the third term does not let the posterior ω diverge from the prior $p(\omega)$, and the last term enforces the model to not collapse into a single Gaussian but uses the mixture model. For the architectural choice, seven convolutional layers for the encoder and seven transpose layers for the decoder are used. The latent variables z and ω are 2D structures with size (32, 42 1) and nine clusters, c.

7.2.7 Context-Encoding Variational Autoencoder

The core idea of Context-Encoding Variational Autoencoder (ceVAE) [298] is to combine the reconstruction term with a density-based anomaly scoring. This allows using the model-internal latent representation deviations and an expressive reconstruction term on the samples and individual pixels. The overall architecture comprises fully convolutional encoders f_u, f_σ and a decoder unit g, where the context encoding part used the encoder module for a data sample. ceVAE can be divided into two branches: VAE branch and CE branch.

VAE Branch: The deviations of the latent from the CE branch is further inspected. The encoders f_u, f_σ, the decoder g, and a standard diagonal Gaussian prior p(z) are used in the VAE module with an objective function:

$$L_{VAE} = L_{KL}(f_\mu(x), f_\sigma(x)^2) + L_{recVAE}(x, g(z)) \tag{7.13}$$

where $z \sim \mathcal{N}(f_\mu(x), f\sigma(x)^2)$ with the reparameterisation trick and L_{KL} loss with a standard Gaussian yield a comparable anomaly score. KL-loss is used to analyse the deviations of posterior from the prior.

CE branch: sample x is subjected to context encoding noise by masking specific regions. Training occurs by reconstructing the perturbed input $\hat{x}$ using the encoder f_μ and the decoder g, formally, $L_{recCE}(x, g(f_\mu(\hat{x})))$. The results show CEs are more discriminative and semantically produce rich representations, which positively influences the expressiveness of model variations.

ceVAE: On combining above modules, the aim is to capture both the effects, i.e., a better calibrated reconstruction loss and model-internal variations. The objective functions is restructured as the following:

$$\begin{aligned} L_{ceVAE} &= L_{KL}(f_\mu(x), f_\sigma(x)^2) \\ &+ L_{recVAE}(x, g(z)) + L_{recCE}(x, g(f_\mu(\hat{x}))) \end{aligned} \tag{7.14}$$

where L_{KL} is the KL-loss, z is sampled with the reparameterisation method, and $\hat{x}$ is perturbed by masking regions similar to CEs. During the training process L_{recCE} has no constraints for normality on the prior $p(z \mid x)$. The combination of CE and VAE has a regularising effect that prevents the posterior from collapsing and better represents the semantics in the input data.

7.2.8 Open Questions

Several methods have been proposed over the years for the task of unsupervised anomaly detection—some of them are presented here. Existing research spanned from model architecture development to new learning techniques. However, many of them fail to perform well when it comes to clinical datasets, resulting in Dice scores less than or close to 50% [298, 347]. This can be attributed to the fact that the anomaly-free dataset used as the training set might be significantly different from the clinical dataset in terms of image properties (e.e. resolution, contrast, brightness, etc.). One possible strategy could be to simplify the input to the model—making the model less dependent on the image properties and more on the structures themselves. Nevertheless, this has not been explored.

7.3 Small Vessel Segmentation

Small vessels in the brain, such as the Lenticulostriate Arteries (LSA), which supply blood to the basal ganglia [76, 83], are the terminal branches of the arterial vascular tree. Pathology of these small vessels is associated with ageing, dementia and Alzheimer's disease [159, 248]. Segmentation and quantification of these small vessels are critical steps in the study of Cerebral Small Vessel Disease or CSVD [348]. 7-Tesla (7T) Time of Flight MRA is capable of depicting such small vessels non-invasively [98, 109]. Manual segmentation of small vessels in 7T MRA is reliable, but also time-consuming and laborious. Non-Deep Learning (non-DL) approaches [43, 254], which are capable of extracting vessels based on structural properties, can detect medium to large vessels but detecting small vessels can be challenging [183]. Manual and semi-automated techniques involve manual efforts in parameter tuning or annotating vessels, making it a time-consuming task requiring experience to achieve the required sensitivity to achieve robust detection of small vessels. Some of the semi-automated techniques use machine learning algorithms to extract features based on the annotations of observers and classify the non-annotated pixels of the image. Another common problem of semi-automatic vessel segmentation [131] is generating spurious points in the segmentation result [258]. This has been discussed earlier in [pub:11].

Different non-deep learning based techniques used for vessel segmentation, including semi-automated techniques and also deep learning based techniques, have been proposed over the years. Some of them are discussed here.

7.3.1 Vessel segmentation using non-DL techniques

The Hessian-based Frangi vesselness filter [43] is one of the most common vessel enhancement approaches and is usually combined with a subsequent, empirically tuned thresholding to obtain the final segmentation. The multi-scale properties of this method enable small vessel segmentation, but considerable parameter fine-tuning might be required to detect the small vessels of interest. [66] proposed a Vesselness Enhancement Diffusion (VED) filter, which is a combination of the

Frangi filter with an anisotropic diffusion scheme. [78] extended the VED filter by enforcing smoothness of the tensor/vessel response function. [205] suggested a two-step approach for the segmentation of small vessels. The first step is to detect all the major vessels and most parts of thin vessels using a model-based approach. In the second step, minimal path approaches are used to fill gaps in the initial segmentation and segment-missing low-contrast vessels. However, the minimal path approach involves the manual intervention of observers for key-point detection. Hence, the accuracy of segmentation depends on the information provided by the observer, and this step is time-consuming. [226] proposed an automatic vessel filtering pipeline to enhance the contrast of the images. They used the Cardano formula to assign each voxel a value between zero and one as the probability of being a vessel. This technique requires input filter parameters such as sphericity, tubeness, background suppression, and scale parameter. The output response shows the technique has improved the contrast of small vessels in comparison to the original image and also delineated them well with respect to the background.

[254] proposed the MultiScale Frangi Diffusion Filter (MSFDF) pipeline to segment cerebral vessels from Susceptibility Weighted Imaging (SWI) and ToFMRA datasets. The MSFDF pipeline initially pre-selects voxels as vessel or non-vessel by performing a binary classification using a Bayesian Gaussian mixture. This image is further enhanced using the Frangi filter [43] and followed by the VED filter [78]. Finally, the output is intensity normalised and thresholded to obtain the final segmentation.

The aforementioned methods require manual fine-tuning of the parameters for each dataset, or even for each volume, to obtain the best possible result. Furthermore, these methods require different pre-processing techniques, such as bias field correction, making the execution of the pipeline a time-consuming task.

7.3.2 Deep Learning techniques

In recent years, Convolutional Neural Network (CNN) models have outperformed traditional image processing techniques in terms of image classification and image segmentation. These CNNs are extended to more layers to form deep CNNs to achieve further improved performance. This makes the networks huge and requires a larger dataset for training. On the contrary, UNet [186, 239, 255, 263] can be fully trained over a smaller set of labelled images. This has made UNet [186, 194] a popular architecture in many biomedical image segmentation tasks. [277] added attention gates to the UNet architecture, to construct attention UNet, implemented for pancreas segmentation. It was observed that the attention UNet improves the segmentation accuracy with less computational effort. [268] have shown that the UNet is relatively insensitive to jagged and noisy segmentation boundaries. [314] proposed an architecture for joint classification and segmentation of retinal arteries from fundus images. [375] combined a UNet with a fully convolutional network to perform vascular segmentation on photo-acoustic imaging. [334] used UNet for vessel segmentation in X-ray coronary angiography. [323] used UNet to perform vessel segmentation on ToF-MRA images of patients with cerebrovascular disease. They have used volumes acquired at 3T for training and then validated their approach on images acquired at 7T with a resolution of 600 μm. Motivated by these studies, UNet was chosen as one of the baseline architectures.

7.3.3 Open Questions

Deep learning models typically require large labelled training datasets for supervised learning. Publicly available datasets like IXI [url:12] which can be used for training vessel segmentation models, only offer 1.5T and 3T MRAs. 7T MRAs, which can show small vessels better, are scarce. Hence, a network model is required which is capable of learning from small datasets. By employing other learning paradigms like semi-supervised learning, it might be possible to improve the performance further. Moreover, annotating large 7T MRA volumes is also a challenging and time-consuming task. Annotations created in a semi-automated fashion will be faster and easier to create, but might be noisy. Using these noisy annotations as labels might hinder the performance of the model—an important point for evaluation. Finally, whether it is possible to learn from 7T MRAs and apply that knowledge to lower field strengths (e.g. 1.5T and 3T) is also another interesting research question that has not been answered yet.

7.4 Image Registration

Image registration is the process where two given images are aligned to an identical geometrical coordinate system [345]. This process has been widely used in satellite imagery, computer vision, military applications, etc. One of these fields is medical imaging [289]. This has become a cornerstone for many other image processing tasks in the field of medicine. Medical image registration tasks can be divided into two categories: intramodal—registering images of the same modality (e.g. MRI-T1 or MRI-T2 or CT) of different subjects into a standard coordinate system, and intermodal—registering images of different modalities (e.g. MRI-T1 to MRI-T2, or MRI to CT) of the same subject. In image registration terminology, the image which will be registered or "moved" to the coordinate of a second image is known as the moving image (or source image), while the second image is known as the fixed image (or destination image). There are various ways to perform the image registration task. Traditional approaches focus on directly optimising pairs of images based on some criteria or function. More modern approaches include machine learning, deep learning based solutions that try to understand the problem as a whole rather than focusing on individual pairs of images. Such approaches have far-reaching advantages, e.g. they can be very fast, and they can have the capability of accommodating a wide array of deformations. Furthermore, some of those solutions do not even require manually delineated training data, meaning that some of those algorithms are unsupervised in nature, thus reducing the human effort and possibility of human errors in data preparation. This has been discussed earlier in [pub:5].

Traditional approaches to perform medical image registration usually solve some optimisation function for each image pair. This optimisation function is often defined using a similarity metric that allows the non-linear mapping of apparently similar voxels while maintaining local smoothness, etc. constraints [295]. This is commonly known as a non-learning based solution. Because such solutions require optimising the same criteria from scratch for each image pair, it requires an extensive amount of time and computational effort to solve such an optimisation problem. To speed up this process, supervised machine/deep learning methods can be used. Here, the optimisation does not have to take place for every pair of images every time. Instead, the learnt parameters of the registration model can be used to transform a newly obtained image. But such methods require a considerable amount of manually delineated training data that is difficult to obtain and, most importantly, more vulnerable to human error.

Deep learning based solutions try to map the moving images to fixed images by learning a generalisable representation of the displacement vector field that represents the change in the two images. Both the images are fed into the network, and the network attempts to learn the parameters to best transform the image from one to another. Plenty of deep learning based solutions have been proposed, and most of these solutions use unsupervised learning [289, 303, 354]. Some of these approaches are based on Generative Adversarial Network (GAN) [273, 324]—which is a class of self-criticising neural networks. But, owing to the fact that GANs are prone to generate unrealistic imagery and the difficulty of explaining them might not always be a preferred solution. However, there are some intuitive yet efficient ways of regularising these GAN-based networks for producing realistic images, such as cycle consistency [295, 354], bio-mechanics informed regulariser [363], etc. However, the existing research either avoided considering or failed to incorporate information about structural connectivity inside the brain.

Advanced Normalization Tools (ANTs) [100] is regarded as one of the state-of-the-art medical image registration and segmentation toolkit and is considered to be a gold standard in comparative studies. ANTs, a non-deep learning framework, offers several types of linear and non-linear transforms and different choices of optimisation metrics. The typical optimisation metrics include mutual information, cross-correlation, mattes and demons. Several deep learning based techniques have been proposed over the years to perform inter- and intramodal image registration. [289] focused on unsupervised learning whereby registration is done by calculating a dense Deformation Vector Fields (DVF) between the moving and the fixed image. In order to achieve this DVF, CNNs weights are used as parameters of the DVF and optimise those to calculate the transformation parameters. ICNet [295] employed inverse consistency and prepared a pipeline in an unsupervised fashion for deformable image registration, where both fixed and moving images are symmetrically deformed toward each other. In order to facilitate the production of realistic images, they introduced an anti-folding constraint and smoothness constraint that restricts the amount of "folds" in the warped image and ensures the smoothness of the produced image, respectively. FIRE [312] focuses on unsupervised image registration for intermodality. To achieve this, it uses cycle consistency loss and inverse-consistent property. Voxelmoroph [303] works with concatenated fixed and moving images and predicts the DVF of the given image pair, which is then applied to the moving image using a spatial transformer. ADMIR [370]—Affine and Deformable Medical Image Registration is an end-to-end method for affine and deformable image registration utilising CNNs. This method does not require the images to be pre-aligned, which in turn helps to do image registration quickly with good accuracy. Among all these deep learning based image registration techniques, ICNet [295], ADMIR [370], and Voxelmorph [303] are discussed here in detail.

7.4.1 ICNet

[295] performs image registration both ways—moving to fixed and fixed o moving. Let's assume one image as A and another as B. with ICNet architecture, image A is registered to image B and called $warped\_A$ and image B registered to image A is called as $warped\_B$. When image A gets registered to image B, in the process, a flow or displacement vector field $Flow\_AB$ is generated. Similarly, when image B gets registered to image A, flow $Flow\_BA$ is generated.

ICNet [295] proposed three types of constraints, inverse-consistent constraint, anti-folding constraint, and smoothness constraint. Inverse consistent is a regularisation term used for penalising the deviation of the transformation fields (flow) from their corresponding opposite or inverse mappings, i.e. inverse flow. The second constraint is the anti-folding constraint to prevent folding from happening in the flow generated. The last constraint used is the smoothness constraint. This constraint encourages the local smoothness of transformation fields. For the objective function, the authors used mean squared distance to measure the similarity between the fixed image and the registered image. So, the final objective function becomes the sum of similarity, inverse consistent, anti.-folding, and smoothness constraint, where the constraints are multiplied by hyperparameters a, b, and c respectively to balance out their values.

ICNet [295] takes image A and B as input, and results in $warped\_A$ and $warped\_B$ by modelling $Flow\_AB$ and $Flow\_BA$. Authors have used two UNet [186, 194] models for registering, one for image A to B, B being the target image and another one for image B to A, A being the target image. These two models share parameters and have the same structure. The ICNet [295] pipeline consists of a grid sampler that takes as input image A and flow $Flow\_AB$ and gives $warped\_A$. Similarly, image B and flow $Flow\_BA$ as inputs result in $warped\_B$ as the output. A grid sampling network is nothing but a spatial transformer network [202] that is fully differentiable. It consists of a generator that generates a spatial grid and a sampler. This spatial grid gets converted to a sampling grid by using the DVF or flow obtained from each of the UNet models. Finally, the source image is warped using bilinear interpolation by using the sampling grid by the sampler.

7.4.2 ADMIR

ADMIR [370] (Affine and Deformable Medical Image Registration) is an affine and deformable image registration that works end to end. This method does not require the images to be pre-aligned, which in turn helps to do image registration quickly with good accuracy.

An affine registration network, a deformable registration network, and a spatial transformer are the three fundamental components of ADMIR. Both Affine and Deformable networks were trained at the same time in the original publication. The traditional image registration techniques improve the registration by iteratively optimising the similarity function.

ADMIR [370] has two sub-networks; one focuses on affine registration, while the other one focuses on deformable registration. The fixed and moving pictures are concatenated and sent into the Affine ConvNet, which predicts 12 affine transformation parameters (rotation, translation, scaling, and shearing) used to calculate the DVF u_a, which is then used by the spatial transform to coarsely warp the moving image. The coarsely warped and fixed images are then combined and fed into the Deformable ConvNet, which calculates the DVF u_d. To get the final registration result, the final DVF u_f is calculated by aggregating DVF u_a and DVF u_d with the help of a spatial transformer, which warps the moving image to fully register the moving and fixed images. Gradient loss has been employed for smoothness in addition to Normalised Cross-correlation (NCC) as the similarity loss.

7.4.3 Voxelmorph

Voxelmorph [303] is a deep learning based, unsupervised, pairwise, deformable registration framework. The framework also incorporates an additional supervised segmentation training regimen that could help further improve the performance on the dataset. Traditional iterative approaches have the disadvantage of optimising the deformation field for each pair of images independently of others and taking minutes to compute on GPU. Voxelmorph takes in two n-D volumes and computes the deformation field in a few seconds. This reduces computation time whilst maintaining performance—the main motivation for using Voxelmorph as the main inspiration of this current paper.

Voxelmorph is based on UNet [194] with LeakyReLU with parameter 0.2 as the activation function that takes a single input of concatenated affinely registered moving and fixed image. In the original work, the scans were resampled to standard $256 \times 256 \times 256$ with 1mm isotropic size. The network then predicts the DVF for the supplied pair of fixed-moving images, which is then used to transform moving images using a spatial transformer [181]. The warped image is then compared against the fixed image using a similarity metric such as MSE and NCC. The deformation field is regularised using a smoothness loss function that penalises unrealistic sharp deformations. Equation 7.15 shows the overall loss function where f is fixed image, m is moving image, ϕ is the deformation field and $\mathscr{L}_{sim}$ is the similarity loss, $\mathscr{L}_{smooth}$ is the deformation field smoothness loss, λ acts as hyperparameter for regularisation term.

$$\mathscr{L}_{us}(f, m, \phi) = \mathscr{L}_{sim}(f, m \circ \phi) + \lambda \mathscr{L}_{smooth}(\phi) \tag{7.15}$$

If additional segmentation is considered for training then the loss function gets modified as below equation 7.16 where $\mathscr{L}_{seg}$ is the segmentation loss

$$\mathscr{L}_a(f, m, s_f, s_m, \phi) = \mathscr{L}_{us}(f, m, \phi) + \gamma \mathscr{L}_{seg}(s_f, s_m \circ \phi) \tag{7.16}$$

7.4.4 Open Questions

Deformable image registration techniques, such as Voxelmorph, have been successful in capturing finer changes and providing smoother deformations. However, they do not explicitly encode global dependencies (i.e. the overall anatomical view of the supplied image) and therefore cannot track large deformations. To improve the performance in case of small as well as large deformations, supervision of the model at different resolutions might be useful, while to support the network to learn and encode the minute structural co-relations of the given image-pairs, a complete Graph Convolution Network (GCN) or a Self-Constructing Graph (SCG) might be suitable. These research questions have not been explored so far.

7.5 General Open Questions

With an ever-growing field of deep learning and its application in various fields, there is one common question ever-lurking—how robust they are when they encounter noise. In the field of medical imaging, like in MRI, there can be several sources of artefacts in the image (discussed in detail Sect. 2.5). It is quintessential to evaluate the deep learning based pipelines and how they perform when they encounter imaging artefacts when they are predominately trained on artefact-free images (*[RQ:8]*). Moreover, as typical deep learning based models require large training datasets—acquisition of such large MRI datasets is very time-consuming. Undersampling can reduce the scan time—effectively reducing the total amount of time required to create such large datasets. But undersampling might lead to artefacts. Indirectly, undersampling can reduce the probability of motion artefacts as well—the probability of a subject moving reduces with a reduction in scan-time for each subject. These make undersampling useful in the case of these automatic image processing pipelines—making this evaluation an important task. If an undersampled reconstruction technique is employed as a precursor to such pipelines, the performance of those pipelines on undersampled data might improve—as the undersampling methods attempt to reduce the artefacts in the image or improve the resolution of the image. This evaluation has not been done in the past and is an interesting research direction (*[RQ:8.1]*).

Part III
Advancing the Field of Undersampled Reconstruction

Summary
Several methods to advance the field of undersampled MRI reconstruction using deep learning have been presented here. Based on their working principles, these methods are divided into three chapters: artefact reduction in the image space, undersampled reconstruction as a super-resolution problem, and methods working in a hybrid space (image + k-space). The first chapter proposes two different methods for artefact reduction in the image space—undersampled reconstruction in the image space by proposing the ReconResNet model and the NCC1701 pipeline for both Cartesian as well as radial MRIs and the CT reconstruction-inspired undersampled radial MR reconstruction technique PDUNet, which optimises in both sinogram space (obtained by applying preprocessing on radial MRI k-space) and image space. The second chapter proposes super-resolution techniques to perform undersampled MRI reconstruction, which is divided into two sections based on the type of MRI used in the research: static and dynamic. First, for static MRIs, the ShuffleUNet model has been proposed, and its applicability while working with static diffusion-weighted MRIs has been shown. The next section presents two different techniques for the super-resolution of undersampled dynamic MRIs—by exploiting the available prior knowledge to improve the quality of super-resolution and by proposing a model to learn the spatiotemporal relationship to improve the super-resolution reconstruction. The third and final chapter presents techniques for working in the k-space and hybrid space. This chapter starts by presenting some preliminary comparisons of CNNs and CV-CNNs. Then proposes KSPResNet, which is a modified version of ReconResNet, as a CV-CNN model working directly with the undersampled k-space data. The final section presents two new models, Fourier-PD and Fourier-PDUNet, working in the hybrid space, by extending the research of PDUNet—by modifying the original models with complex-valued functions—optimising in both k-space and image space.

Artefact Reduction in the Image Space 8

One of the most popular techniques for undersampled MRI reconstruction with deep learning is to first reconstruct the undersampled data with inverse fast Fourier transform, which results in artefacts in the image—followed by artefact reduction using deep learning models in the image space. This chapter aims to improve deep learning based artefact reduction methods working in the image space with better model architectures.

8.1 NCC1701: Artefact Correction in the Image Space using ReconResNet

This study proposes an MR reconstruction framework for undersampled Cartesian and radial data, which utilises a ResNet-based model with modified residual blocks as the network backbone. Apart from the network, the framework uses an existing data consistency step for Cartesian sampling and introduces one for radial sampling to replace all the real acquired data points from the undersampled k-space in the network's output to generate the final output of the framework. The work presented in this section has been published in [pub:8]. The initial findings were previously presented in [pub:48] and a short summary in [pub:25]. The code of this project is available publicly on GitHub [code:8]. The undersampling pipeline and a collection of MRI related code-snippets for PyTorch developed as a part of this work are also available on GitHub [code:7] and [code:13]. Moreover, some initial experiments

---

Supplementary Information The online version contains supplementary material available at https://doi.org/10.1007/978-3-658-50741-1_8.

S. Chatterjee, *Deep Learning Innovations in MRI Reconstruction and Analysis*,
https://doi.org/10.1007/978-3-658-50741-1_8

were performed to filter the undersampled images in the wavelet space with the help of trainable wavelets and CNNs, that was presented in [pub:42].

8.1.1 Methodology

The proposed framework, codenamed NCC1701, contains two main components: the network backbone architecture—ReconResNet and the data consistency step. This section presents the initial assumptions taken while selecting ResNet-based model for the task of undersampled MR reconstruction, then explains the architecture of the proposed ReconResNet, elucidates the different data consistency steps for Cartesian and radial sampling, elaborates on the implementation, presents the data used in this research, discusses the undersampling patterns and finally describes the evaluation techniques.

8.1.1.1 Model Ansatz

According to the universal approximation theorem [32], it should be possible to find a nonlinear function $\Psi(x)$ that maps the corrupted input x to the artefact-free output with a stack of multiple nonlinear layers. However, looking only at the corrupted image data x this can also be seen as a superposition of the ground truth and the artefact. In other words the artefact is simply the residual between the input and output $\Phi(x) = \Psi(x) - x$. Due to strong similarities in the artefact structures of different images, it is more reasonable to force the network to approximate the residual function $\Phi(x)$ and learn the artefact. By reformulating the problem, the initially desired mapping can then be obtained through $\Psi(x) = \Phi(x) + x$.

8.1.1.2 Network Architecture

The proposed network backbone, ReconResNet, has been created using the residual learning approach—ResNet [200]. This research proposes a modified regularised version of the Residual Block by adding a Spatial Dropout [401], to randomly drop or zero out feature maps between the two convolution layers of the Residual Block; this results in better accuracy on the validation set as shown in Fig. 8.1. In this model, first, the input is down-sampled with two down-sampling blocks; each decreases the input size by half in all dimensions while increasing the number of feature maps by two (starting with 64). Next, the network contains 14 modified Residual Blocks. The input to each residual block is added to its output and forwarded to the next residual block—this is the reason behind the nomenclature—residual learning, as residue from the input is forwarded with the output to the next layer. After the residual blocks, the network contains two up-sampling blocks, each doubles the input size and reduces the number of feature maps by half to obtain the original image size back. Finally, a fully-connected convolution layer is added, followed by Sigmoid as the final activation function. Furthermore, Parameterised Rectified Linear Unit (PReLU) has been used as the activation function (except for the final activation), instead of the commonly used ReLU. The graphical representation of the network architecture is shown in Fig. 8.2.

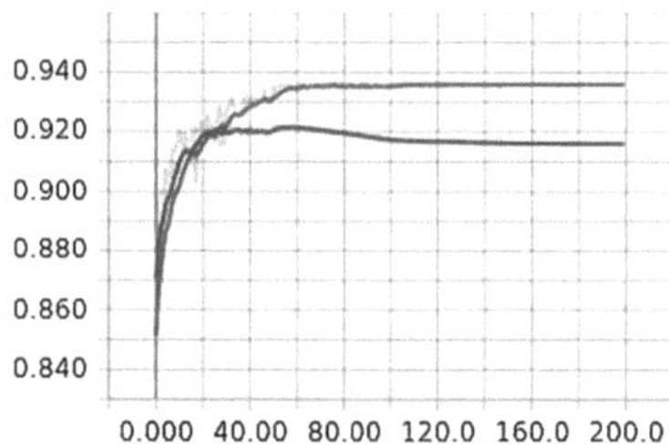

Fig. 8.1 Average accuracy on the validation set over the epochs while training on 1D Varden sampling—Pink shows the model with dropout (proposed model) and blue shows without dropout [pub:8]

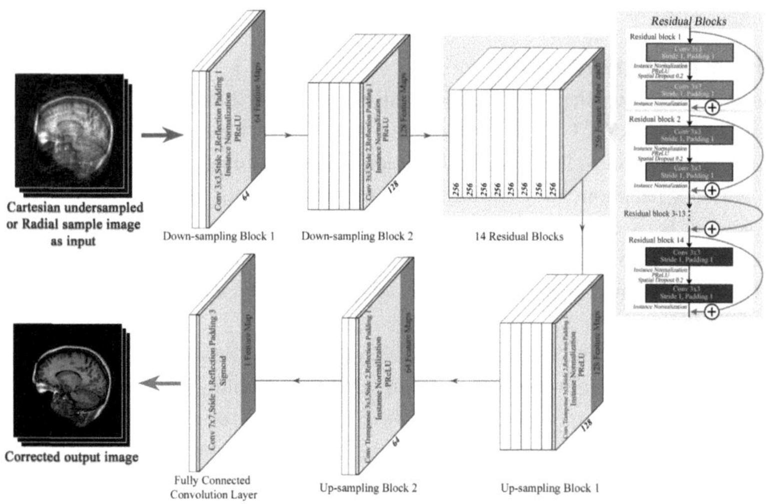

Fig. 8.2 ReconResNet: Architecture of the proposed network backbone [pub:8]

8.1.1.3 Data Consistency Step

The data consistency step replaces the actual acquired undersampled data in the network's output. In this way, the final output is not entirely dependent on the network. The network only helps to fill in the data which were ignored during the undersampled acquisition.

Cartesian sampling For undersampled Cartesian data, data consistency was performed following [269]. First, FFT was performed on the output image to get its corresponding k-space. An inverted sampling mask was applied to this to identify the k-space values, which were not acquired. Then the measured data were combined with the missing k-space data estimated by the network. Eventually, iFFT was applied to this combined k-space to obtain the final output.

Radial sampling This research introduces a data-consistency step for the reconstruction of undersampled radial data. A sampling pattern was generated for a fully sampled image (referred to as Ω_{FS}), considered to have twice the number of spokes than its pixel resolution, and a NUFFT [94] object was created from it. Then using that object, a forward transform was performed on the output image of the network to obtain its fully sampled radial k-space. The measured spokes were then inserted into the output k-space. A Density Compensation Function (DCF) was applied, followed by an adjoint NUFFT using the same NUFFT object was performed to obtain the final output.

8.1.1.4 Implementation

Figure 8.3 exhibits the working mechanism (including the dataflow) of the proposed framework NCC1701, comprising a network backbone: ReconResNet, and data consistency steps for Cartesian and radial undersampling reconstructions. Only the network backbone is used during the training process, whereas the complete framework is used during inference. To train the ReconResNet, the loss has been calculated with the help of the Structural Similarity Index (SSIM) [74] (range 0 to 1), where higher SSIM means closer image similarity, which can be calculated as Eq. 4.35.

The negative of the SSIM value has been used as the loss value (Eq. 4.29), and it was then minimised using Adam Optimiser (Initial learning rate 0.0001, decayed by 10 after every 50 epochs; $\beta_1 = 0.9$, $\beta_2 = 0.999$, $\epsilon = 1e - 09$), and was trained for 50 (for multi-slice models using all the slices), 100 (for multi-slice models, using only the central slices) or 200 epochs (for single-slice models) with a batch size of one. This network was implemented using PyTorch [327] and was trained using Nvidia GTX 980 and 1080Ti GPUs. The code of NCC1701 is available on GitHub [code:8].

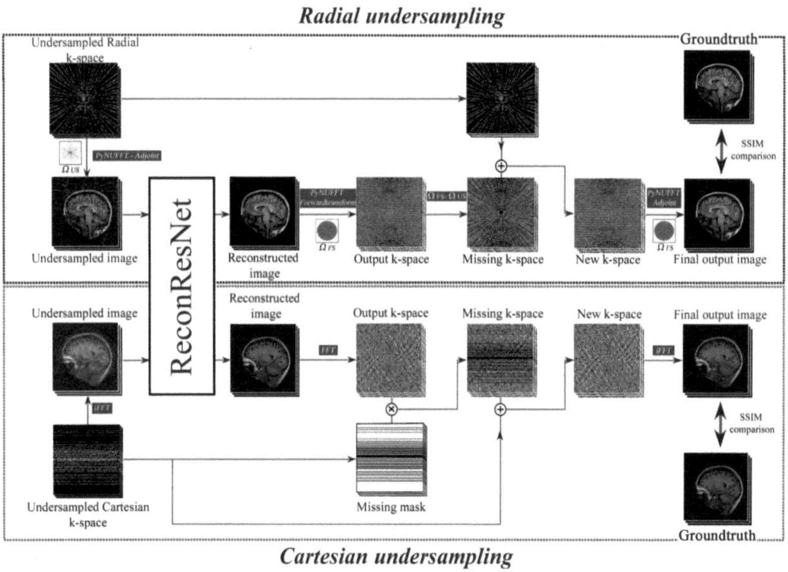

Fig. 8.3 Working Model [pub:8]

The proposed ReconResNet is created using 2D convolutions, yet a 3D version using 3D convolutions, ReconResNet3D, was also created for comparison purposes. Due to the high memory requirement of such a model, ReconResNet3D was trained using an Nvidia Tesla V100 GPU. However, the number of starting feature maps had to be reduced to 32 (from 64 in the 2D version) due to GPU memory limitations.

8.1.1.5 Dataset

Primary Dataset In this work, two different benchmark datasets were used. Firstly, the OASIS-1 [96] dataset, which contains brain MRIs of 416 subjects acquired at 1.5T (details are in Table 8.1), among which 150 were used for training and 100 were used for testing, selected randomly. 20% of the training set were used as the validation set and the remaining 80% as the actual training set. Only the first scan of each subject was used. Secondly, the IXI-dataset [url:12] was used, which contains nearly 600 Brain MRIs of normal healthy subjects of different contrasts (T1, T2, PD, MRA etc.) from three different hospitals and from two different field strengths (1.5T and 3T); T1 weighted volumes (see Table 8.2 for MR acquisition information)

Table 8.1 OASIS-1 Dataset [96]: MR acquisition information

Field strength	1.5T
Sequence	MP-RAGE
Repetition time (TR)	9.7 ms
Echo time (TE)	4.0 ms
Flip angle	10°
Inversion time (TI)	20 ms
Dead time (TD)	200 ms
Orientation	Sagittal
Voxel size	$1 \times 1 \times 1.25\,mm^3$

Table 8.2 IXI Dataset [url:12]: MR acquisition information

	Guy's Hospital	Hammersmith Hospital
Field strength	1.5T	3T
Repetition time (TR)	9.81 ms	9.60 ms
Echo time (TE)	4.60 ms	4.60 ms
Flip angle	8°	8°
Orientation	Sagittal	Sagittal
Voxel size (in mm^3)	$0.94 \times 0.94 \times 1.20$	$0.94 \times 0.94 \times 1.20$

were used for training validation and testing, by randomly taking 100 volumes for each. Only healthy subjects were used for training to avoid any kind of pathology-related bias in the model. As the vast variability of the possible brain pathologies might be difficult to represent in the training set, training set was created only with pathology-free data.

The proposed architecture works with 2D images. Therefore, preliminary tests were performed on the OASIS-1 dataset by taking slice number 50 out of all the MR volumes. Further experiments with 3D volumes were performed by creating a subset of each volume consisting of 30 slices (slice numbers 51 to 80). Each slice of the 3D volumes was sent individually to the framework as 2D slices, and the final outputs were later combined as a 3D volume.

Tests on the IXI dataset were performed separately on 1.5T and 3T volumes by using a subset of each volume containing 30 slices (slice number 61 to 90) to check the framework's performance at different field strengths. Further tests were performed by taking all the slices of each 3T volume in two different manners: a 2D approach, where all the slices were supplied to ReconResNet separately and then combined to form 3D volumes; and a 3D approach, where the volumes were directly supplied to ReconResNet3D.

Dataset for additional testing For training, only data from healthy subjects were used. To check the boundaries and validity of the approach in clinical settings, additional tests were performed on data of subjects with pathology. To test whether the proposed approach can reconstruct undersampled brain MRIs of Alzheimer patients, T1 weighted sagittal MPRAGE images from the ADNI dataset [url:11] were used, and for testing on brain MRIs with tumours, T1 contrast-enhanced images from the BraTS 2019[169] dataset was used. From ADNI, data of patients with Mild Cognitive Impairment (MCI), Significant Memory Concern (SMC), and Alzheimer's disease (AD) were included. From both datasets, only T1 weighted images were taken, as the network was trained on a similar type of acquisition.

8.1.1.6 Undersampling

Images from all the datasets were treated as fully sampled images, for both Cartesian and radial sampled experiments. The datasets don't contain any raw MR data. Thus, simulated single-channel fully sampled raw data, as well as various undersampled datasets, were artificially generated using MRUnder pipeline [url:8, code:7].

Cartesian sampling Artificial undersampling of the Cartesian raw data was achieved using a k-space sampling pattern (sampling mask) [95], which was generated by selecting fully sampled readout lines randomly in the phase encoding direction following a one-dimensional normal distribution (Fig. 8.4a) with the centre of the distribution matching the centre of the k-space (referred to as 1D Varden). Also, a densely sampled centre consisting of eight lines was added. For the two-dimensional variable density mask [95] (referred to as 2D Varden), randomly distributed k-space points were selected following a two-dimensional normal distribution (Fig. 8.4b) with a densely sampled centre covering 2.5% of the k-space. Furthermore, a uniform sampling mask was also generated by taking every n^{th} readout line from, where n equals to the step size, and a densely sampled centre consisting of nine lines were added.

The initial set of experiments was performed with three different Cartesian undersampling patterns. 1D and 2D Varden masks were generated by taking 30% of the k-space, and a uniform sampling mask was generated with a step size of 4, resulting in approximately 28% of the k-space. These patterns resulted in an acceleration factor of approximately 3.5. Further experiments were performed to check the performance of the framework while reconstructing higher acceleration factors, by taking 15%, 10% and 5% of the k-space—resulting in approximate acceleration factors of 7, 10 and 20, respectively.

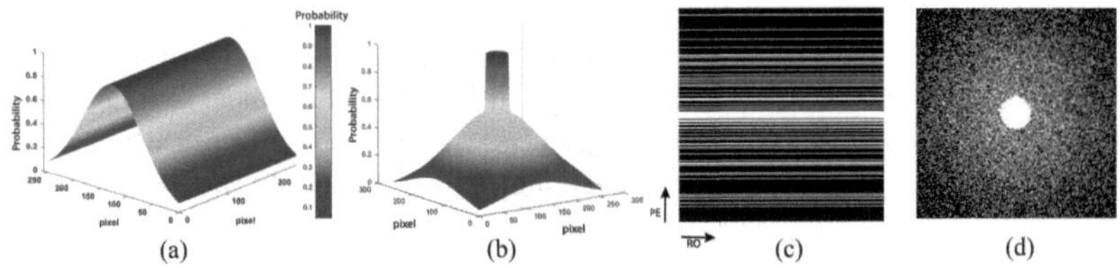

Fig. 8.4 From left to right (a) Probability density function for 1D Varden Mask (b) Probability density function for 2D Varden Mask (c) 1D Varden Mask and (d) 2D Varden Mask. All of them are for image size 256x256, taking 30%of the k-space [pub:8]

Radial sampling Artificially radially undersampled data were generated with the help of PyNUFFT [274]. The sampling pattern Ω_{US} was defined with the golden angle of 111.25° [163] [162] between the radial acquisitions (spokes). A forward NUFFT using a NUFFT object following Ω_{US} was performed on the fully sampled image data to obtain an undersampled radial k-space, then a Density Compensation Function (DCF) was applied to compensate for the over-sampled centre. Finally, an adjoint NUFFT [274] was used to bring it back to the image space with radial undersampling artefacts.

In this research, two different levels of radial undersampling by taking 60 and 30 spokes were evaluated. Considering the fully-sampled data to have twice the number of spokes than its pixel resolution (here 256), the fully-sampled data had 512 spokes. Hence, k-space with 60 and 30 spokes correspond to approximate acceleration factors of 8.5 and 17, respectively.

8.1.2 Results

Results have been evaluated quantitatively with the help of Structural Similarity Index (SSIM, Eq. 4.35), Mean-Squared Error (MSE, Eq. 4.30) and Normalised Root Mean-Squared Error (NRMSE, Eq. 4.32), where the normalisation factor was the averaged Euclidean norm of the fully sampled image. All the similarity measures were calculated by comparing the fully-sampled image with the under-sampled image (zero-filled k-space) or the output image. For quantitative evaluation of the reconstruction of images with Alzheimer's disease, the Percentage Brain Volume Change (BVD, Eq. 8.1) was calculated from the normalised brain volume (NBV) obtained using FSL SIENAX [73, 113, 138]. The BVD values were used to confirm brain volume preservation. Reconstruction performance of the models while reconstructing brain MRIs with tumours from the BraTS dataset was evaluated qualitatively by inspecting the tumour in the reconstruction.

$$\begin{aligned} BVD_{(fully-recon)} &= \frac{\left|NBV_{fully} - NBV_{recon}\right|}{\overline{NBV_{all}}} * 100 \\ BVD_{(fully-under)} &= \frac{\left|NBV_{fully} - NBV_{under}\right|}{\overline{NBV_{all}}} * 100 \end{aligned} \tag{8.1}$$

Various sets of experiments were performed using both the training datasets. OASIS-1 dataset was used initially to compare the framework's performance when trained using one slice and then by training on multiple slices by choosing centre slices of the brain, the framework's performance when multiple sampling patterns are trained together, and to evaluate the limits of the framework in terms of undersampling factor—known herewith as the "Limbo" experiment. Then experiments were performed with the IXI dataset for different undersampling patterns and different field strengths—by taking the central slices, by using all the slices, and also by supplying the 3D volumes to a 3D variant of the ReconResNet model. OASIS-1 dataset was also used to perform the additional experiments on the ability of the network while reconstructing orientations different than the training orientation and reconstructing different undersampling patterns than the training, and IXI dataset trainings were used to evaluate pathological scenarios. The various experiments with the different datasets are summarised in Fig. 8.5.

Training OASIS-1 single slice models took trained on slice number 50 of each volume for 200 epochs took around 4 hours, training the OASIS and IXI multi-slice models (3D volumes, supplied as 2D slices to the models) trained on the central slices for 100 epochs took around 45–50 hours each, while the IXI models trained on the whole volumes (as 2D slices) for 50 epochs took around 95–100 hours each. The reconstruction time using this framework for the Cartesian undersampling patterns took only a fraction of a second for each slice, while the radial undersampling patterns took around a second. It is worth mentioning that the timings may differ depending upon the system configuration (GPU, CPU, etc.) and the current workload. Finally, before going to the results, it is worth mentioning that the training and validation curves were monitored to detect any possible overfitting, and none of

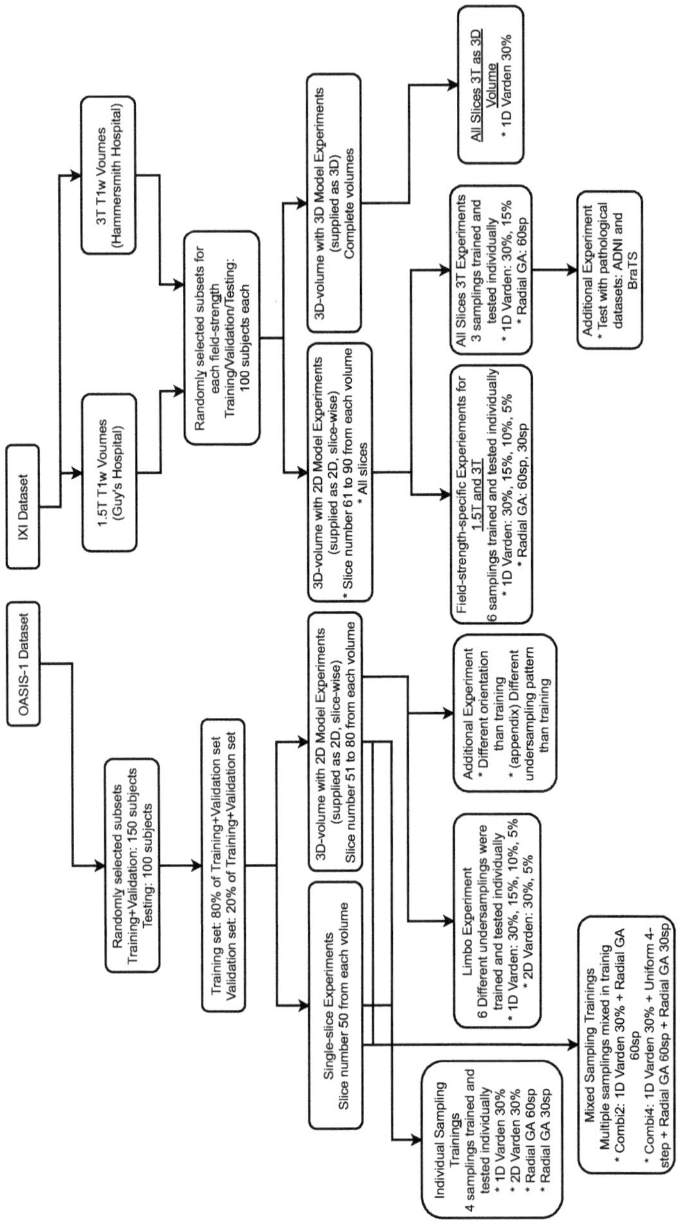

Fig. 8.5 Experiment design [pub:8]

the trainings showed any overfitting behaviour. Examples of training and validation curves have been provided in the Appendix.

8.1.2.1 OASIS-1 dataset

Various undersampling patterns trained independently For the first series of tests, various Cartesian and radial sampling patterns were trained and tested individually (results in Table 8.3 and 8.4). Two different types of trainings were performed: single slice training on slice 50 and trainings with 3D volumes (slice 51 to 80), supplied as 2D slices. While reconstructing 1D Varden 30%, 2D Varden 30%, Radial 60 spokes and Radial 30 spokes, the models yielded 0.986±0.005, 0.986±0.003, 0.981±0.004 and 0.947±0.011 SSIMs respectively. An example set of outputs for different sampling patterns has been shown in Fig. 8.7 for qualitative evaluation.

Cartesian and Radial undersampling trained together To test the robustness of the framework further, different sampling patterns were combined together while training. For the first experiment, each image in the training and test set was undersampled using one Cartesian (1D Varden 30%) and one radial (60 spokes golden-angle) sampling pattern (referred to as Combi2, results in Table 8.5). For the second experiment, each image was undersampled using four different sampling patterns, two Cartesian (30% 1D Varden and uniform sampling with step size 4) and two radial (30 and 60 spokes golden-angle) (referred to as Combi4, results in Table 8.6).

In Combi2, a decrease of 0.20% in SSIM can be observed for 1D Varden 30% (0.984±0.004 for Combi2, 0.986±0.005 for independent) but an increase of 0.10% for Radial 60 spokes (0.982±0.004 and 0.981±0.004). In Combi4, decrease of 0.81% and 0.10% can be observed for 1D Varden 30% and Radial 60 spokes respectively, but an increase of 1.58% for Radial 30 spokes (0.962±0.012 and 0.947±0.011).

Further experiments were performed to test the capabilities of the model while reconstructing different undersampling patterns than the training pattern, which are discussed in the Appendix.

Limbo: How low can it go? The limitations of the network in terms of acceleration factors were evaluated using six different Cartesian undersampling patterns. Four levels of 1D Varden (30%, 15%, 10% and 5% of the k-space) and two levels of 2D Varden (30% and 5%) undersampling were used in this OASIS Limbo experiment. The quantitative results are been portrayed with the help of violin plots in Fig. 8.6. For 30% of the k-space, 1D Varden and 2D Varden sampling resulted in similar SSIM values: 0.986±0.005 and 0.986±0.003 respectively. 15% and 10% 1D Varden

Table 8.3 Result of Cartesian undersampling patterns, while being trained separately

		1D Varden 30%			2D Varden 30%		
		SSIM	*MSE*	*NRMSE*	*SSIM*	*MSE*	*NRMSE*
Single Slice	*Under*	0.336±0.019	0.0235 0.0036	0.80±0.10	0.410±0.022	0.0103 0.0020	0.53±0.07
(Slice 50)	*Output*	0.966±0.006	0.0006 0.0002	0.12±0.02	0.971±0.003	0.0005 0.0002	0.11±0.03
3D Volume	*Under*	0.338±0.020	0.0234 0.0034	0.88±0.12	0.427±0.022	0.0102 0.0019	0.58±0.07
(Slice 51–80)	*Output*	0.986±0.005	0.0003 0.0002	0.09±0.04	0.986±0.003	0.0003 0.0002	0.10±0.03

Table 8.4 Result of Radial undersampling patterns, while being trained separately

		Radial 60sp			Radial 30sp		
		SSIM	*MSE*	*NRMSE*	*SSIM*	*MSE*	*NRMSE*
Single Slice	*Under*	0.513±0.026	0.0024 0.0004	0.25±0.02	0.375±0.019	0.0067 0.0010	0.42±0.02
(Slice 50)	*Output*	0.960±0.007	0.0006 0.0002	0.12±0.02	0.927±0.013	0.0010 0.0003	0.16±0.02
3D Volume	*Under*	0.539±0.024	0.0022 0.0003	0.27±0.02	0.413±0.017	0.0061 0.0008	0.45±0.03
(Slice 51–80)	*Output*	0.981±0.004	0.0004 0.0002	0.11±0.03	0.947±0.011	0.0010 0.0002	0.18±0.03

Table 8.5 Result of Cartesian 1D Varden mask and Radial 60 spokes, while being trained together

		1D Varden 30%			Radial 60sp		
		SSIM	*MSE*	*NRMSE*	*SSIM*	*MSE*	*NRMSE*
Single Slice	*Under*	0.336±0.019	0.0235 0.0036	0.80±0.10	0.513±0.026	0.0024 0.0004	0.25±0.02
(Slice 50)	*Output*	0.964±0.006	0.0006 0.0003	0.13±0.03	0.963±0.007	0.0005 0.0002	0.12±0.03
3D Volume	*Under*	0.338±0.020	0.0234 0.0034	0.88±0.12	0.540±0.024	0.0022 0.0003	0.27±0.02
(Slice 51–80)	*Output*	0.984±0.004	0.0003 0.0002	0.10±0.03	0.982±0.004	0.0003 0.0002	0.10±0.03

Table 8.6 Result of Cartesian 1D Varden sampling, Uniform sampling and Radial 60, 30 spokes, while being trained all four of the sampling pattern together

		Cartesian				Radia			
		1D Varden 30%		Uniform 4-step		60 spokes		30 spokes	
		SSIM	*NRMSE*	*SSIM*	*NRMSE*	*SSIM*	*NRMSE*	*SSIM*	*NRMSE*
Single Slice	*Under*	0.336±0.019	0.80±0.10	0.282±0.018	0.93±0.12	0.513±0.026	0.25±0.02	0.375±0.019	0.43±0.02
(Slice 50)	*Output*	0.957±0.008	0.14±0.02	0.947±0.012	0.15±0.02	0.961±0.008	0.12±0.02	0.935±0.011	0.15±0.02
3D Volume	*Under*	0.345±0.020	0.88±0.12	0.289±0.018	1.02±0.14	0.540±0.024	0.27±0.02	0.413±0.018	0.45±0.03
(Slice 51-80)	*Output*	0.978±0.005	0.12±0.03	0.968±0.010	0.14±0.03	0.980±0.005	0.11±0.03	0.962±0.012	0.15±0.03

resulted in 0.975±0.007 and 0.963±0.008 respectively. When it comes to the highest acceleration factor evaluated in this research, 5% of the k-space, 2D Varden resulted in 0.968±0.005 SSIM, but 1D Varden resulted in 0.906±0.017. Looking at the distribution in Fig. 8.6, it can be said that the 1D Varden 5% cannot yield reliable results (while still resulting in aesthetically good but inaccurate results, further discussed in the Appendix) and other of the undersampling patterns resulted in similar performances.

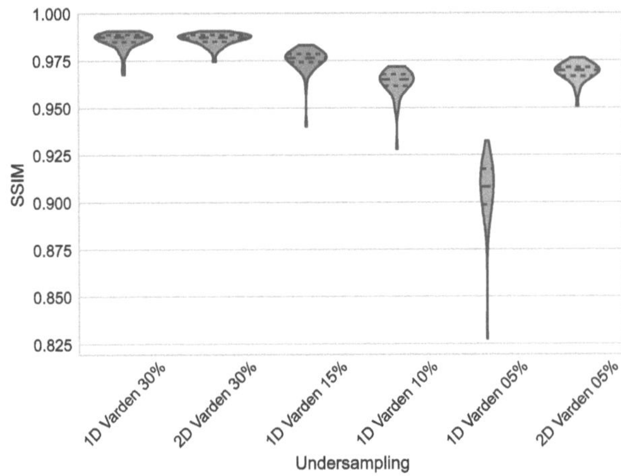

Fig. 8.6 OASIS Limbo [pub:8]

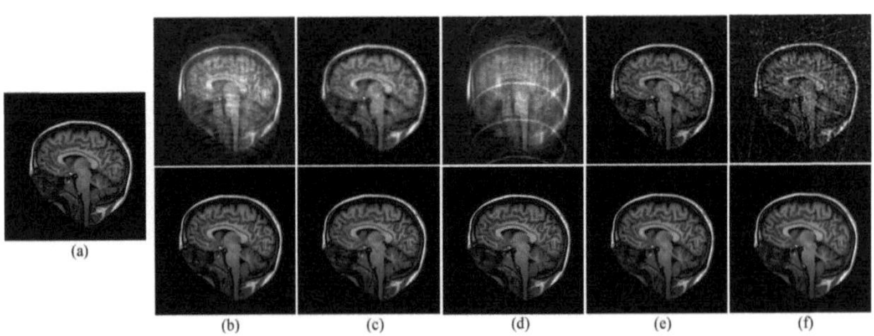

Fig. 8.7 Pipeline's example output (OASIS-1 dataset). (a) Fully sampled. From b to f—top undersampled, bottom output of the framework. Undersampling patterns: (b) Cartesian 1D Varden 30%, (c) Cartesian 2D Varden 30%, (d) Cartesian Uniform sampling step size 4, (e) Radial 60 spokes, and (f) Radial 30 spokes [pub:8]

8.1.2.2 IXI dataset

The experiments with the IXI dataset were separated based on the field strengths (1.5T and 3T). Furthermore, the performance of the proposed framework was compared against compressed sensing (CS) reconstruction. Initially, two different CS reconstruction algorithm were evaluated for 1D Varden 30% (3T IXI)—L1 Wavelet regularised reconstruction [95] and total variation (TV) regularised reconstruction [92], resulted in 0.732±0.036 and 0.701±0.04 SSIM values respectively. Given that the L1 wavelet resulted in higher SSIM values in this experiment and since it has been used successfully in various studies [352, 396], L1 wavelet was used for all the remaining experiments. Both L1 and TV reconstructions were performed using SigPy [326] for 100 iterations with the regularisation parameter $\lambda = 0.01$. IXI dataset contains already coil-combined images, and to assist the CS algorithms, coil images were simulated using a four-channel birdcage coil profile before undersampling. These coil-simulated images were then supplied to the CS algorithms. However, for RecoResNet, the original coil-combined images were used. Four different levels of 1D Varden sampling (30%, 15%, 10% and 5% of the k-space) and two different levels of radial sampling (60 and 30 spokes) were evaluated on T1w MRIs acquired at 3T and 1.5T scanners. This set of experiments was performed using a subset of the slices (slices 61 to 90) from each volume. The quantitative results are shown in Table 8.7. Figure 8.8 and Fig. 8.9 portray the distribution of the resultant SSIM values. It can be observed that the proposed method was able to reconstruct both 1.5T and 3T images significantly better than the L1 wavelet regularised reconstruction for all the undersampling patterns. It can be further said that the proposed network was able to reconstruct with high SSIM values even for higher undersampling factors, for 1D Varden Cartesian sampling acceleration factor 10 resulted in 0.939±0.012 and for radial sampling acceleration factor 17 resulted in 0.932±0.013 SSIM values for 3T IXI dataset. However, 1D Varden with acceleration factor 20 resulted in unreliable results (0.860±0.021 for 3T and 0.828±0.025 for 1.5T). This is in line with the findings of the OASIS Limbo experiment discussed earlier.

Whole volume experiments Further experiments were performed using the whole volumes for training and testing. These experiments were performed using the 3T volumes, by supplying them as 2D slices to the network, with three different undersampling patterns: 1D Varden 30%, 1D Varden 15% and Radial 60 spokes. The results of the proposed ReconResNet were compared against the UNet [269] results, and hypothesis testing was performed using the independent two-sample t-test. The proposed model performed successfully while achieving statistically significant improvement over UNet (p-values always less than 10^{-14}), for all three

Table 8.7 Quantitative results while reconstructing different undersampling patterns, of 1.5T and 3T volumes from IXI dataset

		SSIM		MSE		NRMSE	
		CS:L1Wavelet	*ReconResNet*	*CS:L1Wavelet*	*ReconResNet*	*CS:L1Wavelet*	*ReconResNet*
1D Varden 30%	*1.5T*	0.806±0.024	0.973±0.006	0.0066±0.0026	0.0006±0.0002	0.34±0.08	0.10±0.02
(AF: 3.5)	*3T*	0.732±0.036	0.976±0.007	0.0125±0.0037	0.0004±0.0003	0.62±0.11	0.11±0.03
1D Varden 15%	*1.5T*	0.694±0.031	0.948±0.014	0.0097±0.0031	0.0012±0.0010	0.42±0.09	0.14±0.05
(AF: 7)	*3T*	*0.651±0.033*	0.954±0.011	0.0171±0.0041	0.0008±0.0004	0.73±0.11	0.15±0.03
1D Varden 10%	*1.5T*	0.690±0.028	0.929±0.014	0.0099±0.0032	0.0016±0.0008	0.42±0.09	0.17±0.04
(AF: 10)	*3T*	0.639±0.030	0.939±0.012	0.0181±0.0038	0.0009±0.0005	0.76±0.11	0.17±0.03
1D Varden 5%	*1.5T*	0.526±0.028	0.828±0.025	0.0112±0.0023	0.0041±0.0012	0.45±0.06	0.27±0.05
(AF: 20)	*3T*	0.587±0.027	0.860±0.021	0.0142±0.0033	0.0020±0.0006	0.67±0.10	0.25±0.03
Radial 60 spokes	*1.5T*	0.652±0.034	0.959±0.012	0.0061±0.0021	0.0009±0.0008	0.33±0.07	0.13±0.05
(AF: 8.5)	*3T*	0.608±0.040	0.965±0.009	0.0119±0.0035	0.0006±0.0004	0.61±0.11	0.13±0.03
Radial 30 spokes	*1.5T*	0.514±0.030	0.920±0.016	0.0080±0.0010	0.0017±0.0009	0.38±0.05	0.17±0.04
(AF: 17)	*3T*	0.462±0.035	0.932±0.013	0.0119±0.0022	0.0010±0.0004	0.62±0.11	0.11±0.03

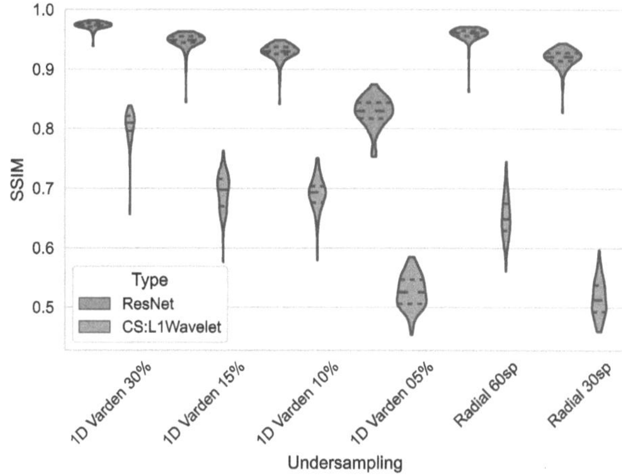

Fig. 8.8 Violin plots showing the resultant SSIM values on the IXI 1.5T dataset [pub:8]

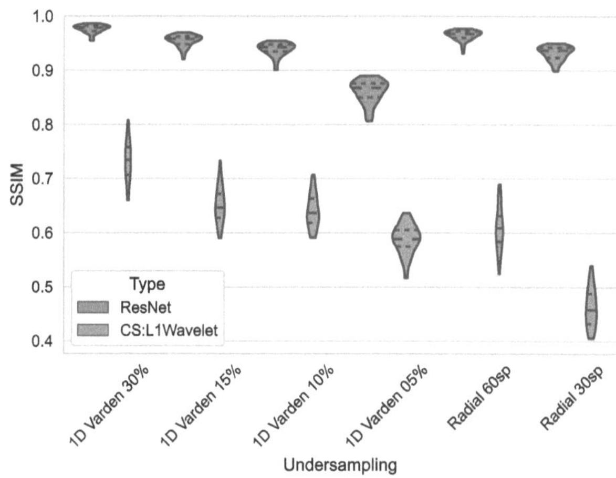

Fig. 8.9 Violin plots showing the resultant SSIM values on the IXI 3T dataset [pub:8]

sampling patterns and ReconResNet resulted in 0.990±0.006, 0.983±0.008 and 0.989±0.005 SSIM values respectively. Fig 8.10 shows the quantitative results with the help of violin plots.

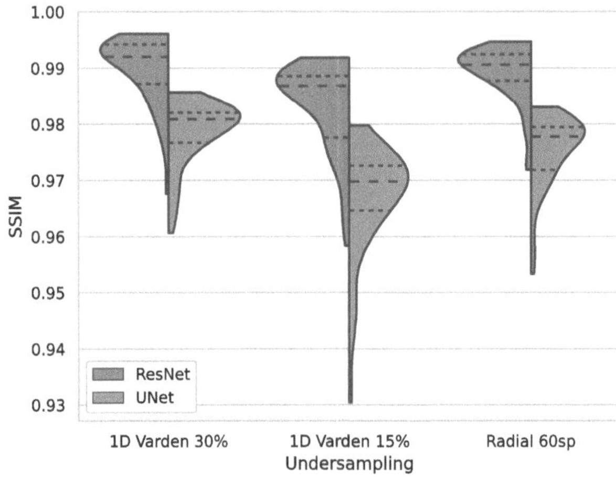

Fig. 8.10 Violin plots showing the resultant SSIM values on the IXI 3T dataset, full volumes [pub:8]

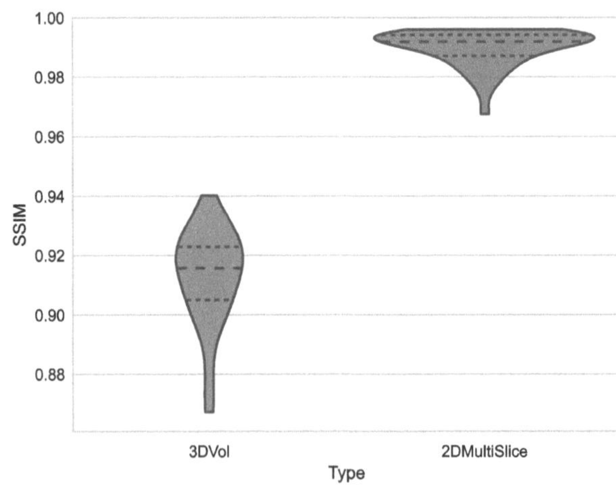

Fig. 8.11 Resultant SSIM values on IXI 3T dataset, full volume—ReconResNet 3D vs 2D [pub:8]

One additional experiment was performed by directly supplying the 3D 3T volumes undersampling using 1D Varden 30% to the framework using ReconResNet3D as the backbone network. It was observed that the performance of the network was 7.88% lower in terms of SSIM values than the 2D version, resulting in 0.912±0.017. Figure 8.11 shows the comparative results of both the models.

8.1.2.3 Additional experiments

Reconstructing images in different orientations than the training The models trained using the OASIS-1 dataset were used in testing the performance of the network while reconstructing different orientations (coronal and axial) from the one used during training (sagittal). As observed in Fig. 8.13, testing the models using the coronal orientation yield higher SSIM values than the axial orientation. The models trained on a combination of undersampling patterns (Combi4) did better in reconstructing different orientations scoring higher SSIM values than the specific models trained on one pattern. As illustrated in Fig. 8.14 reconstructions of coronal orientations are better than reconstructions of axial orientations, where some artefacts are still present. It also shows a better reconstruction in the case of radial undersampling patterns in comparison to Cartesian for both axial and coronal orientations.

Reconstructing brain MRIs with pathological abnormalities For testing the network's applicability on clinical data, MR volumes of Alzheimer's disease patients and patients with brain tumours were used from the ADNI and BraTS datasets, respectively.

For the ADNI dataset, the performance was measured quantitatively by measuring the difference in brain volume (BVD, Eq. 8.1), to evaluate the framework's ability to preserve the brain volume. Fig. 8.15 shows that the BVD values between fully-sampled-reconstructed and fully-sampled-undersampled for both radial and Cartesian undersampling patterns. Fig. 8.12 shows the reconstructed output, which is free from artefacts and highly similar to the fully-sampled image.

As for testing the clinical tumour data, the performance was evaluated using qualitative assessment. Fig. 8.16 shows visual examples of the reconstructed images, where the tumour seems to be preserved, and most of the artefacts are gone.

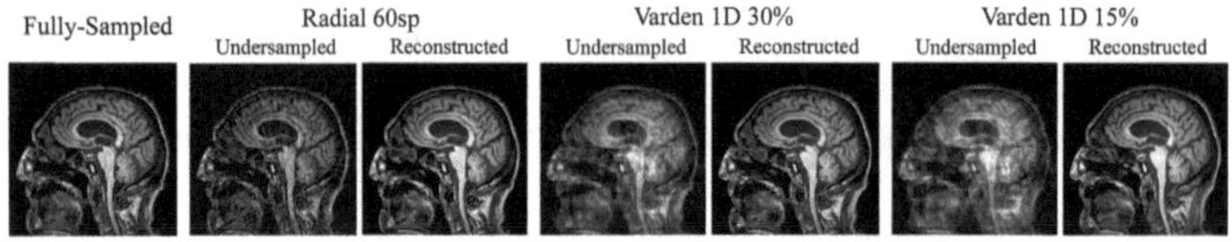

Fig. 8.12 Examples of undersampled reconstruction of ADNI images: Radial 60sp, 1D Varden 30% and 15% [pub:8]

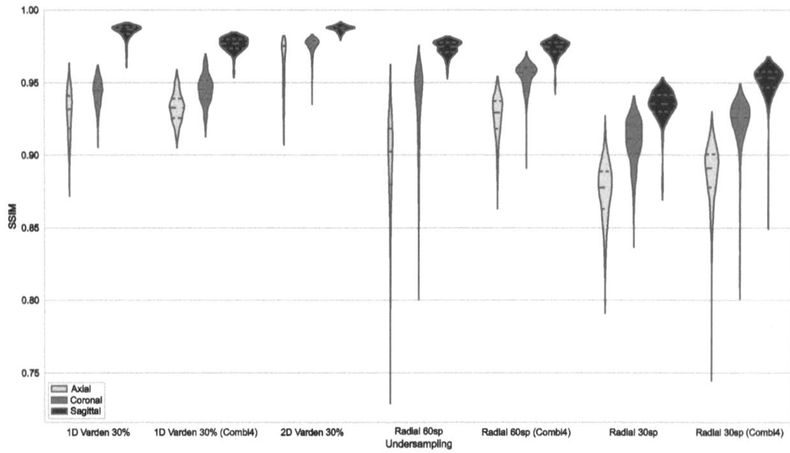

Fig. 8.13 SSIM values for 2D Models Tested on Original orientation (Sagittal in blue) or different orientation (Coronal and Axial in orange and green respectively). In Combi4, the model was trained on Radial 60sp, Radial 30sp, Uniform4step and Varden 1D 30% [pub:8]

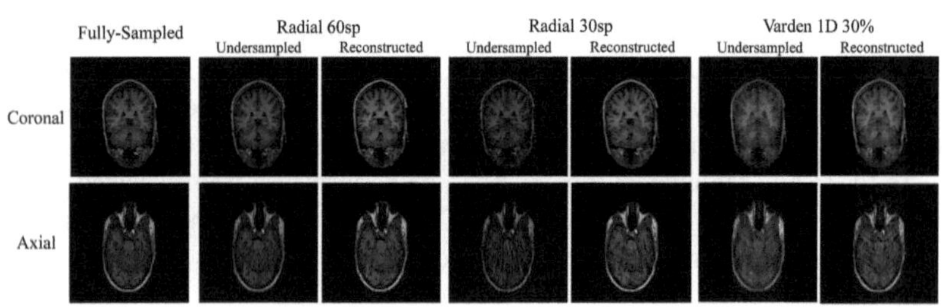

Fig. 8.14 The top row is an example of the reconstruction of the coronal orientation and the axial orientation is shown in the bottom row: For Radial 60sp, 30sp and 1D Varden 30% [pub:8]

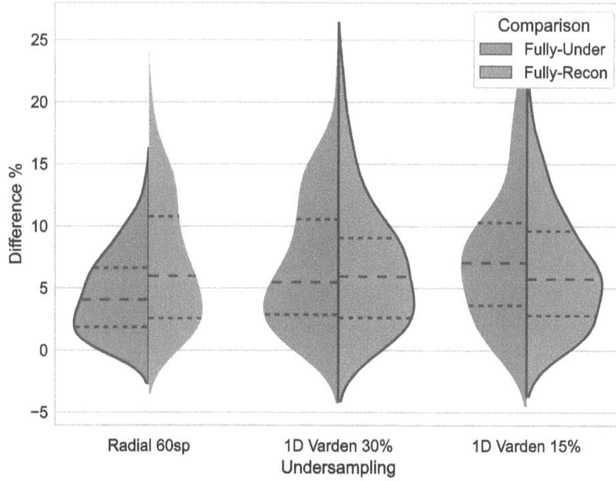

Fig. 8.15 Paired comparison of brain-volumes (BVD) values for (fully-sampled-reconstructed) in green and (fully-sampled-under) in orange or each tested model on Alzheimer's dataset ADNI [pub:8]

8.1.3 Discussion

The evaluation of the proposed framework for undersampled MRI reconstruction has proven its applicability based on different datasets. It can be observed that the framework seems to be robust against various undersampling patterns. When different sampling patterns were trained together, the accuracy was a little bit lower. For example, for 3D volumes undersampled using 1D Varden 30%, when trained only on this sampling pattern resulted in 0.986±0.005, while being trained along with radial 60 spokes resulted in 0.984±0.004, and while being trained with three other sampling patterns resulted in 0.978±0.005 SSIM. For radial 60, the resultant SSIM values are similar for all three cases. For radial 30 spokes, the result improved when the training was performed in combination with other patterns (Combi4: 0.962±0.012, independent: 0.947±0.011). So, it can be said that the network can be trained with various types of sampling patterns all at the same time to make the trained network work with various sampling patterns.

OASIS Limbo experiments have shown that the framework was able to reconstruct undersampling factors as high as 20 for 2D Varden sampling and 10 for 1D Varden sampling, yielding 0.968±0.005 and 0.963±0.008 respectively. However, the network struggled to reconstruct an acceleration factor of 20 for 1D Varden sampling (resultant SSIM: 0.906±0.017). With the help of the IXI dataset, it was shown

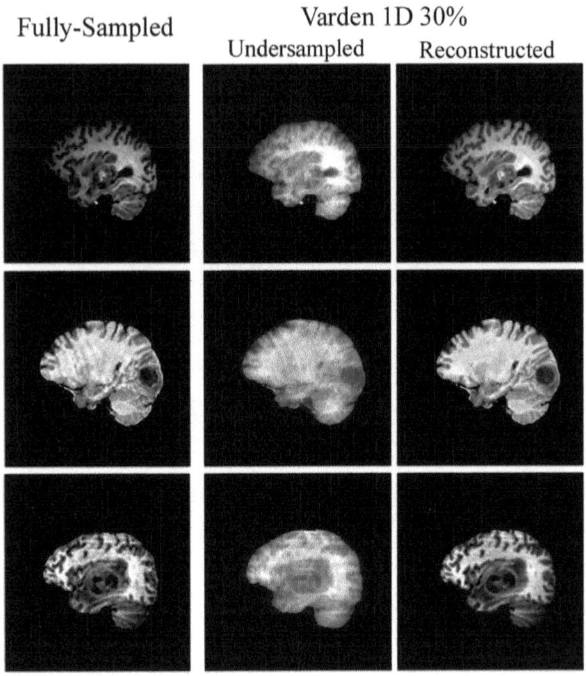

Fig. 8.16 Examples of the BraTS 2019 dataset outcome tested on Varden 1D 30% model. It can be seen that the tumours are preserved and most of the artefacts are gone [pub:8]

that the network was able to perform well for both 1.5T and 3T MRIs, for up to an acceleration factor of 17 for radial sampling and an acceleration factor of 10 for 1D Varden sampling, resulting in 0.932±0.013 and 0.939±0.012 respectively for 3T, 0.920±0.016 and 0.929±0.014 respectively for 1.5T. It was further observed that the proposed ReconResNet performed significantly better than L1-wavelet regularised compressed sensing reconstruction and UNet.

While experimenting with the whole 3D volume, it was observed that the 2D version of the model trained on the slices separately worked superior to the ReconResNet3D working with the whole volume as 3D, resulting in 0.990±0.006 and 0.912±0.017 SSIM values respectively for 3T volumes undersampled with 1D Varden 30%. It should be taken into account that these two models are not directly comparable, as the number of starting features was cut by half (from 64 to 32) for the 3D model due to GPU memory limitations, even while using a more powerful GPU than the 2D model (discussed in Sect. 8.1.1.4). Moreover, it is to be noted that the 3D model received a significantly less number of data points than the 2D model, albeit each data point is larger: because the 3D model received the slices combined together as a volume, whereas the 2D model received them individually. This might also have a negative impact on the results of the 3D model. Given these results and also taking into account the GPU memory demand of ReconResNet3D, it can be said that the 2D model might be sufficient to perform this task of undersampled MRI reconstruction. However, this demands further experimentation to reach a definitive conclusion.

The experiments performed while reconstructing images in different orientations than the training orientation showed that the reconstruction of images in coronal orientation was more successful than in the axial orientation, while the network was trained on sagittal images. One possible explanation is that sagittal and coronal orientations share more structural similarities (for example, the resemblance of the cerebellum and brainstem) than sagittal and axial images. Besides, the eyes' optical nerve structure in axial images is more pronounced and differs from sagittal and coronal orientation. As for the difference in the performance between the network trained on multiple patterns and on a specific one, one can argue that it is better for the network to see more different undersampling patterns, which also serves as data augmentation and helps to produce more accurate results. Trainings performed with different orientations mixed together might be able to improve the model's ability to reconstruct different orientations. Furthermore, the ReconResNet3D might perform better handling different orientations than the training orientation—which should be investigated in the future.

The models were only trained with healthy subjects. So, testing the reconstruction performance with volumes containing pathology is an important experiment. In this research, the framework's performance for two different types of pathology was evaluated to judge the clinical applicability. The first type of pathology was Alzheimer's disease, which was evaluated using the ADNI dataset and yielded close to zero change in BVD (fully-sampled vs reconstructed), which is a good indication and it confirms the network's ability to preserve the original volume when reconstructing undersampled data. The BVD (fully-sampled vs undersampled) for the Radial 60 spokes was close to the BVD (fully-sampled vs reconstructed) results because, unlike Cartesian, the radial under-sampling pattern does not distort the image in a way that leads to a change in brain segmentation performed by SIENAX, resulting in the change of brain volume. The second type of pathology that was evaluated here was brain tumours. It was observed qualitatively that the network was able to preserve the tumours while removing most of the artefacts while reconstructing undersampled brain MRIs with tumours from the BraTS dataset. It is to be noted that the resolution of the BraTS dataset is different from the resolution of the training set. Furthermore, the volumes were already brain-extracted while the training set volumes were not. These might have impacted the results. Further experiments should be performed by training the network with healthy brain-extracted volumes and/or experimenting with different brain tumour datasets where brain extraction has not been performed. It is also worth mentioning that an in-depth clinical applicability study, including evaluation by radiologists, was not performed under the scope of this research and will be performed in the future.

Another interesting thing to point out is that all the trainings and evaluations in this research were conducted using T1 weighted MRIs. However, other MR sequences, such as T2-weighted, FLAIR, etc, are interesting for brain pathologies. It can be hypothesised that the framework should be able to work in a similar fashion with other types of sequences—a hypothesis that will be investigated in the future.

On a closing note, it is to be taken into account that the proposed framework currently works with coil-combined images, and the artificial undersampling technique used here only simulates single-channel data. Real data with multiple channels might produce a better result and can further be accelerated using parallel imaging—this open question is to be investigated in the future.

8.1.4 Summary and Outlook

A residual learning based framework for undersampled MR data, codenamed NCC1701, has been proposed here. The framework uses the proposed ReconResNet as the network backbone and applies data consistency after reconstructing using the backbone network—making the network model only responsible for predicting the missing data from the undersampled k-space. Evaluation using multiple datasets has shown that the proposed framework can efficiently work with both Cartesian and radial undersampled data, even while trained together, and provides results with high accuracy (SSIM value as high as 0.99) and achieved statistically significant improvements over the baseline methods (***[RQ:2]***, ***[RQ:5]***). Moreover, the proposed framework was able to reconstruct MRIs with pathology (Alzheimer's disease and brain tumours), while being trained using MRIs of only healthy subjects. Experiments presented here have shown that the framework was able to reconstruct properly for undersampled data with an acceleration factor of 20 for Cartesian (2D Varden 5%) and an acceleration factor of 17 for radial (60 spokes) acquisitions (***[RQ:6]***). The acquisition time can further be reduced using techniques such as parallel imaging. Given the fast inference speed of a trained network and the high acceleration factors that can be achieved with the proposed framework, this can be a candidate to be used for real-time MRIs like interventional MRIs.

8.2 Primal-Dual UNet: Radial MRI Reconstruction using Sinogram Upsampling

Radial MRI and computed tomography are very similar mathematically—a common solution might be possible to find for the problems of undersampled radial MRI and sparse CT reconstructions. However, the deep learning research community did not try to unify these problems before. This study aims at unifying these problems and attempts to improve the existing deep learning based CT reconstruction techniques which can be applied for both undersampled radial MRI and sparse CT reconstruction. The work presented in this section has been performed in collaboration with Philipp Ernst and has been published in [pub:6], while some of the initial results were presented in [pub:33]. The code of this project is available publicly on GitHub [code:9].

8.2.1 Methodology

8.2.1.1 Fourier Slice Theorem

CT and radial MRI measurements are mathematically very similar. The Fourier slice theorem tells that the one-dimensional Fourier transform ($\mathscr{F}_1$) of a parallel-beam projection for a certain angle (p_θ) is the same as extracting a one-dimensional central slice with the same angle (S_θ) from the two-dimensional Fourier transform ($\mathscr{F}_2$) of the image function (f):

$$\mathscr{F}_1[p_\theta](u) = (S_\theta \circ \mathscr{F}_2[f])(u) \tag{8.2}$$

This means that every parallel-beam CT projection fills up the 2D Fourier space with one spoke. The spokes acquired in radial MRI do exactly the same, which means that conversely, applying the one-dimensional inverse Fourier transform on these spokes results in values similar to CT projections, such that it is possible to use CT reconstruction algorithms like FBP to ultimately reconstruct MRI images. This was the earliest method to reconstruct such MRIs [234].

Though not directly applicable to the more commonly used fan- and cone-beam projections, the Fourier slice theorem has been used in reconstruction algorithms for these beam types with slight modifications [39].

8.2.1.2 Proposed Architecture

The network architecture proposed in this research is based on the Primal-Dual Network [249], which achieves superior reconstruction quality compared to other deep learning based reconstruction algorithms [389], and is combined with UNet [186], which is well-known and often used in medical image processing.

The Primal-Dual Network can be interpreted as an unrolled iterative reconstruction algorithm that optimises in both the sinogram and image space using blocks of fully convolutional layers. In each iteration, the processed sinograms are reconstructed using FBP, combined with the processed images, and reprojected to be combined with the previous sinograms. This does not only improve the quality of the final reconstructions, but also ensures data consistency with the original sinogram. As with many iterative algorithms, the quality of the reconstructions of the Primal-Dual Network depends on the number of iterations, i.e. for a fixed number of parameters in an iteration, an optimal (usually minimal) number of iterations has to be found for the network to converge. If the number of parameters in the convolutional blocks is low, more iterations are needed for convergence. This, however, increases the processing time, with the reconstruction and projection operators being the bottleneck. On the other hand, if the number of parameters in the convolutional blocks increases, much fewer iterations are needed for convergence. The processing time is not much different though, since the convolutions are the bottleneck in this case.

The idea of the network architecture proposed in this research tackles this trade-off by (1) keeping the number of iterations low, i.e. the reconstruction and projection operators are not a bottleneck, and (2) replacing the convolutional blocks in the image space with a UNet, to get a high number of parameters while keeping the processing time low. This architecture will be referred to as *PD-UNet*, and it is presented in Fig. 8.17.

8.2.1.3 Baselines

The performance of the proposed model was compared against three deep learning based methods. First, a UNet [186] was applied on reconstructed Undersampled CT images [227] and MRIs [269], referred to hereafter as *Reconstruction UNet*. The CT images were reconstructed using FBP, and the undersampled MRIs were reconstructed using adjoint NUFFT [274]. The undersampled images were supplied as input to the Reconstruction UNet model, and the outputs of the model were compared against the ground-truth fully-sampled images to calculate the loss while training.

The second baseline was the *Sinogram UNet* [271]. For this method, the sparsely sampled sinograms were upsampled using bilinear interpolation before supplying them to the UNet model as input, and the loss was computed by comparing the outputs of the model against the corresponding fully-sampled sinograms.

The final deep learning baseline was the *Primal-Dual Network* [249], where the sparsely sampled sinograms and zero-initialised reconstructions were supplied as input to the model, and the outputs from the model were compared against the ground-truth fully-sampled images to calculate the loss.

Lastly, the performance of the models was compared with FBP reconstructions after upsampling the sparse-sampled sinograms using bilinear interpolation. For MRI, one additional comparison was performed against images obtained by applying adjoint NUFFT [274] on the undersampled radial k-space (referred in this research as *Undersampled (NUFFT)*). It is noteworthy that the majority of the baselines (Sinogram UNet, Primal-Dual Network, and Sinogram Bilinear) were never used before for undersampled MR reconstruction.

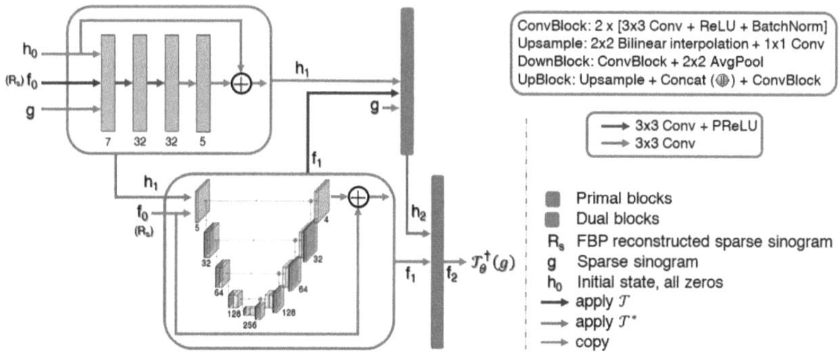

Fig. 8.17 Primal-Dual UNet (PD-UNet): Architecture of the proposed network. The orange and the green boxes signify the primal and dual iterates, respectively. The proposed primal block uses a UNet model instead of a fully convolutional network. In contrast, the dual block is the same fully-convolutional network as the original primal-dual network. It is to be noted that the second orange and green blocks have the same architecture as the first one. The initial dual block receives the image reconstructed by applying FBP on the sparse sinogram as f_0 (unlike the original primal-dual network, which receives all zeros), all-zeros as the initial-state h_0, and the sparse sinogram g. The output of this block, along with the same f_0 is given as input to the first primal block [pub:6]

8.2.1.4 Data Normalisation

Data in the image space was normalised by dividing each slice by the 99th percentile of the intensity values present in the total training and validation sets. The sinograms were normalised using the Z-score normalisation method—by applying the following equation on each sinogram:

$$S_N = \frac{S - \mu_S}{\sigma_S} \tag{8.3}$$

where S_N is the normalised sinogram, S is the original sinogram, and μ_s and σ_s are the mean and standard deviation of the values present in the sinogram.

For the Reconstruction UNet and Sinogram UNet, the input was normalised using the image space normalisation and sinogram normalisation methods, respectively. The output given by those models were "de-normalised" to obtain the final output—for image space, multiplying with the 99th percentile of the intensity values; for sinogram, by applying the following equation:

$$S = (S_N \times \sigma_s) + \mu_s \tag{8.4}$$

For the Primal-Dual Network and the proposed PD-UNet, several normalisation and de-normalisation steps were performed. Initially, μ_s and σ_s were calculated from the sparsely sampled input sinogram, and all the sinogram normalisations and de-normalisations were performed using these values. Each time before a sinogram or an image was given as input to any block, they were normalised using sinogram and image normalisation methods as discussed earlier, and after receiving output from that block, they were de-normalised. This was performed to preserve the relationship between image and sinogram values while using two different types of normalisation techniques for two different data spaces. Before providing the final output of these models, the values were also de-normalised using the image space technique.

8.2.1.5 Implementation and Training

The models (the proposed model and the baseline models) were trained for 151 epochs with an effective batch size of 32, the best epoch was chosen based on the validation loss and was used for inference on the test set. The memory requirements of the proposed model and the different baseline models are not the same, making it impossible to have the same batch size for the different models. To achieve a constant effective batch size for all the models, instead of the conventional "forward-pass then backward-pass" technique for each training step, multiple forward passes were performed, the gradients were summed up, and finally, an accumulated backward-pass was performed. The number of forward passes to be performed before an accumulated backward pass was calculated as: (32 $\div$ actual batch size of the model). To train the models, the loss was calculated using mean absolute error (L1-Loss, Eq. 4.28), and it was minimised using Adam optimiser (learning rate $=$ 1e $-$ 3, $\beta_1 = 0.9$, $\beta_2 = 0.999$, $\epsilon =$ 1e $-$ 08). The models were implemented using PyTorch [327], with the help of PyTorch Lightning [311]; and were trained with mixed precision [236] using Nvidia RTX 2080Ti and Nvidia RTX A6000 GPUs. The code of this project is available on GitHub [code:9].

8.2.1.6 Dataset

This section furnishes the details regarding the datasets used for the MRI experiments, while the CT datasets are explained in the Appendix (Sect. A.2.2). Two different publicly available benchmark datasets of two different organs were used in this research: IXI dataset [url:12] for the brain and CHAOS challenge dataset [pub:16] for the abdomen. The IXI dataset contains nearly 600 brains MRIs of normal healthy subjects, acquired using different MRI protocols (T1w, T2w, PDw, MRA, and DWI), collected from three different hospitals at two different field strengths (1.5T and 3T). 30 central slices from 100, 35 and 50 T1w volumes acquired at 3T were used in this study as the training, validation and test set, respectively. The CHAOS challenge dataset contains abdominal MRIs of 40 healthy subjects, acquired using two different MR sequences: T1-Dual (In-phase and Opposed-phase) and T2-SPIR. All the slices from 24, 6 and 10 subjects (three volumes each: T1-in, T1-opposed, and T2) were used in the training, validation, and test set, respectively. All the images were interpolated with sinc interpolation to have an in-plane matrix size of 256x256.

The datasets do not contain any raw MRI data, only the magnitude images, which were treated as the fully-sampled ground-truth. The corresponding single-coil radial k-spaces of those magnitude images were generated using NUFFT (implemented in PyNUFFT [274]). The fully-sampled raw data was considered to have the number of spokes (radial acquisitions) as twice the base resolution, which was 512 for this dataset. The sampling was performed following the equidistant radial sampling scheme, where the angle between the spokes, calculated as $\Delta\phi = \pi \div n_{Sp}$ for $n_{Sp} \in \mathbf{N}$ spokes, was $\Delta\phi = 0.3515625$. A NUFFT was performed on each slice of the MRI volumes, following $\Delta\phi$, to obtain the equivalent fully-sampled radial k-space.

Pre-processing The fully-sampled sinograms of the respective radial k-spaces were obtained by applying the one-dimensional inverse Fourier transform on each spoke of the k-space. To keep the setup similar to the one described in the Appendix (Sect. A.2.2), the spokes were shifted by half a detector pixel using sinc interpolation and were cropped to the central 363 detector pixels. This step was necessary because each spoke extracted by NUFFT contained 512 frequency components, which corresponds to a detector pixel number of 512, as well. In contrast to the simulated CT projections in the Appendix (Sect. A.2.2), the MR sinograms correspond to parallel-beam projections according to the Fourier slice theorem (cf. Sect. 8.2.1.1).

Undersampling To simulate the undersampled datasets, the sinograms were made sparse by retaining only every n^{th} projection (spoke), where n denotes the level of sparsity. For CT, three different sparsity levels were experimented with: 4, 8 and 16 (referred to herewith as Sparse 4, Sparse 8 and Sparse 16, respectively). In the case of MRI, two levels were used: Sparse 8 and Sparse 16.

8.2.2 Results

The performance of the models was evaluated and compared quantitatively with the help of Root Mean-Squared Error (RMSE, Eq. 4.31) and Structural Similarity Index (SSIM, Eq. 4.35) [74]. Moreover, the statistical significance of the improvements observed was evaluated by the Mann-Whitney U test [7]. Finally, they were also compared qualitatively for selective slices with the help of difference images and SSIM maps. Slices that resulted in SSIM values identical to the median value up to three decimal points for Primal-Dual UNet (the proposed method) and Primal-Dual Network (main baseline) were chosen for qualitative portrayal for each experiment—to be able to choose slices which are representative of the results for each of the experiments.

This section presents the results of the MRI experiments and the CT results are presented in the Appendix (Sect. A.2.2). The proposed Primal-Dual UNet was compared (see Sect. 8.2.1.3) against the undersampled radial k-space reconstruction using PyNUFFT [274] (referred to as *Undersampled (NUFFT)*), by applying FBP on the corresponding sinograms (obtained by applying 1D-iFFT on each spoke, explained in Sect. 8.2.1.6), referred to here as *Sinogram Bilinear*, and finally, against three deep learning baseline models: Reconstruction UNet [269], Sinogram UNet [271], and Learned Primal-Dual Network [249]—for two different publicly available benchmark datasets for two different organs: IXI for T1w brain MRIs and CHAOS for T1-Dual and T2w abdominal MRIs. Experiments were performed for two different levels of undersamplings: with an acceleration factor of 8 and 16—which, in terms of the sparsity of the corresponding sinograms for equidistant radial samplings (see Sect. 8.2.1.6 and 13), are referred to here as Sparse 8 and 16—to have the same terminology for both CT and MRI.

Table 8.8 Resultant metrics for MRI for the IXI dataset (mean±std)

Method	SSIM		RMSE	
	Sparse 8	Sparse 16	Sparse 8	Sparse 16
Undersampled (NUFFT)	0.595±0.027	0.410±0.021	0.046±0.009	0.085±0.016
Bilinear Sinogram	0.819±0.026	0.682±0.035	0.058±0.017	0.067±0.016
Sinogram UNet	0.860±0.044	0.782±0.032	0.058±0.021	0.073±0.023
Reconstruction UNet	0.948±0.011	0.877±0.025	0.021±0.006	0.037±0.012
Primal-Dual Network	0.947±0.012	0.867±0.025	0.021±0.006	0.041±0.015
Primal-Dual UNet	**0.965±0.008**	**0.903±0.019**	**0.017±0.005**	**0.034±0.011**

IXI Dataset Qualitative results of SSIM and RMSE are shown in Table 8.8, whilst the range of the resultant SSIM values with the help of box plots is shown in Fig. 8.18 and 8.19 for Sparse 8 and 16, respectively. It can be observed that the proposed model outperformed all the baseline methods in terms of both SSIM and RMSE, and the statistical tests revealed that these improvements were significant. In terms of average SSIM values, the Primal-Dual UNet achieved improvements of 1.9% and 4.15% over the main baseline Primal-Dual Network for Sparse 8 and 16, respectively. Qualitative comparisons of the results using difference images and SSIM maps are shown in Figures of 8.20 and 8.21 for Sparse 8 and 16, respectively.

CHAOS Dataset The final set of experiments was performed on the CHAOS MRI dataset. Quantitatively, the proposed method outperformed all the baselines for both acceleration factors with statistical significance when being compared using SSIM and RMSE, reported in Table 8.9. Figure 8.22 and 8.23 portray the range of resultant SSIM values for Sparse 8 and 16, respectively. The proposed method improved the average SSIM values by 0.41% and 0.84% over the baseline Primal-Dual Network, respectively, for Sparse 8 and 16.

The CHAOS dataset includes three different types of MRIs: T1 in-phase, T1 opposed-phase, and T2, acquired using two different sequences for two different contrasts: T1-Dual and T2-SPIR (explained in 8.2.1.6) and all three types were combined during training. The metrics that have been shown so far also included all these three types of MRIs together. However, in addition, they were also evaluated separately. Table 8.10 shows the resultant SSIM values for each of the three types

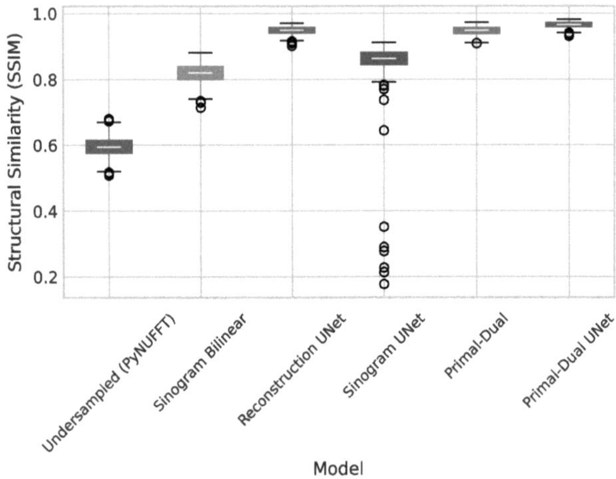

Fig. 8.18 Box-plots of the resultant SSIM values for MRI Sparse 8 for the IXI dataset [pub:6]

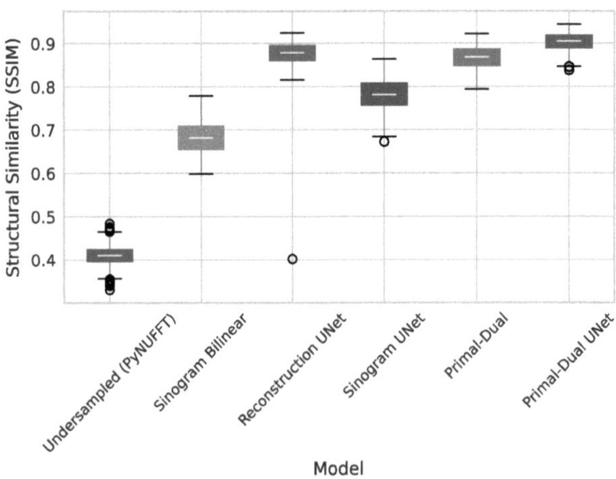

Fig. 8.19 Box-plots of the resultant SSIM values for MRI Sparse 16 for the IXI dataset [pub:6]

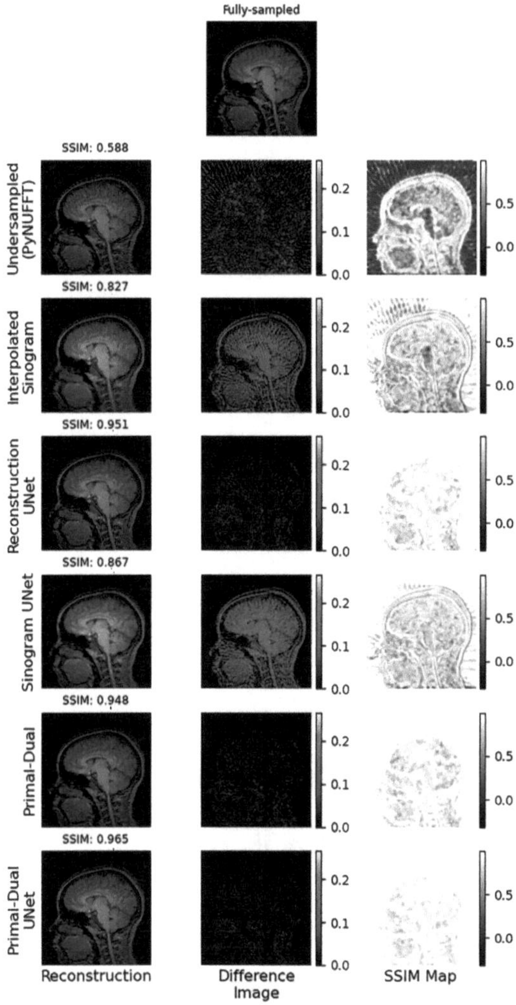

Fig. 8.20 Qualitative comparisons of the reconstructions of MRI Sparse 8 for the IXI dataset [pub:6]

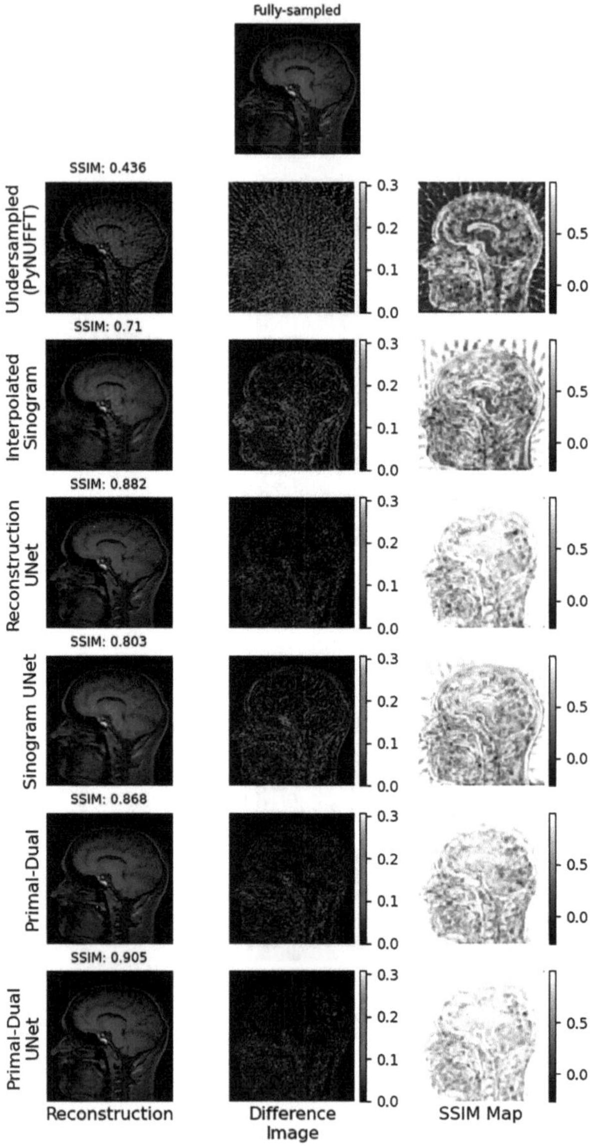

Fig. 8.21 Qualitative comparisons of the reconstructions of MRI Sparse 16 for the IXI dataset [pub:6]

Table 8.9 Resultant metrics on the CHAOS dataset (mean±std)

Method	SSIM		RMSE	
	Sparse 8	Sparse 16	Sparse 8	Sparse 16
Undersampled (NUFFT)	0.528±0.092	0.359±0.060	0.034±0.015	0.068±0.027
Bilinear Sinogram	0.915±0.033	0.843±0.060	0.035±0.013	0.044±0.016
Sinogram UNet	0.926±0.026	0.897±0.038	0.032±0.011	0.036±0.014
Reconstruction UNet	0.981±0.013	0.943±0.029	0.011±0.007	0.022±0.014
Primal-Dual Network	0.982±0.012	0.949±0.025	0.010±0.005	0.021±0.010
Primal-Dual UNet	**0.986±0.011**	**0.957±0.023**	**0.009±0.006**	**0.018±0.011**

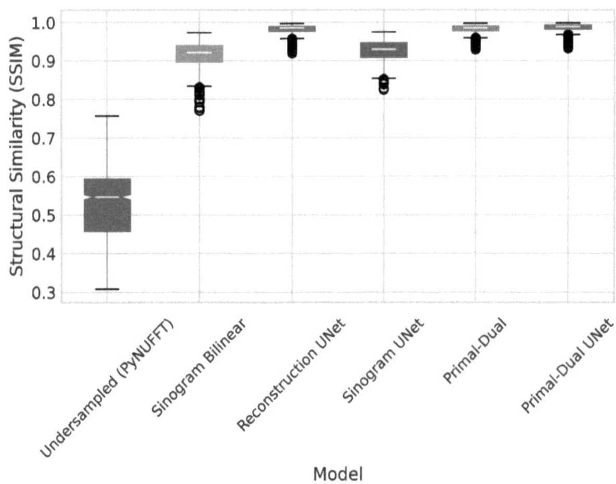

Fig. 8.22 Box-plots of the resultant SSIM values for MRI Sparse 8 for the CHAOS dataset [pub:6]

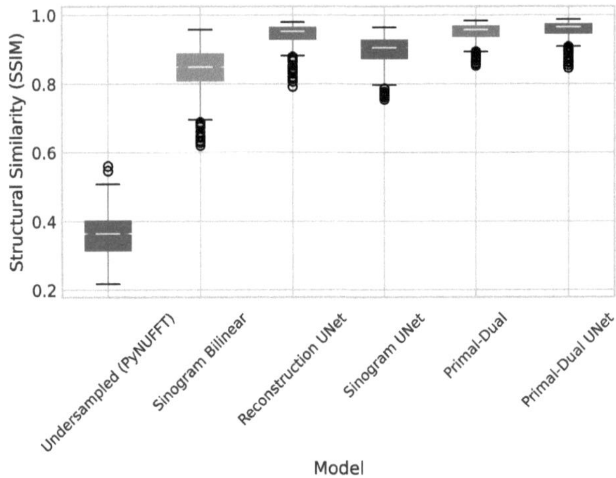

Fig. 8.23 Box-plots of the resultant SSIM values for MRI Sparse 16 for the CHAOS dataset [pub:6]

of MRIs separately, as well as for both types of acquisition sequences T1-Dual (all the results of T1 in-phase and opposed-phase) and T2-SPIR. The proposed method achieved statistically significant improvements over all the baselines in every scenario. Among the different types of MRIs, T1 opposed-phase achieved the best score, while T1 in-phase achieved the worst. When only the contrasts/sequences are taken into account, then T2-SPIR performed better than T1-Dual. These observations hold true for both levels of sparsity. However, in terms of the percentage of improvement of average SSIM values achieved by the proposed method against the primary baseline model (Primal-Dual Network), the amounts are different for the different levels of sparsity. Both T2-SPIR and T1-Dual improved 0.41% for Sparse 8, and for Sparse 16 obtained improvements of 0.63% and 1.06%, respectively. Considering the three types of MRIs separately, it is noteworthy that even though the T1 in-phase was the worst-performing, it managed to get the most amount of improvement, 0.51% and 1.18% for Sparse 8 and 16. In contrast, the best-performing type of MRI achieved the least amount of improvements: 0.30% and 0.73% for Sparse 8 and 16.

Finally, the reconstructions were compared qualitatively with the help of difference images and SSIM maps for Sparse 8 are shown in Figs. 8.26, 8.28, and 8.24, for T1 in-phase, T1 out-phase, and T2-SPIR; while the Sparse 16 are shown in Figs. 8.27, 8.29, and 8.25, respectively.

Table 8.10 Resultant SSIM values for different types of MR acquisitions from the CHAOS dataset (mean±std)

Method	Sparse 8				Sparse 16			
	T2-SPIR	T1-Dual	T1 In-phase	T1 Opposed-phase	T2-SPIR	T1-Dual	T1 In-phase	T1 Opposed-phase
Undersampled (NUFFT)	0.599±0.060	0.493±0.084	0.430±0.058	0.557±0.053	0.394±0.051	0.342±0.057	0.306±0.043	0.378±0.045
Bilinear Sinogram	0.944±0.017	0.901±0.029	0.887±0.029	0.916±0.021	0.896±0.035	0.817±0.052	0.796±0.051	0.838±0.044
Sinogram UNet	0.950±0.014	0.915±0.023	0.904±0.023	0.925±0.017	0.927±0.021	0.882±0.036	0.867±0.036	0.898±0.027
Reconstruction UNet	0.983±0.009	0.980±0.015	0.975±0.014	0.985±0.013	0.950±0.019	0.939±0.033	0.926±0.031	0.952±0.029
Primal-Dual Network	0.983±0.008	0.981±0.013	0.976±0.013	0.986±0.011	0.955±0.017	0.945±0.028	0.934±0.028	0.957±0.022
Primal-Dual UNet	**0.987±0.007**	**0.985±0.012**	**0.981±0.011**	**0.989±0.011**	**0.961±0.015**	**0.955±0.026**	**0.945±0.026**	**0.964±0.023**

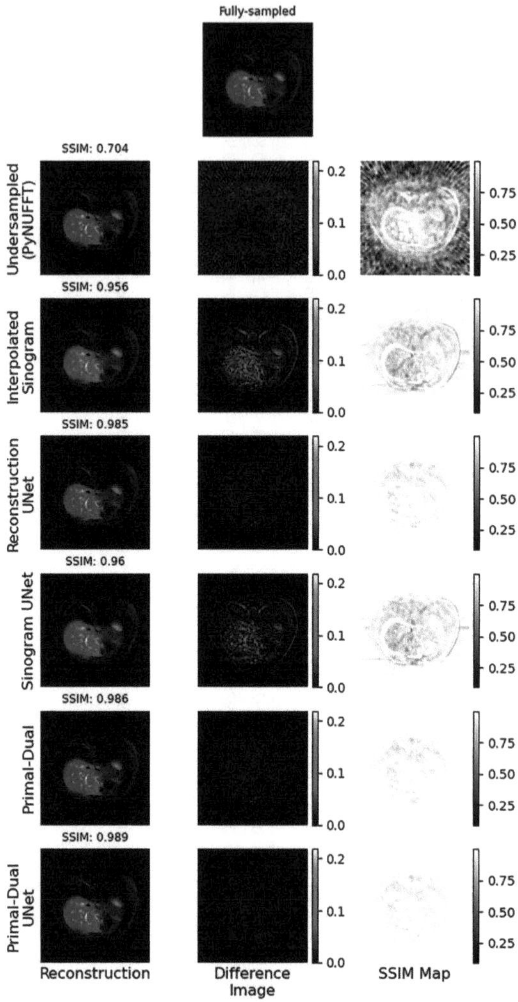

Fig. 8.24 Qualitative comparisons of the reconstructions of CHAOS T2w Sparse 8 [pub:6]

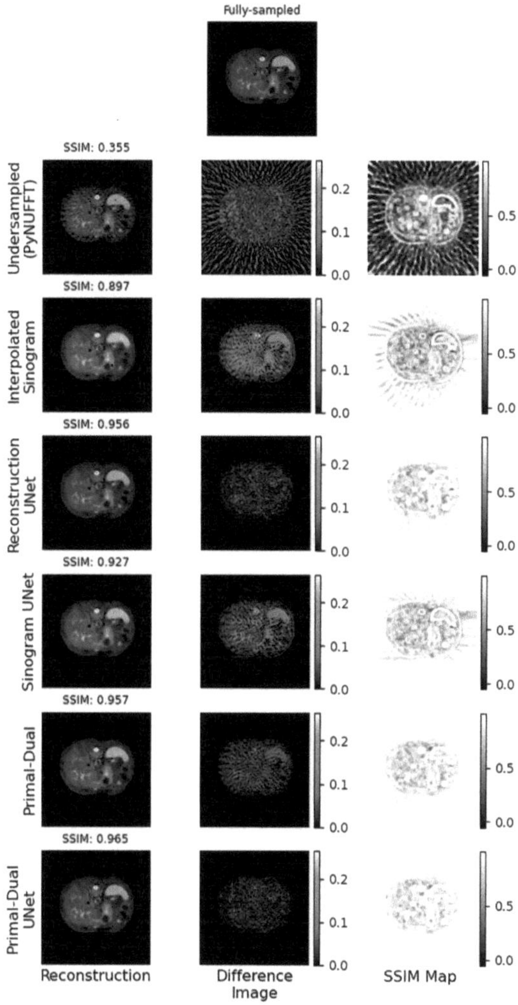

Fig. 8.25 Qualitative comparisons of the reconstructions of CHAOS T2w Sparse 16 [pub:6]

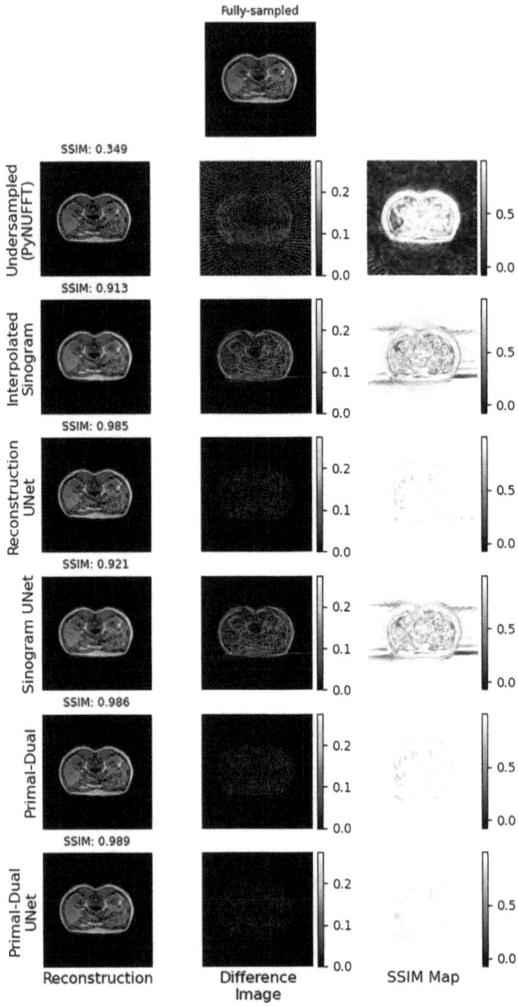

Fig. 8.26 Qualitative comparisons of the reconstructions of CHAOS T1w In-phase Sparse 8 [pub:6]

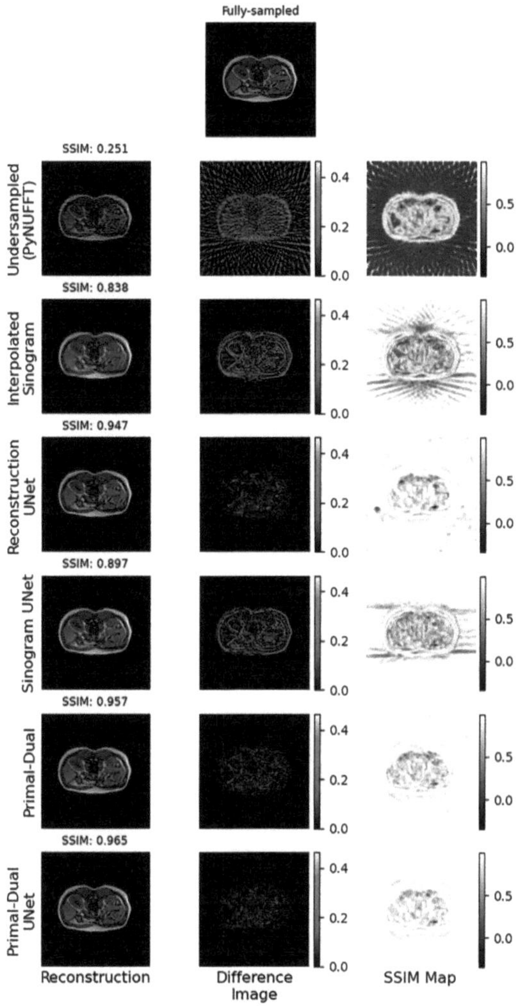

Fig. 8.27 Qualitative comparisons of the reconstructions of CHAOS T1w In-phase Sparse 16 [pub:6]

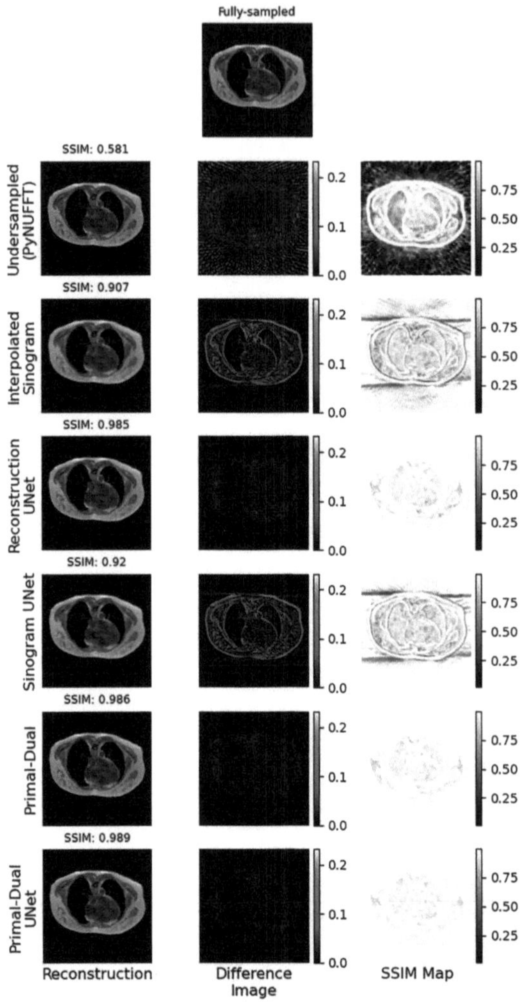

Fig. 8.28 Qualitative comparisons of the reconstructions of CHAOS T1w Opposed-phase Sparse 8 [pub:6]

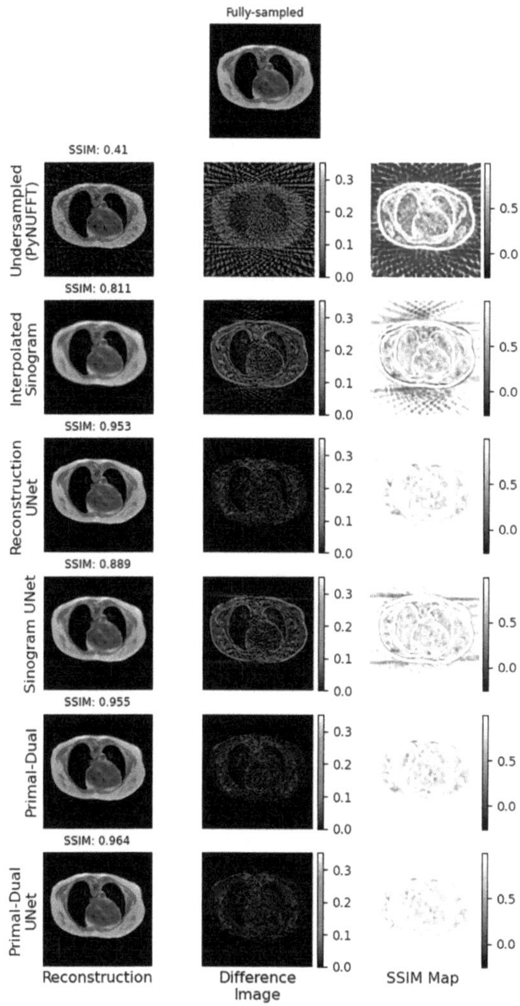

Fig. 8.29 Qualitative comparisons of the reconstructions of CHAOS T1w Opposed-phase Sparse 16 [pub:6]

Evaluation for Region Of Interest Further evaluations were performed to compare the performance of the proposed model against the other methods for different Region Of Interest (ROI). For this purpose, the images were segmented into three different regions: liver, kidneys (both left and right), and spleen, with the help of the available segmentation labels from the CHAOS dataset with the images. Then the images were cropped to have only the ROI. Those segmented-cropped images obtained from the results of different methods were then compared against the segmented-cropped version of the ground-truth images. Tables 8.11, 8.12, 8.13 show the quantitative results for liver, kidneys, and spleen. Statistical tests revealed that the proposed model achieved statistically significant improvements over all the other methods. Figure 8.30 shows a qualitative comparison of the reconstruction quality for the liver for Sparse 8.

Table 8.11 Resultant metrics for liver from the CHAOS dataset (mean±std)

Method	SSIM		RMSE	
	Sparse 8	Sparse 16	Sparse 8	Sparse 16
Undersampled (NUFFT)	0.831±0.083	0.656±0.102	0.028±0.014	0.065±0.026
Bilinear Sinogram	0.883±0.046	0.788±0.076	0.037±0.016	0.048±0.021
Sinogram UNet	0.893±0.042	0.826±0.063	0.032±0.014	0.038±0.016
Reconstruction UNet	0.951±0.029	0.864±0.061	0.013±0.007	0.026±0.014
Primal-Dual Network	0.951±0.030	0.866±0.059	0.012±0.005	0.025±0.011
Primal-Dual UNet	**0.963±0.026**	**0.888±0.054**	**0.011±0.006**	**0.021±0.010**

8.2.2.1 Comparison of Execution Speeds

Sparse sampling can reduce the speed of acquisition, which is an essential factor when it comes to MR imaging. However, the time required for reconstruction can be an additional overhead—increasing the total time for imaging. For this reason, the execution speed of the proposed Primal-Dual UNet was compared against the main baseline model—Primal-Dual Network. Figure 8.31 shows the required amount of time to reconstruct one slice for these two methods and how much they change with a change in the number of projections. It can be observed that the proposed method is faster than the baseline Primal-Dual Network. Moreover, it can be observed that

Table 8.12 Resultant metrics for both left and right kidneys from the CHAOS dataset (mean±std)

Method	SSIM		RMSE	
	Sparse 8	Sparse 16	Sparse 8	Sparse 16
Undersampled (NUFFT)	0.877±0.180	0.695±0.168	0.027±0.014	0.070±0.028
Bilinear Sinogram	0.866±0.083	0.759±0.112	0.051±0.023	0.065±0.031
Sinogram UNet	0.881±0.079	0.809±0.095	0.045±0.020	0.054±0.025
Reconstruction UNet	0.962±0.054	0.862±0.086	0.016±0.008	0.035±0.018
Primal-Dual Network	0.963±0.057	0.866±0.077	0.016±0.008	0.039±0.023
Primal-Dual UNet	**0.971±0.049**	**0.892±0.073**	**0.013±0.007**	**0.030±0.014**

Table 8.13 Resultant metrics for spleen from the CHAOS dataset (mean±std)

Method	SSIM		RMSE	
	Sparse 8	Sparse 16	Sparse 8	Sparse 16
Undersampled (NUFFT)	0.848±0.149	0.699±0.155	0.024±0.010	0.059±0.021
Bilinear Sinogram	0.878±0.114	0.817±0.130	0.045±0.018	0.055±0.022
Sinogram UNet	0.885±0.121	0.840±0.127	0.039±0.017	0.045±0.019
Reconstruction UNet	0.953±0.104	0.895±0.117	0.011±0.005	0.023±0.010
Primal-Dual Network	0.950±0.106	0.894±0.116	0.012±0.006	0.025±0.013
Primal-Dual UNet	**0.957±0.118**	**0.907±0.128**	**0.009±0.004**	**0.019±0.009**

with the increase in the number of projections, the required reconstruction time increases for both models, but also the difference between the models increases constantly.

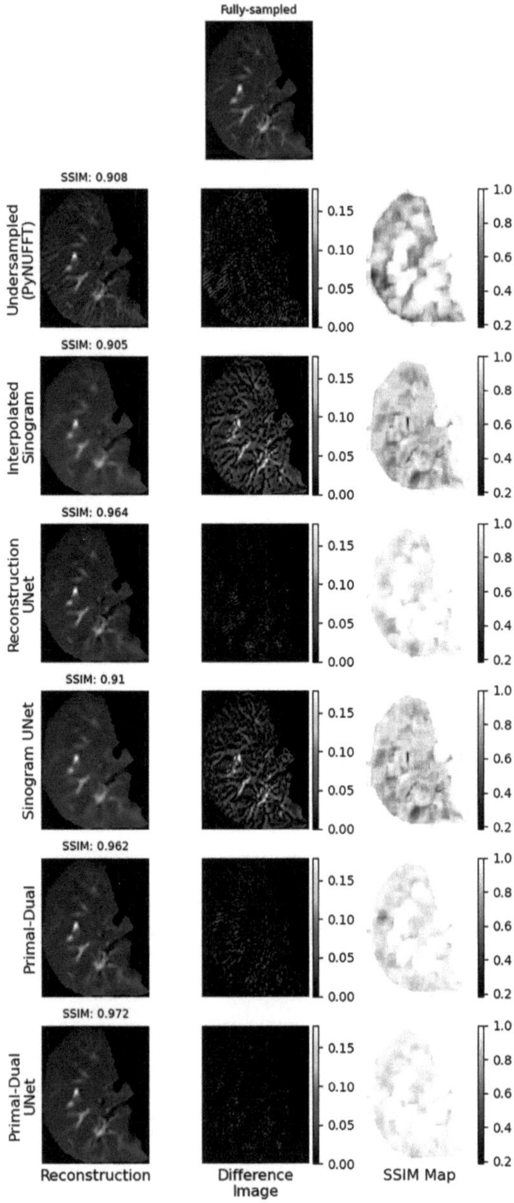

Fig. 8.30 Qualitative comparisons of the reconstructions of CHAOS T2w Sparse 8: Liver ROI [pub:6]

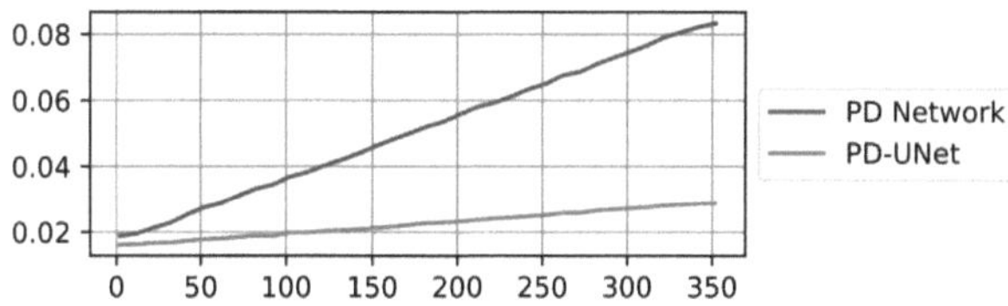

Fig. 8.31 Comparison of the inference times for Primal-Dual Network (PD Network) and Primal-Dual UNet (PD-UNet). Number of projections on horizontal axis. Average inference time for one batch (batch size: 4) in seconds on vertical axis [pub:6]

8.2.3 Discussion

Regarding CT-inspired techniques for performing radial MRI reconstruction, an interesting fact to be noted is that the first-ever 2D MRI was also produced using a back-projection algorithm [15, 84]. The only difference between the sparse CT reconstruction and undersampled radial MRI reconstruction using the proposed model (as well as the CT-inspired baselines) is the 1D inverse Fourier transform as pre-processing. The experiments performed as a part of this research show the possibility of using sinogram upsampling techniques combined with FBP to reconstruct undersampled radial MRI.

It was observed that converting the undersampled radial k-space into the corresponding sinogram, then applying bilinear interpolation before finally performing FBP, already results in scores better than the undersampled radial MRIs, which were reconstructed with the conventional NUFFT. Sinogram UNet, which aims to improve the quality of the bilinearly upsampled sinograms, improves the results even further in terms of SSIM. Even though these two methods perform better than the traditional reconstruction of the undersampled radial MRIs using NUFFT, the quantitative and qualitative evaluations yield that they are much inferior to the other three models. It is to be noted that Reconstruction UNet—a well-established model for deep learning based undersampled MRI reconstruction, which is the only model out of the four models which is not CT-inspired and works directly with the reconstructed images, performed very similar to the main baseline model of this research: Primal-Dual Network. Reconstruction UNet performed better for the brain MRI reconstruction task, whereas the Primal-Dual Network performed better while reconstructing abdominal MRIs. As noted earlier, the proposed method Primal-Dual UNet performed better than all the other methods for all the tasks.

8.2.4 Summary and Outlook

This research presents a novel Primal-Dual UNet model for performing sinogram upsampling, which improves upon the Primal-Dual Network in terms of accuracy and speed of reconstruction. The proposed method was applied for the task of the undersampled brain and abdominal MRI reconstruction with radial acquisitions for acceleration factors 8 and 16 (*[RQ:6]*), and the proposed method outperformed all the baseline methods with statistical significance (*[RQ:5]*). This research shows the potential of using sinogram upsampling techniques, such as deep learning based techniques, for radially-acquired undersampled MRIs (*[RQ:5.1]*). Needle insertion simulation experiment (on CT) showed that the proposed network not only yields better image reconstruction, but also generalises better than the other networks—reconstructs images with a needle significantly better than the rest while being trained on a dataset without any needle (results are presented in the appendix, Sect. A.2.2). Moreover, Region Of Interest based evaluations showed significant improvements over the baselines not only for the overall images, but also for the regions-of-interest liver, kidneys, and spleen (*[RQ:5]*). Future research directions are towards applying the proposed method to cone-beam CT projections [pub:28] and 3D radial MRI (unlike stack-of-stars radial MRI used in this research, e.g. [128]), thus yielding true 3D reconstructions and incorporating known information about the volume for facilitating partial volume updates to further reduce memory and time requirements of the reconstruction task. Moreover, data augmentation techniques were only used minimally here, but could have a big impact on the generalisability and overall quality of the networks. Using data from different acceleration/sparsity factors during training could not only eliminate the need for one model per factor, but even help the network learn how subsampling the data influences the creation of artefacts and ultimately learn how to suppress these artefacts independent of the factor, which might make the model applicable to unseen subsampling patterns—will also be evaluated.

9 Undersampled Reconstruction as Super Resolution

The task of undersampled MRI reconstruction in the image space can also be treated as a Super-Resolution (SR) problem. The so-called SR algorithms aim to restore images with high spatial resolution from the corresponding low-resolution images. This chapter starts by exposing the mathematical representation of Super-Resolution reconstruction and then presents three different methods—one model—ShuffleUNet—has been proposed and employed while working with static MRIs, and then two different methods working with dynamic MRIs.

9.1 Mathematical Representation

The reconstruction of the high-resolution image from the corresponding low-resolution image can be modelled as:

$$\hat{I}_{HR} = \mathcal{F}(I_{LR}; \theta) \tag{9.1}$$

where I_{LR} is the low-resolution image, $\hat{I}_{HR}$ is the super-resolved image, $\mathcal{F}$ is the mapping function which models the spatial super-resolution relationship between the corresponding low- and high-resolution images using a given set of parameters θ [373]. The SR image reconstruction is an ill-posed problem to approximate the super-resolved image from a given low-resolution counterpart, where the network model is trained to optimise the objective function:

$$\hat{\theta} = arg\,min \mathcal{L}(\hat{I}_{HR}, I_{HR}) + \lambda R(\theta) \tag{9.2}$$

S. Chatterjee, *Deep Learning Innovations in MRI Reconstruction and Analysis*,
https://doi.org/10.1007/978-3-658-50741-1_9

The operator $\mathscr{L}(\hat{I}_{HR}, I_{HR})$ describes the loss function between the predicted super-resolved image $\hat{I}_{HR}$ and a ground-truth image I_{HR}, $R(\theta)$ is a regularisation term and λ denotes the regularisation parameter [pub:13].

9.2 Super-Resolving Static MRI

9.2.1 ShuffleUNet: Super Resolution of Diffusion-Weighted MRIs

The work presented in this section has been performed in collaboration with Alessandro Sciarra and has been published in [pub:18]. The code of the proposed model is available publicly on GitHub [code:11].

9.2.1.1 Methodology

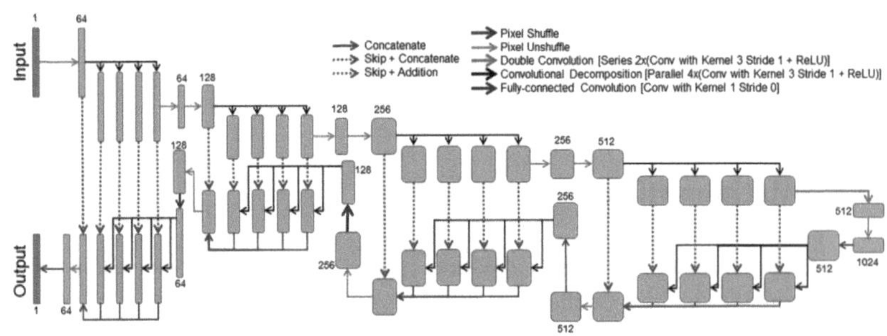

Fig. 9.1 Proposed network architecture: ShuffleUNet [pub:18] © 2021IEEE

Network Architecture The proposed ShuffleUNet consists of four blocks in the contraction path, each of them downsamples the input by half in all dimensions. Each block consists of three sub-blocks: double convolution, convolutional decomposition, and pixel unshuffle. The input goes to double convolution, and its output serves as the input of the convolutional decomposition sub-block. The input of this sub-block is provided to each convolution of this block, and four different outputs are obtained—these outputs are referred to as convolutional decomposition of the input of this sub-block. Pseudo-loss-less downsampling operation, pixel unshuffle (see Eq. 9.3), is applied on the fourth output, which down-samples the input by a factor of two in all dimensions, and the rest of the outputs are directly forwarded as skip-connection to the expansion path.

$$I^D = f^l(I^L) = \boldsymbol{PU}(W^l \times f^{(l-1)}(I^L) + b^l) \quad (9.3)$$

where ***PU*** denotes the pixel-unshuffling operation, which rearranges all the elements of a tensor of dimension (n, c, rH, rW, rD) tensor to a tensor of dimension (n, $r^3 \times$ c, H , W , D), r is the scaling factor. After the contraction path, one double convolution sub-block is applied to the output of the final pixel unshuffle as the latent convolution. The output of this latent convolution is passed to the expansion path (Fig. 9.1).

Similar to the contraction path, the expansion path also contains four blocks—each of them upsamples the input by a factor of two in all dimensions. Each of these blocks contains three sub-blocks: pixel shuffle, convolutional decomposition, and double convolution. Pixel shuffle sub-blocks upscale the given input by a factor of k (here $k = 2$) in all dimensions by reducing the number of filters by a factor of $1/k^3$. A tensor of dimensions (n, $r^3 \times$ c, H, W, D) is upsampled with periodic shuffling operation in feature space as explained by the following equation:

$$I^S = f^l(I^L) = \boldsymbol{PS}(W^l \times f^{(l-1)}(I^L) + b^l) \quad (9.4)$$

where ***PS*** denotes pixel-shuffling operation which shuffles and arranges all the pixel elements of a tensor with dimensions (n, $k^3 \times$ c, H, W, D) to a tensor with dimensions (n, c, kH Œ kW $\times$ kD), k is the scaling factor. W^l, b^l denotes the weights and bias matrices, and $f^{(l-1)}$ represents feature maps from low-resolution feature space from the previous layer, and f^l represents upsampled feature space after pixel shuffling. The output of the pixel shuffle sub-block is forwarded to the convolutional decomposition sub-block to obtain four different outputs, which are then added with the incoming skip-connections from the same level of the contraction path, and finally, these four results are concatenated together. Then, this is further concatenated together with the output of the skip-connection coming from the pixel unshuffle operations of the contraction path, which is then forwarded to the double convolution sub-block. The final output of the model is obtained after the fully-connected convolution layer.

The initial number of filters of the network (first convolution layer) is 64, and the number of features is doubled by each of the contraction path blocks after every down-sampling step and is reduced by half in every up-sampling process in the expansion path. Also, it is noteworthy that here both pixel shuffle and pixel unshuffle operators have learnable parameters. Furthermore, the weight initialisation (convolution, pixel shuffle and pixel unshuffle layers) was performed using a normal distribution (also known as Kaiming Normal) [179], rather than the typical choice of uniform distribution (Kaiming Uniform), to help the convergence and to avoid degrading effects arising from random weight initialisation [215].

Implementation The implementation was done using PyTorch and was trained using an Nvidia V100 GPU. The volumes were divided into patches with a patch size of $96 \times 96 \times 48$ with the help of TorchIO [394]. The network was trained for 80 epochs (when it converged) with a batch size of four. The loss was calculated using L1 loss (mean absolute error, Eq. 4.28) and was optimised using the Adam optimiser with a learning rate of 1e–4.

Dataset For this research, DWIs from the IXI dataset [url:12] were used, which were acquired with 1.5T and 3T MRI scanners with 16 diffusion directions for 399 healthy subjects. The volumes in the dataset are originally of anisotropic spatial resolutions $1.75 \times 1.75 \times 2.35\ mm^3$ and $1.75 \times 1.75 \times 2\ mm^3$. The 3D volumes were under-sampled to simulate the low-resolution dataset, by a factor of two in all dimensions using FSL [138] to a spatial resolution of $3.5 \times 3.5 \times 4.7\ mm^3$, $3.5 \times 3.5 \times 4.0\ mm^3$ respectively, which is 12.5% of the original data. UNet-based models require the same size for input and output. Hence the low-resolution dataset was then interpolated using sinc interpolation to achieve the same pixel dimensions as the high-resolution volumes before supplying as input. The dataset was split into three subsets: training, test and validation with 300, 50 and 49 subjects, respectively.

9.2.1.2 Results

The results have been quantitatively and qualitatively compared initially using the actual structural output of the models and then by comparing the derived data obtained. The results of the proposed model and the baselines (Sinc interpolation, tri-linear interpolation and UNet model) were compared quantitatively using three metrics: Structural Similarity Index (SSIM, Eq. 4.35) [74], Root Mean-Squared Error (RMSE, Eq. 4.31), and Universal Quality Index (UQI, Eq. 4.34) [64]. Furthermore, derived voxel-wise data, such as Axial Diffusivity (AD), Fractional Anisotropy (FA), Mean Diffusivity (MD), and diffusion tensors (E) [243], were obtained using DIPY [164]. These derived data can be utilised to generate the fibre tracts and assess the integrity of white-matter microstructure [243]. The derived data from the ShuffleUNet results and baseline results were compared against the derived data obtained from the ground-truth volumes quantitatively using RMSE and UQI.

Output Evaluation Fig. 9.2 portrays the resultant SSIM, RMSE and UQI values for the three baselines (trilinear, sinc and UNet) and for ShuffleUNet, while comparing the structural results against the high-resolution ground-truth volumes. It can be observed that trilinear resulted in the lowest values for all three metrics (0.788±0.052, 0.076±0.024, 0.568±0.049), while ShuffleUNet outperformed all the baselines (0.913±0.045, 0.017±0.007, 0.702±0.060). It is to be noted that for two of the metrics (RMSE and UQI), UNet achieved the second position after ShuffleUNet, but for SSIM, UNet achieved the third position behind sinc interpolation.

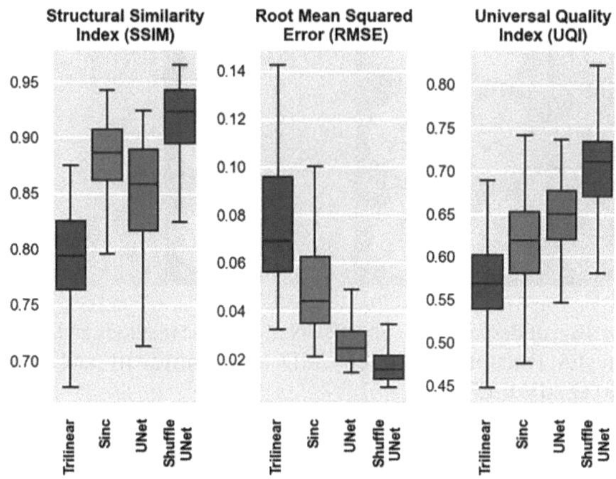

Fig. 9.2 Metrics comparison of the baseline methods and the proposed ShuffleUNet [pub:18] © 2021IEEE

Evaluation of the Derived Data Fig. 9.3 shows the results of the derived data for qualitative analysis, and Fig. 9.5 shows the resultant UQI and RMSE while comparing the derived data obtained from the baseline and ShuffleUNet results against the ground-truth. It can be observed from both metrics that the proposed ShuffleUNet outperformed all the baselines for all four types of derived data. Moreover, Fig. 9.4 shows an example of obtained fibre tracks from a ShuffleUNet result.

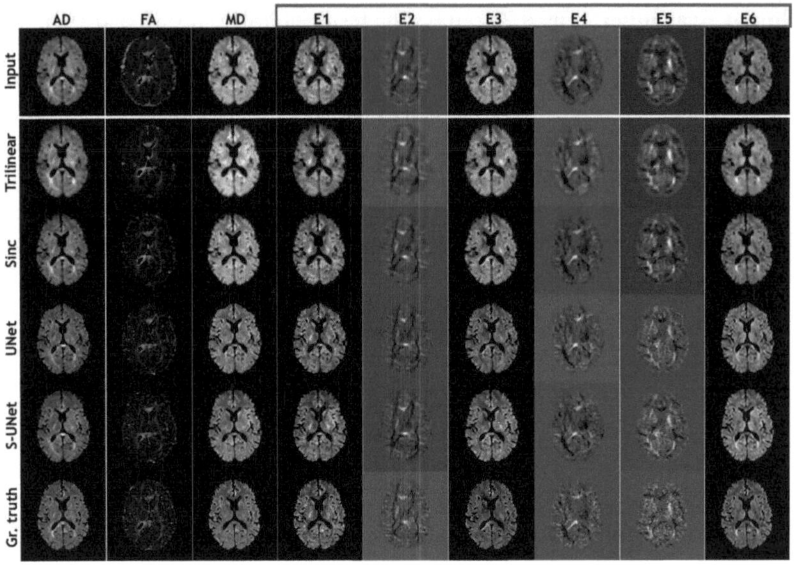

Fig. 9.3 Comparison of derived data (ShuffleUNet, baseline methods and ground-truth). AD: Axial Diffusivity, FA: Fractional Anisotropy, MD: Mean Diffusivity, and E1 to E6: Diffusion Tensors [pub:18] © 2021IEEE

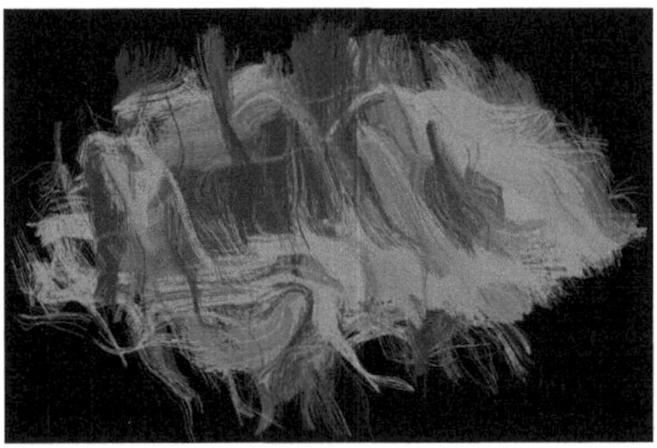

Fig. 9.4 Three dimensional visualisation of the fibre tracts, generated from the derived data obtained from a ShuffleUNet result [pub:18] © 2021IEEE

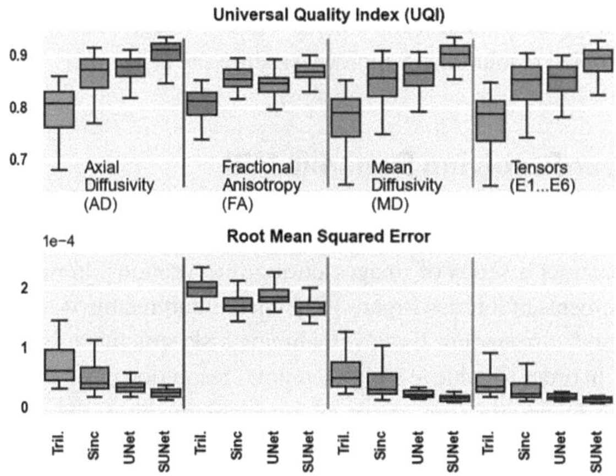

Fig. 9.5 Quantitative evaluation of the derived data obtained from the ShuffleUNet and the baseline methods [pub:18] © 2021IEEE

Statistical Hypothesis Testing The significance of the improvements observed by the ShuffleUNet was analysed using the independent two-sample t-test, on the basis of the resultant UQI values. Initially, the structural output of the methods was compared and then the derived data. The resultant p-values obtained while comparing the UQI values of ShuffleUNet against the values of the three baseline methods were always < 0.01. Hence, it can be said that ShuffleUNet achieved significant (p-values < 0.05) improvements over all the baseline methods explored here—for the structural output and also the derived data.

9.2.1.3 Summary and Outlook

This work presents a modified tight-frame UNet architecture, ShuffleUNet, by incorporating pixel unshuffle and pixel shuffle operations for improved down- and upsampling capabilities of the network, thereby reducing artefacts introduced by UNet-based processing pipelines. The method has been evaluated for the task of patch-based single-image super-resolution of diffusion-weighted MRIs. Images from the IXI dataset were downsampled by a factor of two in all dimensions, resulting in a theoretical MRI acceleration factor of eight (***[RQ:6***). The proposed model successfully super-resolved such low-resolution images and achieved statistically significant improvements over the baselines, resulting in an average SSIM value of 0.913 ± 0.045 (***[RQ:3***). The evaluation metrics obtained on the derived data indicate that after super-resolving the artificially downsampled images, similar fibre tracts can be derived as the ground-truth images. Hence, images can be acquired with those downsampled resolutions, which will reduce the scan time.

9.3 Super-Resolving Dynamic MRI

MR-guided interventions, such as liver biopsy, it is necessary to continuously acquire data and reconstruct a series of images during interventions, in order to examine dynamic movements of internal organs [69]. Clear interpretable visualisation of the target lesion and surrounding tissues, including risk structures, is crucial during interventions. In order to achieve a high temporal resolution during dynamic MRIs, because of the inherently slow speed of image acquisition, the amount of data to be acquired has to be reduced—which may result in loss of spatial resolution. Although there are a number of techniques dealing with this spatio-temporal trade-off [68, 85, 95, 108], their speed of reconstruction creates a hindrance for real-time or near real-time imaging. Therefore, a compromise between spatial and temporal resolution is inevitable during real-time MRIs and needs to be mitigated. The work presented in this section has been performed in collaboration with Chompunuch Sarasaen. Before moving towards deep learning based super-resolution approaches, initial experiments were performed by estimating the deformation field from low-resolution dynamic MRI and applying them to the high-resolution prior scans—to obtain high resolution dynamic MRI. This work has been published in [pub:21].

9.3.1 Fine-Tuning With Prior Knowledge

This study proposes a patch-based super-resolution reconstruction, aiming at addressing the problem of the lack of large abdominal datasets for training. This research intends to improve deep learning based super-resolution of dynamic MR images by incorporating prior images (planning scan), by training the network first on a publicly available abdominal dataset of 40 subjects, acquired using different sequences than the dynamic MR that is to be reconstructed, and then the network is fine-tuned using a high resolution prior planning scan of the same subject as the dynamic acquisition. The initial findings of this work were presented in [pub:45], and [pub:41]. The final results have been published in [pub:13]. The code of this project is available publicly on GitHub [code:5].

9.3.1.1 Methodology

This research proposes a framework of patch-based MR super-resolution based on UNet, incorporating prior knowledge. The framework can be divided into three stages: main training, fine-tuning and inference. The UNet model was initially trained with a benchmark dataset for main-training and then was fine-tuned using a subject-specific prior static scan. Finally, in the inference stage, high-resolution dynamic MRIs were reconstructed from low-resolution scans. This chapter starts with the description of various datasets used in this research, followed by the network architecture, and finally exposes the details regarding implementation and training.

Data In this work, 3D abdominal MR volumes were artificially downsampled in-plane using MRUnder pipeline [url:8, code:7] to simulate low-resolution datasets. The low-resolution data was generated by performing undersampling in-plane (phase-encoding and read-out direction) by taking the centre of k-space without zero-padding. The CHAOS challenge dataset [pub:16] (T1-Dual; in- and opposed-phase images) was used for the main training. During this phase, the dataset was split into training and validation sets, with a ratio of 70:30. High-resolution 3D static (breath-hold) and 3D "pseudo"-dynamic (free-breathing) scans for 10 time-points (TPs) using T1w FLASH sequence were acquired for fine-tuning and inference, respectively. Each time-point of the dynamic acquisition was treated as a separate 3D volume. Three healthy subjects were scanned with the same sequence but with different parameters on a 3T MRI (Siemens Magnetom Skyra). This aims to test the generalisation of the network. For each subject, the 3D static and the 3D dynamic scans were acquired in different sessions using the same sequences and parameters. The sequence parameters of the various datasets have been listed in Table 9.1. The CHAOS dataset (for main training), the 3D static scans (for fine-tuning) and the 3D dynamic scans (for inference) were artificially downsampled for three different levels of image resolution, by taking 25%, 10% and 6.25% of the k-space centre, resulting in MR acceleration factor of 2, 3 and 4 respectively (considering undersampling only in the phase-encoding direction). This can be accelerated theoretically to a factor of 4, 9 and 16, respectively, considering the amount of data used for the SR reconstruction. The effective resolutions and the roughly estimated acquisition times of the low-resolution images were calculated from the corresponding high-resolution images, which are reported in Table 9.2. The acquisition times were calculated as:

$$AcqTime = PE_n \times TR \times S_m \tag{9.5}$$

Table 9.1 MRI acquisition parameters CHAOS dataset and subject-wise 3D dynamic scans. Static scans were performed using the same subject-wise sequence parameters as the dynamic scans for one Time-point (TP), acquired at a different session

	CHAOS (40 Subjects)	Subject 1	Subject 2	Subject 3
Sequence	T1 Dual In-Phase & Opposed-Phase	T1w Flash 3D	T1w Flash 3D	T1w Flash 3D
Resolution	$1.44 \times 1.44 \times 5$ – $2.03 \times 2.03 \times 8\ mm^3$	$1.09 \times 1.09 \times 4\ mm^3$	$1.09 \times 1.09 \times 4\ mm^3$	$1.36 \times 1.36 \times 4\ mm^3$
FOV x, y, z	$315 \times 315 \times 240$ – $520 \times 520 \times 280\ mm^3$	$280 \times 210 \times 160\ mm^3$	$280 \times 210 \times 160\ mm^3$	$350 \times 262 \times 176\ mm^3$
Encoding matrix	$256 \times 256 \times 26$ – $400 \times 400 \times 50$	$256 \times 192 \times 40$	$256 \times 192 \times 40$	$256 \times 192 \times 44$
Phase/Slice oversampling	–	10/0 %	10/0 %	10/0%
TR/TE	110.17 – 255.54 ms / 4.60 –4.64 ms (In-Phase) 2.30 ms (Opposed-Phase)	2.34/0.93 ms	2.34/0.93 ms	2.23/0.93 ms
Flip angle	80°	8°	8°	8°
Bandwidth	–	975 Hz/Px	975 Hz/Px	975 Hz/Px
GRAPPA factor	None	2	None	None
Phase/Slice partial Fourier	–	Off/Off	Off/Off	Off/Off
Phase/Slice resolution	–	75/65 %	75/65 %	50/64 %
Fat suppression	–	None	On	On
Time per TP	–	5.53 sec	11.76 sec	8.01 sec

Table 9.2 Effective resolutions and estimated acquisition times (per TP) of the dynamic and static datasets after performing different levels of artificial undersampling

	Subject 1		Subject 2		Subject 3	
	Resolution (mm^3)	Acq. Time (sec)	Resolution (mm^3)	Acq. Time (sec)	Resolution (mm^3)	Acq. Time (sec)
High Resolution Ground-truth	$1.09 \times 1.09 \times 4$	4.81	$1.09 \times 1.09 \times 4$	9.61	$1.36 \times 1.36 \times 4$	6.62
25% of k-space	$2.19 \times 2.19 \times 4$	1.22	$2.19 \times 2.19 \times 4$	2.43	$2.73 \times 2.73 \times 4$	1.65
10% of k-space	$3.50 \times 3.50 \times 4$	0.47	$3.50 \times 3.50 \times 4$	0.94	$4.38 \times 4.38 \times 4$	0.66
6.25% of k-space	$4.38 \times 4.38 \times 4$	0.28	$4.38 \times 4.38 \times 4$	0.56	$5.47 \times 5.47 \times 4$	0.42

where PE_n is the number of phase-encoding lines—which equals to the matrix dimension in this direction, TR is the repetition time—a parameter of an MRI pulse sequence which defines the time between two consecutive radio-frequency pulses, and S_m is the number of slices. Phase/slice oversampling, phase/slice resolutions and GRAPPA factor were taken into account while calculating PE_n. The low resolution images served as the input to the network and were compared against the high resolution ground-truth images.

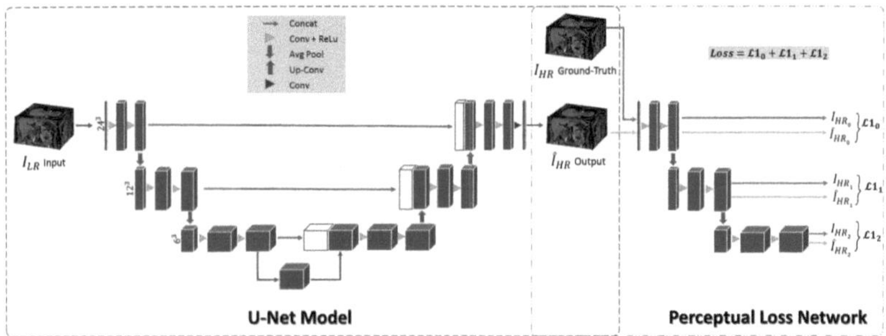

Fig. 9.6 The proposed network architecture [pub:13]

Network Architecture Fig. 9.6 portrays the proposed network architecture. In this work, a 3D UNet based model [186, 194, pub:45] with perceptual loss network [pub:11] was employed for super-resolution reconstruction. The UNet architecture consists of two main paths; contracting (encoding) and expanding (decoding). The contracting path consists of three blocks; each comprises two convolutional layers and a ReLU activation function. The expanding path also consists of three blocks, but convolutional transpose layers were used instead of the convolutional layers. The training was performed using 3D patches of the volumes, with a patch size of 24^3.

The UNet model requires the same matrix size for input and output, therefore, the input low-resolution image patches were interpolated using trilinear interpolation before supplying them to the network. The interpolation factor was determined based on the undersampling factor—thus, the low-resolution input is modified to the same matrix size as the target high-resolution ground-truth.

Patch-based training may result in patching artefacts during inference. To remove these artefacts, the inference was performed with a stride of one, and the overlapped portions were averaged after reconstruction.

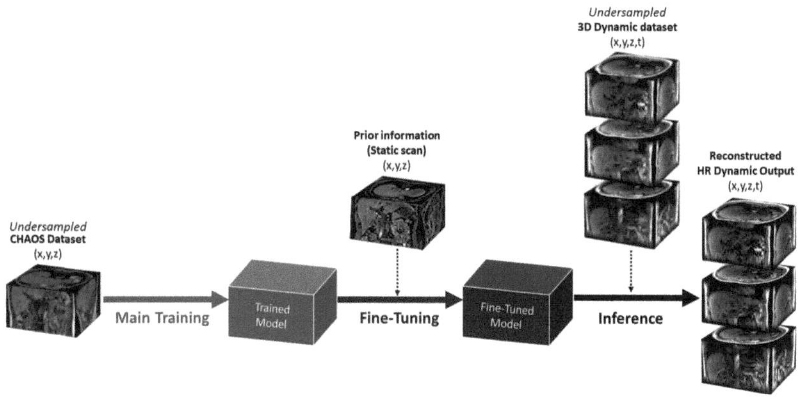

Fig. 9.7 Method Overview [pub:13]

Model Implementation and Network Training Fig. 9.7 shows the method overview. The main training was performed using 3D patches of the 80 volumes (40 subjects, in-phase and opposed-phase for each subject), with a patch size of 24^3 and a stride of 6 for the slice dimension and 12 for the other dimensions. After that, the network was fine-tuned using a single 3D static scan of the same subject from an earlier session, labelled as x,y,z,t in Fig. 9.7. This static scan has the same resolution, contrast and volume coverage as the high-resolution dynamic scan. The static and the dynamic scans were not co-registered to keep it similar to the real-life scenario and to keep the speed of inference fast. Fine-tuning and evaluations were performed with a patch size of 24^3 and a stride of one. The implementation was done using PyTorch [327] and was trained using Nvidia Tesla V100 GPUs. The loss was minimised using the Adam optimiser with a learning rate of $1e-4$. The main training was performed for 200 epochs. The network was fine-tuned for only one epoch, using the planning scan with a lower learning rate ($1e-6$). The code of this research is available online on GitHub [code:5].

Perceptual Loss Loss during the training and fine-tuning of the network was calculated with the help of perceptual loss (Sect. 4.4.3.1) [203]. The first three levels of the contraction path of a pre-trained (on 7T MRA scans, for vessel segmentation) frozen UNet MSS model [pub:11] were used as the Perceptual Loss Network (PLN) to extract features from the final super-resolved output of the model and the ground-truth images (refer to Fig. 9.6). Typically VGG-16 trained on three-channel RGB non-medical images (ImageNet Dataset) is used as PLN, even while working with medical images [313] as the PLN doesn't have to be trained on a similar dataset. In this research, the pre-trained network was chosen because it was originally trained on single-channel medical images, but of different contrast and organ; and it was hypothesised that using a network trained as such will be more suitable than a network trained on three-channel images. The extracted features from the model's output and from the ground-truth images were compared using mean absolute error (L1 loss, Eq. 4.28). The losses obtained at each level for each feature were then added together and backpropagated.

9.3.1.2 Results and Discussion

To evaluate the quality of the reconstructed images against the ground-truth HR images, two of the most commonly used metrics for evaluating image quality were selected. For perceptual quality assessment, the accuracy of the reconstructed images was compared to the ground truth using SSIM [74], which is calculated using Eq. 4.35. Additionally, the performance of the model was measured statistically with PSNR calculated using Eq. 4.33.

The performance of the model was evaluated for three different levels of undersampling: by taking 25%, 10% and 6.25% of the k-space centre. The network was tested before and after fine-tuning using 3D dynamic MRI. The proposed approach was compared against the low-resolution input, the traditional trilinear interpolation, area interpolation (based on adaptive average pooling, as implemented in PyTorch [url:2]) and finally against the most widely used technique in clinical MRIs—Fourier interpolation of the input (zero-padded k-space, also known as the sinc interpolation).

There was a noticeable improvement qualitatively and quantitatively while reconstructing low-resolution data using the proposed method, even for only 6.25% of the k-space. Fig. 9.8 shows the comparison qualitatively for the low resolution images by taking 25%, 10% and 6.25% of the k-space. Fig. 9.9 portrays the comparison of the low resolution input for 6.25% of the k-space, the lowest resolution investigated during this study, with the SR result after fine-tuning over different time-points. The SSIM maps were calculated against the high-resolution ground-truth, and the respective SSIM value can be found on top of the image. Fig. 9.10 illustrates the deviations of an example result from its corresponding ground-truth for two different regions of interest. It can be observed that SR results after fine-tuning could alleviate the undersampling artefacts, which are still present in the SR results of the main training, even for relatively low-resolution images like 10% and 6.25% of the k-space. Consequently, the visibility of small details is improved.

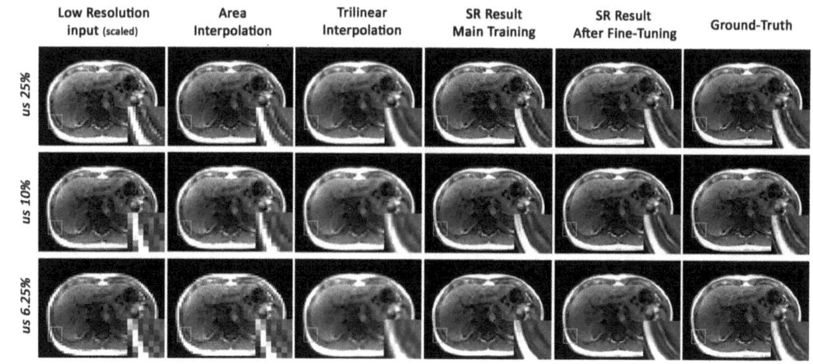

Fig. 9.8 Qualitative comparison of the results of low resolution (25%, 10% and 6.25% of k-space) 3D Dynamic data of the same slice. From left to right: low resolution images (scaled-up), Interpolated input (Trilinear), super-resolution results of the main training (SR Results Main Training), super-resolution results after the fine-tune (SR After Fine-Tuning) and ground-truth images [pub:13]

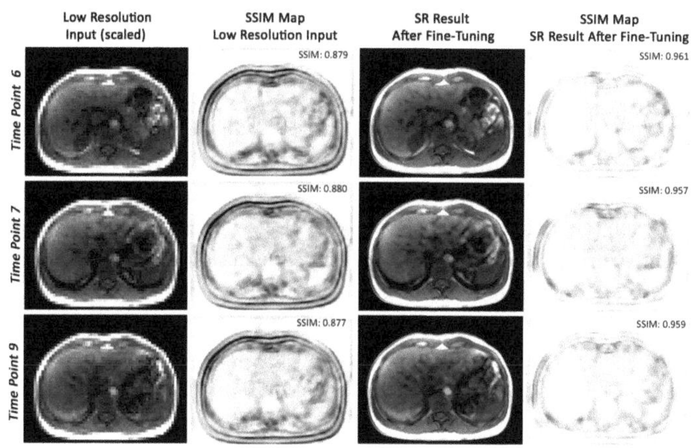

Fig. 9.9 An example comparison of the low resolution input of the 6.25% of k-space with the SR result after fine-tuning over three different time-points, compared against the high resolution ground-truth using SSIM maps [pub:13]

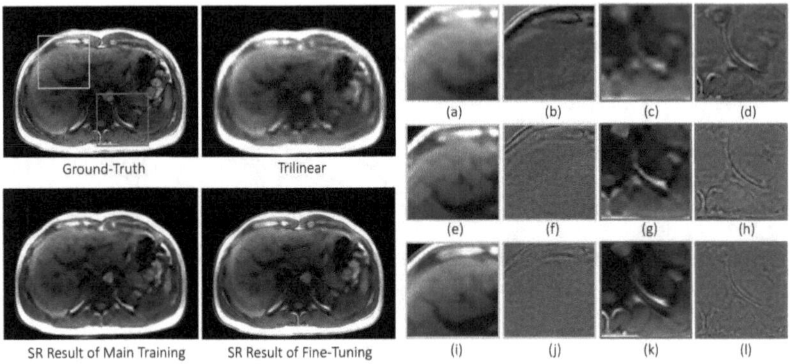

Fig. 9.10 An example from the reconstructed results, compared against its ground-truth (GT) for low resolution images from 6.25% of k-space. From left to right, upper to lower: ground-truth, trilinear interpolation (input of the network), SR result of main training and SR result after fine-tuning. For the yellow ROI, (a–b): trilinear interpolation and the difference image from GT, (e–f): SR result of the main training and the difference image from GT and (i–j): SR result after fine-tuning and the difference image from GT. The images on the right part are identical examples for the red ROI [pub:13]

Additionally, for quantitative analysis, Table 9.3 displays the average and standard deviation (SD) of SSIM, PSNR and the SD of subtracted images for all time-points for the dynamic datasets. Here, each time-point has been considered independent of each other as separate 3D volumes. Fig. 9.11 shows the distribution of the resultant metrics over all resolutions and subjects. The first row portrays the SSIM values, and the second row shows the PSNR values. The columns denote 25%, 10%, and 6.25% of the k-space, respectively. The blue, orange, green, red, and violet lines represent the reconstruction results of trilinear interpolation, area interpolation, zero-padding (sinc interpolation), SR main training, and SR after fine-tuning, whereas the thickness of the lines represents the standard deviation over time-points. It can be observed that the proposed method (SR after fine-tuning) significantly outperformed all the baseline methods experimented with here.

Fine-tuning with the planning scan helped in obtaining sharper images and achieving a better edge delineation. Furthermore, the statistical significance of the improvements in terms of SSIM achieved by the model after fine-tuning was evaluated using paired t-test and Wilcoxon signed-rank test. Separate tests were performed considering all the different resolutions together and also by considering each resolution separately. It was observed that the improvement was statistically significant in every evaluated scenario (p-values were always less than 0.001).

Table 9.3 The average and the standard deviation of SSIM, PSNR, and SD of difference images with ground-truth. The table shows the results of different resolutions

Data	25% of k-space			10% of k-space			6.25% of k-space		
	SSIM	PSNR	diff SD	SSIM	PSNR	diff SD	SSIM	PSNR	diff SD
Area Interpolation	0.911±0.011	29.721±1.948	0.068±0.011	0.814±0.018	26.250±1.867	0.071±0.013	0.723±0.031	24.092±1.964	0.080±0.013
Trilinear Interpolation	0.964±0.005	37.680±1.770	0.013±0.002	0.906±0.007	33.148±1.780	0.022±0.004	0.872±0.011	31.504±1.786	0.026±0.005
Zero-padded	0.977±0.013	37.980±4.078	0.064±0.011	0.926±0.009	31.844±2.260	0.067±0.013	0.888±0.012	29.803±2.147	0.069±0.015
SR Main Training	0.986±0.007	42.781±2.424	0.009±0.002	0.961±0.009	36.710±1.086	0.014±0.002	0.939±0.008	35.377±1.653	0.017±0.003
SR Fine-tuning	***0.993±0.004***	***45.706±2.169***	***0.005±0.002***	***0.973±0.005***	***39.433±2.144***	***0.007±0.001***	***0.957±0.006***	***37.306±2.357***	***0.014±0.004***

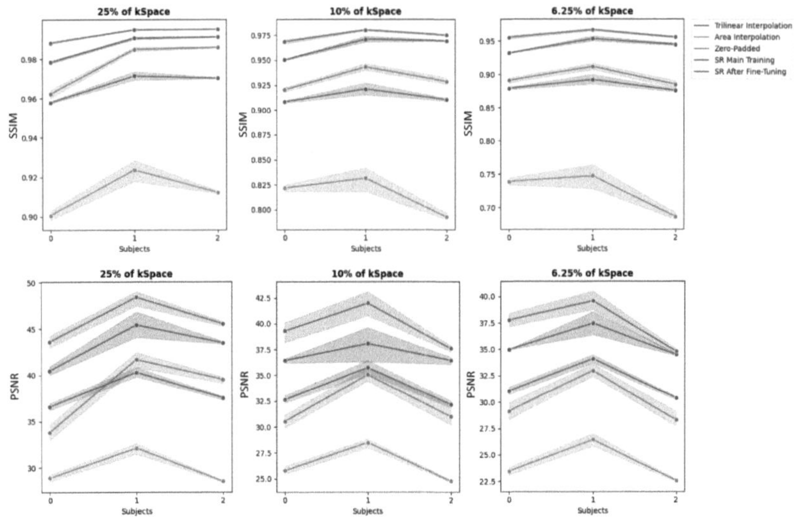

Fig. 9.11 Line plot showing the mean and 95% confidence interval of the resultant SSIM and PSNR over the different time-points for each subject. The blue, orange, green, red, and violet lines represent the reconstruction results of trilinear interpolation, area interpolation, zero-padding (sinc interpolation), SR main training, and SR after fine-tuning, respectively. The upper row shows the SSIM values and the lower row shows the PSNR values [pub:13]

The acquisition time of high-resolution 3D "pseudo"-dynamic reference data without parallel acquisition in this study was ten seconds and five seconds with GRAPPA factor two (Table 9.2). These are not sufficient for real-time or near real-time applications and might lead to blurring in free-breathing subjects. This research shows the potential to acquire such volume with only minimal loss of spatial information in less than half a second.

The fine-tuning process took approximately eight hours to finish for each subject using the earlier mentioned setup (Sect. 9.3.1.1). Super-resolving each time-point took only a fraction of a second. The required time for fine-tuning and inference can further be reduced by reducing the patch-overlap (stride), though that might reduce the quality of the resultant super-resolved images. It can be further perceived that the network was able to produce results highly similar to the ground-truth (SSIM of 0.957) even while super-resolving 6.25% of k-space, which can make the acquisition 16 times faster. Combining this fast acquisition speed with the inference speed of the method, this study can be extended to be used for real-time or near real-time MRIs during interventions.

In the current study, only the centre of the k-space was used during undersampling, which resulted in a loss of resolution without creating explicit image artefacts. Other undersampling patterns, such as variable density or GRAPPA-like uniform undersampling of higher spatial frequencies may be investigated in the future.

It should be noted that the static planning scans and the actual dynamic scans during interventions are typically acquired with different sequences, with planning scans having higher contrast and resolution than the dynamic scans. This study was conducted using the same sequence for static and dynamic scans, but different resolutions and positions (different scan sessions). An additional experiment was performed by fine-tuning using a volumetric interpolated breath-hold examination (VIBE) sequence as a planning scan for one subject. Super-resolving the dynamic low-resolution images from 6.25% of the k-space resulted in 0.032 lower SSIM than using the identical sequence with higher resolution for fine-tuning. This may be a limitation of the current approach but requires further investigation.

9.3.1.3 Summary and Outlook

This research shows that fine-tuning with a subject-specific prior static scan can significantly improve the results of deep learning based Super-Resolution (SR) reconstruction (*[RQ:1]*, *[RQ:1.1]*). A 3D UNet-based model was trained with the help of perceptual loss to estimate the reconstruction error. The network model was initially trained using the CHAOS abdominal benchmark dataset and was then fine-tuned using a static high-resolution prior scan. The model was used to obtain super-resolved high-resolution 3D abdominal dynamic MRI from their corresponding low-resolution images. Even though the network was trained using MRI sequences different than the reconstructed dynamic MRI, the SR results after fine-tuning showed higher similarity with the ground-truth images. The proposed method could overcome the spatio-temporal trade-off by improving the spatial resolution of the images without compromising the speed of acquisition. This approach could be applied to real-time dynamic acquisitions, such as for interventional MRIs, because of the high speed of inference of deep learning models. In the presented approach, a 3D UNet was used as the network model, which needs interpolation as a pre-processing step. Therefore the reconstructed images could be suffering from interpolation errors. As future work, network models such as SRCNN, which do not need interpolation, will be studied. In addition, image resolutions lower than the already investigated ones will be studied to check the network's limitations. Moreover, clinical interventions are performed with devices, such as a needle, which are not present in the planning scan. An interesting future direction of research will be evaluate the performance of this method with such devices.

9.3.2 DDoS: Dynamic Dual-Channel UNet

This research extends the previous research of Single-Image Super-Resolution (SISR) of dynamic MRIs, treating each time-point as individual 3D volumes by incorporating the temporal information into the network model. The proposed method super-resolves the low-resolution dynamic MRI with the help of a static prior scan and by exploiting the temporal relationship between the different time-points. The initial findings of this work was presented in [pub:40] and [pub:34], while the final findings are presented in [pub:23] and the complete work has been published in [pub:4]. The code of this project is available publicly on GitHub [code:3].

9.3.2.1 Methodology

In this work, the dynamic training data was initially generated from the benchmark dataset due to the lack of dynamic abdominal data. After that, it was undersampled, and a modified UNet model was trained on that. The dual-channel input consists of the low-resolution image of the current time-point ($LR_T Pn$) and the super-resolved image of the previous time-point ($HR_T Pn-1$). The network was trained and tested with different levels of undersampling.

Network Architecture The 3D UNet architecture from the previous work [pub:13] was extended using multi-channel for supplying prior-information [pub:27] to create the proposed **D**ynamic **D**ual-channel **o**f **S**uper-resolution UNet architecture (DDoS-UNet, or simply DDoS), as shown in Fig. 9.12. The basic architecture of the UNet is similar to the previous work [pub:13]—except for two differences, having contracting (encoding) and expanding (decoding) paths. The contracting path is made of three blocks; each of the blocks comprises two pairs of 3D convolutional layers (kernel size:3, stride:1, padding:1) and ReLU activation functions, followed by average pool layers (kernel size: 2)—making the output size of the block half the size of the input received by that block. The expanding path also consists of three blocks, each consisting of a pair of trilinear upsampling layers (scale factor:2) and a 3D convolutional layer (kernel size:1, stride:1, padding:0), unlike the original work, which used 3D convolutional transpose layers (first difference with the earlier model); followed by a convolutional block similar to the contracting path, except for the pooling layers. It is noteworthy that initial experiments were performed using 3D convolutional transpose layers similar to the earlier model, but for volumetric super-resolution, this model resulted in checkerboard artefacts [207]. This can be attributed to the fact that overlapped portions of the patches are averaged in the patch-based super-resolution—mitigating the checkerboard problem, but in volumetric super-resolution, there is no averaging operation that could mitigate this effect. Each block of the expanding path increases the size of its input by a factor of

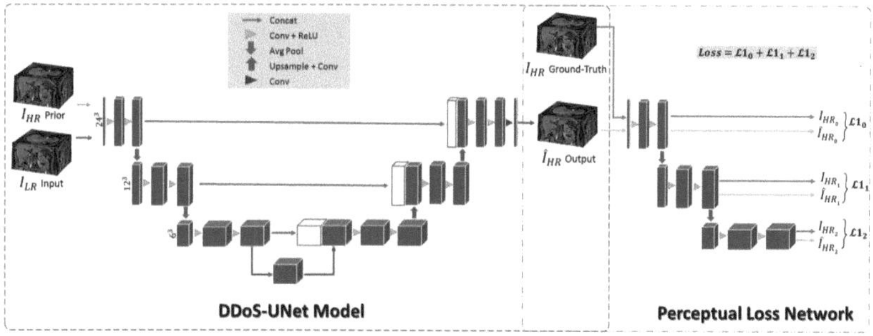

Fig. 9.12 DDoS-UNet: Proposed network architecture [pub:4]

two. Inside these expanding path blocks, after upsampling the input using trilinear-convolution pair, the output is concatenated with the input coming from a similar depth of the contraction path—known as skip-connections. The initial layer of the network provides an output of 64 feature maps. Then, each block of the contraction path increases the number of feature maps by two, whereas each of the expanding path blocks decreases it by two. Finally, a fully connected 3D convolutional layer (kernel size: 1, stride: 1, padding: 0) is applied to merge all the feature maps to generate the final output. The other difference between the earlier UNet [pub:13] and this DDoS-UNet is the fact that the initial layer of the network receives two input channels rather than one.

Since the UNet-like architectures warrant for the matrix size of the input to be the same as the output (ground-truth), the low-resolution input volumes were interpolated using trilinear interpolation with the interpolation factor equivalent to the acceleration factor before providing them as input to the DDoS-UNet model.

DDoS: Working Mechanism and Theory The DDoS-UNet works with dynamic MRIs while using the static planning scan as a prior image. Initially, the network is supplied with a patient-specific fully sampled high-resolution (HR) static prior scan on the first channel and the first time-point (TP0) of the undersampled low-resolution (LR) dynamic MRI on the second channel. It is to be noted that the static planning scan is acquired with the same protocol as the dynamic scan, but they are not co-registered. Given this pair of HR-LR images, DDoS-UNet super-resolves the LR to obtain the TP0 of the super-resolved (SR) HR dynamic MRI. This initial phase is termed here as the "Antipasto" phase as it precedes the main reconstruction phase. The reconstruction phase starts by supplying this SR-TP0 on the first channel, while the LR-TP1 is supplied on the second channel of the network to generate SR-TP1. This process is continued recursively for all the subsequent time-points. This can be formulated by modifying Eq. 9.1 as:

$$\hat{HR}_{TP(n)} = \digamma(LR_{TP(n)}; \hat{HR}_{TP(n-1)}; \theta) \tag{9.6}$$

where $\hat{HR}_{TP(n)}$ is the super-resolved time-point, $LR_{TP(n)}$ is the low-resolution time-point, $\hat{HR}_{TP(n-1)}$ is the super-resolved previous time-point, $\digamma$ is the super-resolution model that maps those three images, and θ is the set of parameters of $\digamma$. It is hypothesised that the network learns two different representations: the temporal relationship between $\hat{HR}_{TP(n)}$ and $\hat{HR}_{TP(n-1)}$, the super-resolution relationship between $LR_{TP(n)}$ and $\hat{HR}_{TP(n)}$. If Ψ is the DDoS operator and Φ is the set of parameters of the DDoS network, this hypothesis can be formulated as:

$$\Psi(\Phi) = \digamma_1(LR_{TP(n)}; \hat{HR}_{TP(n)}; \theta_1) + \digamma_2(\hat{HR}_{TP(n-1)}; \hat{HR}_{TP(n)}; \theta_2) \tag{9.7}$$

where $\digamma_1$ is the super-resolution operator learnt for the relationship between $LR_{TP(n)}$ and $\hat{HR}_{TP(n)}$, whereas θ_1 is its parameters; $\digamma_2$ is the temporal-relationship operator learnt for the relationship between $\hat{HR}_{TP(n-1)}$ and $\hat{HR}_{TP(n)}$, where θ_2 is its parameters.

It is worth mentioning that the patch-based super-resolution idea of the previous work [pub:13] was dropped in this current research due to the working theory of DDoS-UNet. Due to physiological movements, the organs can move in and out of the 24^3 patches (as used in the previous work). Consequently, the supplied $LR_{TP(n)}$ and $\hat{HR}_{TP(n-1)}$ patches might not contain similar organs—making the hypothesis of the temporal-relationship operator $\digamma_2$ of Eq. 9.6 invalid. Hence, this work performs volumetric super-resolution (using complete 3D volumes) instead of 3D patch-based super-resolution.

Data The proposed method was trained using the publicly-available abdominal benchmark dataset: the CHAOS dataset (T1-dual images, in- and opposed phase) [pub:16], comprising 80 volumes (40 subjects, in-phase and opposed-phase for each subject). Dynamic training data was generated artificially by applying random elastic deformation, explained in detail in Sect. 9.3.2.1. The dataset was divided into training and validation sets with a ratio of 70:30. For testing the approach, high-resolution 3D static (breath-hold) and 3D "pseudo"-dynamic (free-breathing) scans for 25 time-points of five healthy subjects were acquired using a 3T MRI (Siemens Magnetom Skyra). Each subject's static and dynamic scans were acquired in different sessions using the same sequence, parameters, and volume coverage. All the datasets (except the high-resolution static scans) were artificially undersampled to simulate the low-resolution datasets. The acquisition parameters of the datasets are listed in Table 9.4.

Table 9.4 MRI acquisition parameters CHAOS dataset and subject-wise 3D dynamic scans. Static scans were performed using the same subject-wise sequence parameters as the dynamic scans for one time-point (SR), acquired at a different session

	CHAOS (40 Subjects)	Protocol 1 (2 Subjects)	Protocol 2 (1 Subject)	Protocol 3 (1 Subject)	Protocol 4 (1 Subject)
Sequence	T1 Dual In-Phase & Opposed-Phase	T1w Flash 3D	T1w Flash 3D	T1w Flash 3D	T1w Flash 3D
Resolution	$1.44 \times 1.44 \times 5$ – $2.03 \times 2.03 \times 8\ mm^3$	$0.90 \times 0.90 \times 4\ mm^3$	$0.90 \times 0.90 \times 4\ mm^3$	$0.90 \times 0.90 \times 4\ mm^3$	$1.00 \times 1.00 \times 4\ mm^3$
FOV x, y, z	$315 \times 315 \times 240$ – $520 \times 520 \times 280\ mm^3$	$300 \times 225 \times 176\ mm^3$	$350 \times 262 \times 176\ mm^3$	$350 \times 262 \times 192\ mm^3$	$350 \times 262 \times 176\ mm^3$
Encoding matrix	$256 \times 256 \times 26$ – $400 \times 400 \times 50$	$320 \times 240 \times 44$	$384 \times 288 \times 44$	$384 \times 288 \times 48$	$352 \times 264 \times 44$
Phase/Slice oversampling	–	10/0 %	10/0 %	10/0 %	10/0 %
TR	110.17 – 255.54 ms	2.37 ms	2.40 ms	2.40 ms	2.31 ms
TE	4.60 –4.64 ms (In-Phase) 2.30 ms (Opposed-Phase)	1.00 ms	1.02 ms	1.02 ms	0.97 ms
Flip angle	80°	8°	8°	8°	8°
Bandwidth	–	920 Hz/Px	930 Hz/Px	930 Hz/Px	950 Hz/Px
GRAPPA factor	None	None	None	None	None
Phase/Slice partial Fourier	–	Off/Off	Off/Off	Off/Off	Off/Off
Phase/Slice resolution	–	50/64 %	50/64 %	50/64 %	50/64 %
Fat saturation	–	On	On	On	On
Time per SR	–	10.52 sec	12.80 sec	13.96 sec	11.36 sec

Dynamic Data Generation Since large dynamic MRI datasets that would be required for training are not available publicly, an artificial dynamic dataset was created. This was achieved by applying random elastic deformation of TorchIO [394] on the volumes from the CHAOS dataset. Random displacement fields were generated using Torchio's random elastic deformation with five control points, 5-20-20 mm of maximum displacements along x-y-z dimensions, respectively, and two locked borders. The displacement files were then applied to the volumes of the CHAOS dataset using cubic B-spline interpolation, considering them as TP0, to generate artificial TP1. Then, a new set of random displacement fields with the same parameters were generated and applied on TP1 to generate TP2. In this manner, 24 artificial time-points (TP1—TP24) were generated for each of the volumes present in the original dataset. The displacement field tries to imitate the movement induced by breathing during a dynamic acquisition. The displacement field was set to expand and/or contract more in anterosuperior (front-back) and superoinferior (up-down) but less in lateral (left-right) direction. However, this manner of generating artificial breathing motion is not equivalent to physiological motion. It is to be noted that the goal of using this kind of artificial motion was to create a dataset from which a network can learn the pseudo-temporal relationship between two subsequent time-points. This process results in an artificially created dynamic dataset—CHAOS dynamic, comprising 25 time-points in total for each volume.

Undersampling The training data—the original CHAOS dataset and the artificially created CHAOS dynamic dataset, as well as the testing data (3D dynamic scans) were artificially undersampled in-plane using MRUnder [url:8, code:7] by taking only 10%, 6.25%, and 4% of the centre k-space, resulting in MR acceleration factors of 3, 4, and 5, respectively (considering undersampling only in the phase-encoding direction). Considering the actual amount of data used during SR-reconstruction, this results in theoretical acceleration factors of 10, 16, and 25, respectively. The effective resolutions and estimated acquisition times for each of the dynamic test datasets are calculated using Eq. 9.8 and shown in Table 9.5

$$T_{acq} = PE_n \times TR \times S_m \tag{9.8}$$

where T_{acq} is the estimated acquisition time, given the number of phase-encoding lines PE_n, the repetition time TR, and the number of slices acquired S_m [pub:13]. During the calculation of Table 9.5, phase/slice resolution and phase/slice oversampling (Table 9.4) were also taken into consideration while calculating PE_n and S_m.

Table 9.5 Effective resolutions and estimated acquisition times (per SR) of the dynamic and static datasets after performing different levels of artificial undersampling

	Protocol 1		Protocol 2		Protocol 3		Protocol 4	
	Resolution (mm^3)	Acq. Time (sec)	Resolution (mm^3)	Acq. Time (sec)	Resolution (mm^3)	Acq. Time (sec)	Resolution (mm^3)	Acq. Time (sec)
High Resolution (Ground-truth)	$0.90 \times 0.90 \times 4$	8.76	$0.90 \times 0.90 \times 4$	10.68	$0.90 \times 0.90 \times 4$	11.76	$1.00 \times 1.00 \times 4$	9.38
10% of k-space	$2.70 \times 2.70 \times 4$	0.88	$2.70 \times 2.70 \times 4$	1.07	$2.70 \times 2.70 \times 4$	1.18	$3.00 \times 3.00 \times 4$	0.94
6.25% of k-space	$3.60 \times 3.60 \times 4$	0.55	$3.60 \times 3.60 \times 4$	0.67	$3.60 \times 3.60 \times 4$	0.74	$4.00 \times 4.00 \times 4$	0.59
4% of k-space	$4.50 \times 4.50 \times 4$	0.35	$4.47 \times 4.47 \times 4$	0.43	$4.47 \times 4.47 \times 4$	0.47	$4.99 \times 4.99 \times 4$	0.38

Implementation, Training, and Inference The proposed model was trained on 3D volumes from the artificially created dynamic version of a publicly available benchmark dataset, as summarised in Fig. 9.13. Figure 9.14 shows an overview of the inference steps. The inference process starts with the Antipasto phase—by supplying the high-resolution patient-specific static scan as a prior image on the first channel of the network (as $\hat{HR}_{TP(n-1)}$ is not yet available), and by supplying $LR_{TP(0)}$ on the second channel of the network.

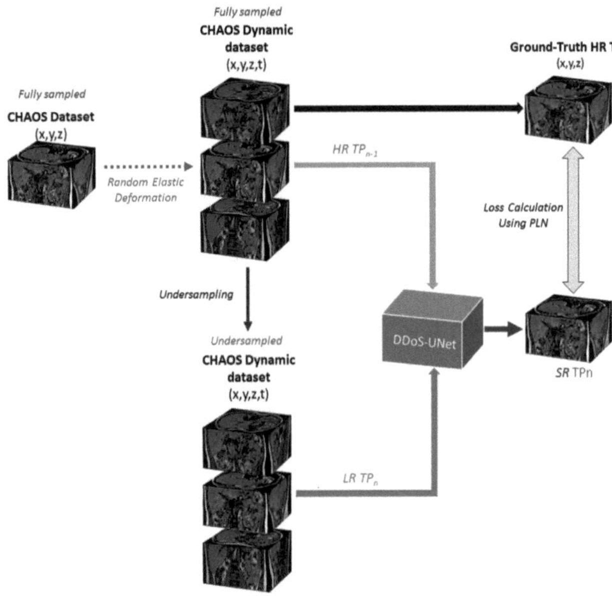

Fig. 9.13 Method Overview: Training. Initially, random elastic deformation is applied on the CHAOS dataset (fully sampled) to generate the artificial CHAOS dynamic dataset. Then the CHAOS dynamic dataset was undersampled to generate the final training dataset. Then the model is trained by providing low-resolution (undersampled) current time-point ($LRTP_n$) along with the high-resolution (fully sampled) previous time-point ($HRTP_{n-1}$) as input and the output is compared against the ground-truth high-resolution current time-point ($HRTP_n$) [pub:4]

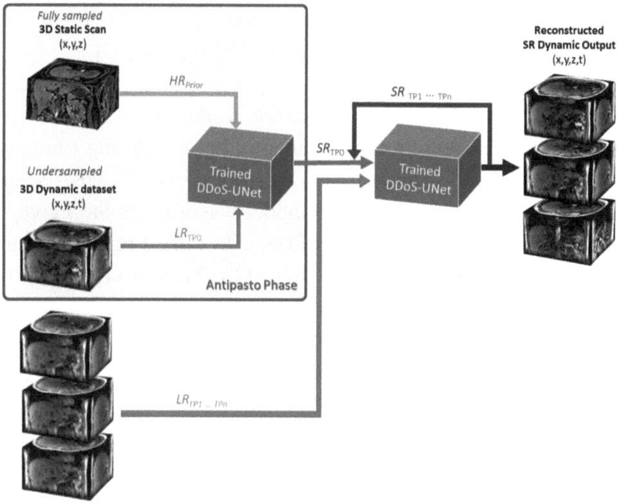

Fig. 9.14 Method Overview: Inference. 3D static subject-specific planning scan (fully sampled) is supplied as the high-resolution prior image (HR_{Prior}) along with the first low-resolution (undersampled) time-point (LR_{TP0}) of the 3D dynamic dataset are supplied as input to the trained DDoS-UNet model, and the model super-resolves LR_{TP0} to obtain SR_{TP0}. This initial phase is called as the "Antipasto" phase. LR_{TP0} is then supplied as input together with the next low-resolution time-point LR_{TP1} to the same trained DDoS-UNet model to obtain SR_{TP1}. This process is continued recursively until all the time-points of the low-resolution (undersampled) 3D dynamic are super-resolved, by supplying pairs of SR_{TPn-1} and LR_{TPn} to obtain each of the SR_{TPn} [pub:4]

It is to be noted that the static scan has the same resolution, contrast and volume coverage as the high-resolution ground-truth dynamic scan. However, to keep the testing environment similar to the real-life scenario and keep a fast speed of inference, the static and dynamic datasets were not co-registered, as registration is typically time-consuming. After this, the network super-resolves $LR_{TP(0)}$ to $\hat{H}R_{TP(0)}$. Now for the next time-point, $\hat{H}R_{TP(0)}$ and $LR_{TP(1)}$ are supplied as input to the network and the network provides $\hat{H}R_{TP(1)}$ as output.

The implementation was done using PyTorch [327], and training-inference were performed using Nvidia Tesla V100 GPUs. Following the hypothesis of using batch size one to be able to learn an exact mapping function between the specific pair of low-high-resolution images [pub:8], batch size during training and inference in this research was also set to one. The loss during training was calculated using perceptual loss [203], with the help of a perceptual loss network [pub:11], and was minimised using the Adam optimiser with a learning rate of 1e-4 for 100 epochs. The code of the implementation is available on GitHub [code:3].

Perceptual Loss Similar to the previous work [pub:13], perceptual loss (Sect. 4.4.3.1) [203] was employed to compute the loss during training. For the same, the initial three blocks of the frozen pre-trained (on 7T MRA scans, for the task of vessel segmentation) UNet MSS model was used as the Perceptual Loss Network (PLN) [pub:11]. The job of this PLN is to extract minute features of different abstraction levels at the different levels of the PLN, from the super-resolved volumes and their corresponding ground-truths. The features generated from the super-resolved output and the ground-truth were compared against each other using mean absolute error (L1 loss, Eq. 4.28). Finally, all these 11 losses were added together and backpropagated.

9.3.2.2 Results

The quality of super-resolution was evaluated quantitatively with the help of the Structural Similarity Index (SSIM, Eq. 4.35)[74]—to evaluate the perceptual quality of the output, the Peak Signal to Noise Ratio (PSNR, Eq. 4.33) and the Normalised Root Mean-Squared Error (NRMSE, Eq. 4.32)—to measure the image quality statistically. The statistical significance of the differences in the quantitative metrics for the proposed method against the other baselines was computed using the Mann-Whitney U-test [7]. Apart from quantitative evaluations, the results were also compared qualitatively.

The performance of the DDoS-UNet was compared for three different levels of undersampling: 10%, 6.25%, and 4% of the centre k-space, against the low-resolution input, traditional trilinear-interpolation, Fourier interpolated input (zero-padded k-space), and finally against two different baseline deep learning models: two UNet models identical to the DDoS-UNet except for the initial layer (unlike DDoS-UNet, these UNets received one input)—one of them trained on the original CHAOS dataset (T1-dual images, in- and opposed phase) [pub:16], and the other one was trained using artificial dynamic CHAOS (see Sect. 9.3.2.1). The training dataset of the second UNet was identical to the training dataset of the DDoS-UNet. The models were evaluated on real dynamic datasets of five subjects, each consisting of 25 time-points (details in Sect. 9.3.2.1). The inference process for the DDoS-UNet was started with the patient-specific prior high-resolution static scan and first low-resolution time-point as input and then continued by supplying the previous super-resolved time-point with the current low-resolution time-point to super-resolve the current time-point (as explained in Sect. 9.3.2.1).

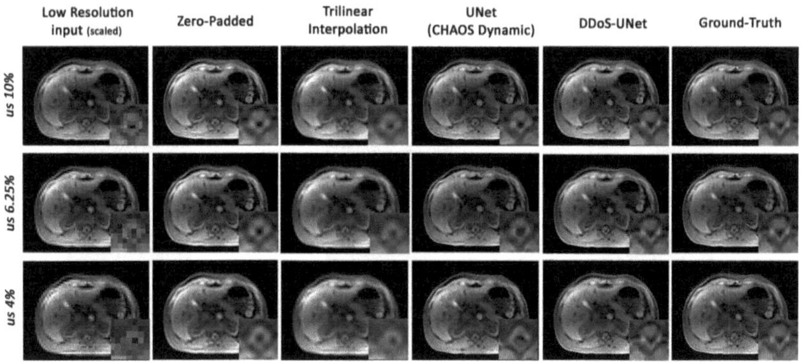

Fig. 9.15 Comparative qualitative results of low resolution (10%, 6.25% and 4% of k-space) 3D Dynamic data of the same slice. From left to right: low resolution images (scaled-up, nearest-neighbour interpolation), Interpolated input (Trilinear), Zero-padded reconstruction, Output of UNet trained on CHAOS dynamic dataset, Output of DDoS-UNet and ground-truth images [pub:4]

Figure 9.15 shows a qualitative comparison of the results obtained by the different methods for different levels of undersampling. It can be observed that the proposed DDoS-UNet managed to restore finer details better than the other methods. Moreover, both the baseline UNet models show better anatomical structures than the zero-padded reconstructions. Furthermore, the comparison with the help of SSIM maps between the input (low-resolution images) and output (super-resolved images) of the DDoS-UNet are shown in Fig. 9.16. It reveals that the reconstruction quality of the initial time-point is not very good, but the network manages to recover from the initial struggle during the Antipasto phase, and manages to reconstruct the subsequent time-points much better and consistently over all the time-points. This can be attributed to the fact that the static and dynamic scans are acquired in two different sessions, and they are not co-registered. Finally, Fig. 9.17 shows the qualitative comparisons of the different methods for two Region Of Interest (ROI). It shows the proposed DDoS-UNet framework results in better reconstruction performance than the baseline UNet models. Between the baseline UNets, UNet trained on the CHAOS dynamic dataset managed to recover finer anatomical details better than the UNet model trained on the CHAOS dataset.

Table 9.6 presents the quantitative results for all the methods. It can be observed that both the baseline UNet models (trained on the original CHAOS dataset and on the CHAOS dynamic dataset) outperformed the non-DL baselines: trilinear interpolation and zero-padded reconstruction (sinc interpolation), and the proposed

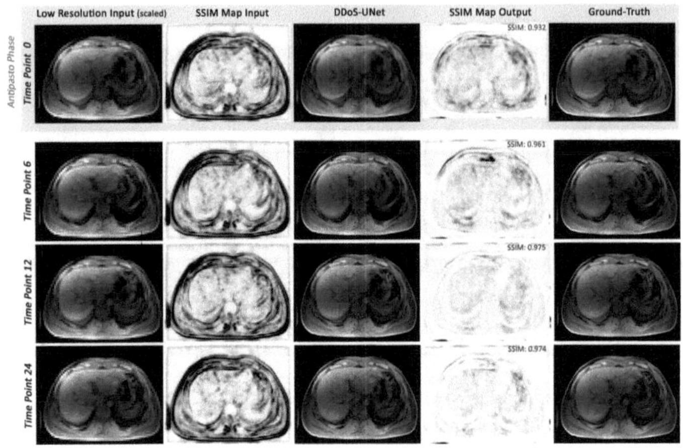

Fig. 9.16 An example comparison of the low resolution input of the 4% of k-space with the SR result of the DDoS-UNet over four different time-points, compared against the high resolution ground-truth using SSIM maps [pub:4]

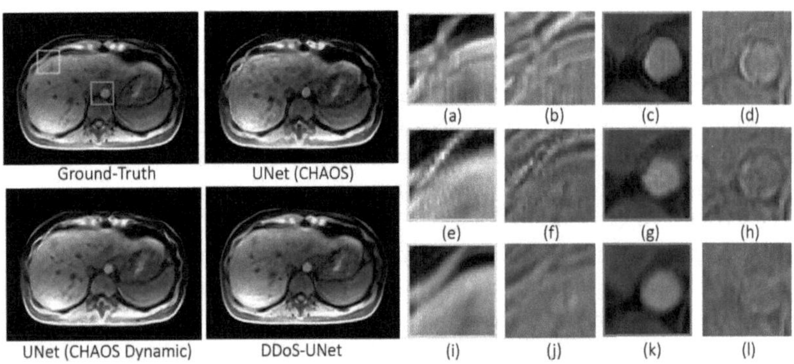

Fig. 9.17 An example of reconstructed results from UNet baselines and DDoS-UNet, compared against its ground-truth (GT) for low resolution images from 4% of k-space. From left to right, upper to lower: ground-truth, SR result of the UNet baseline (UNet CHAOS), SR result of the UNet baseline trained on CHAOS dynamic (UNet CHAOS Dynamic) and SR result of the DDoS-UNet. For the yellow ROI, (a–b): UNet CHAOS and the difference image from GT, (e–f): SR result of UNet CHAOS Dynamic and (i–j): SR result of DDos-UNet and the difference image from GT. The images on the right part are identical examples for the green ROI [pub:4]

Table 9.6 The average and the standard deviation of SSIM, PSNR, and NRMSE. The table shows the results for different resolutions

Data	10% of k-space			6.25% of k-space			4% of k-space		
	SSIM	PSNR	NRMSE	SSIM	PSNR	NRMSE	SSIM	PSNR	NRMSE
Trilinear Interpolation	0.872±0.014	28.631±1.364	0.192±0.023	0.821±0.017	26.770±1.226	0.238±0.024	0.765±0.022	25.248±1.298	0.283±0.025
Zero-padded	0.949±0.013	36.138±1.753	0.082±0.016	0.910±0.018	29.761±1.640	0.124±0.019	0.863±0.021	32.520±1.508	0.170±0.025
UNet (CHAOS)	0.967±0.006	38.359±1.580	0.021±0.004	0.944±0.010	35.623±1.552	0.029±0.005	0.916±0.015	32.658±1.598	0.041±0.007
UNet (CHAOS Dynamic)	0.959±0.012	37.376±1.275	0.024±0.003	0.941±0.012	35.113±1.566	0.031±0.006	0.914±0.012	33.620±1.035	0.036±0.004
DDoS-UNet	**0.980±0.006**	**41.824±2.070**	**0.014±0.003**	**0.967±0.011**	**39.494±2.121**	**0.019±0.005**	**0.951±0.017**	**37.557±2.179**	**0.024±0.006**

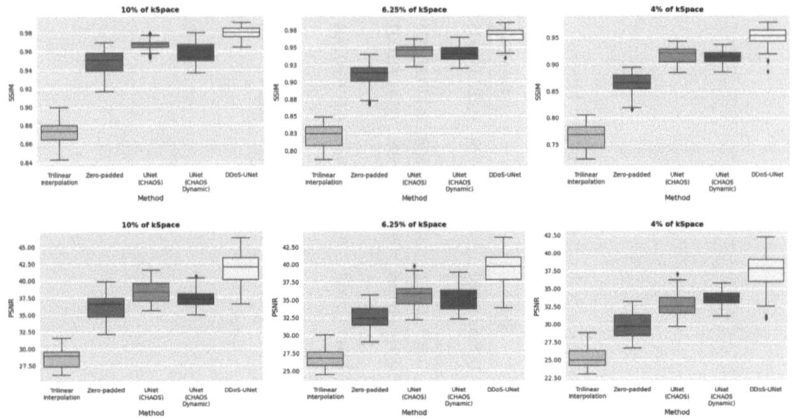

Fig. 9.18 Quantitative comparison of different methods using SSIM and PSNR—for all subjects and time-points combined, for different levels of undersampling [pub:4]

DDoS-UNet method outperformed all the baselines for all three undersampling factors in all three metrics with statistical significance (p-values always less than 0.001). It can be further observed that the UNet trained on the original CHAOS dataset outperformed the UNet trained on the CHAOS dynamic dataset. Figure 9.18 shows the resultant SSIM and PSNR values over all subjects and time-points by means of box plots. It can be observed that the improvements obtained by the proposed method increase with the increase in the undersampling factor. Figure 9.19 portrays the SSIM values over the different time-points averaged for all five subjects. It can be seen that after the initial time-point TP0 (Antipasto phase), the proposed DDoS-UNet achieved consistently better SSIM values compared to all the other methods. Finally, Fig. 9.20 shows the average SSIM values over the different time-points (excluding the Antipasto phase) for each subject. The median values over TP1 to TP24 for each of the subjects resulted in SSIM values in the range 0.988 to 0.975, 0.980 to 0.960, and 0.970 to 0.945, for 10%, 6.25%, and 4% of k-space, respectively. Figure 9.19 and 9.20 show that the proposed DDoS-UNet is able to reconstruct different protocols and subjects efficiently while being stable over different time-points.

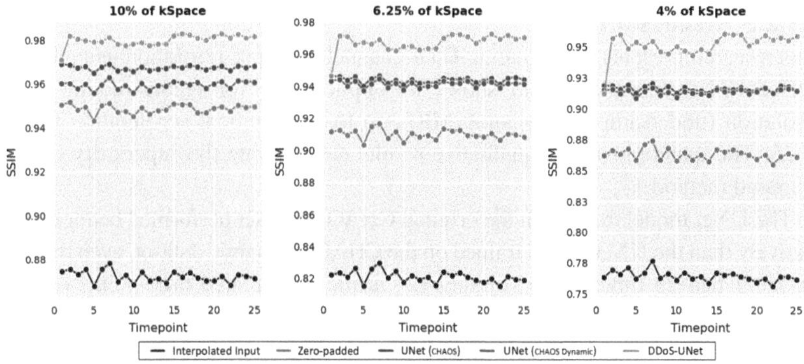

Fig. 9.19 Line plot showing the average SSIM values over each subject for all time-points, for different levels of undersampling. Initial drop can be observed for the first time-point for DDoS-UNet, which is referred here as the Antipasto phase, then the network performs with stability for the rest of the time-points [pub:4]

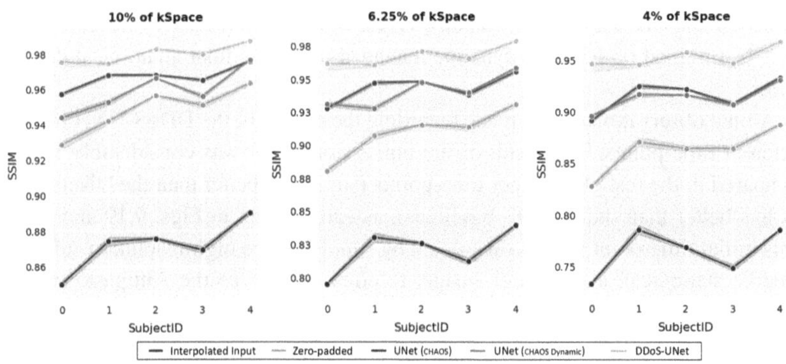

Fig. 9.20 Line plot showing the mean and 95% confidence interval of the resultant SSIM values over the different time-points (excluding the initial one, the Antipasto phase) for each subject. The red, blue, orange, green, and violet lines represent the reconstruction results of trilinear interpolation, zero-padding (sinc interpolation), UNet trained on CHAOS dataset, UNet trained on CHAOS Dynamic dataset, and DDoS-UNet, respectively [pub:4]

9.3.2.3 Discussion

This research presents the **D**ynamic **D**ual-channel **o**f **S**uper-resolution using **UNet** (DDoS-UNet) framework and shows its applicability for reconstructing low-resolution (undersampled) dynamic MRIs up to a theoretical acceleration factor of 25. The quantitative and qualitative results demonstrate the superiority of the proposed method.

The UNet model trained on the original CHAOS dataset performed better quantitatively than the UNet model trained on the CHAOS dynamic dataset, even though the latter had 25 times more volumes (24 artificially created time-points on top of the original one). This can be attributed to the quality of the CHAOS dynamic dataset. Due to the repeated applications of the random elastic deformation on the original dataset, which includes interpolation, the sharpness of the later time-points decreased caused by the accumulated interpolation errors. This might have also negatively impacted the results of the DDoS-UNet. Improving the quality of the artificial dynamic dataset might improve the performance of both of these models. It is worth mentioning, however, that for the highest undersampling factor (4% of the k-space), UNet trained on the CHAOS dynamic dataset resulted in better PSNR than UNet trained on the CHAOS dataset, and also visual comparison (Fig. 9.17) revealed that the UNet trained on CHAOS dynamic managed to restore finer anatomical details better.

A final observation can be made regarding the results of the DDoS-UNet for the different time-points. The result of the initial time-point was considerably worse compared to the rest of the other time-points (similar or better than the UNets, and always better than the non-DL baselines), as can be seen in Figs. 9.19 and 9.16. This initial time-point was reconstructed by supplying the high-resolution subject-specific static scan as the prior image, referred to here as the Antipasto phase, whereas the rest of the time-points were reconstructed by supplying the super-resolved previous time-point as the prior image. The static scan has a big temporal difference from the first time-point of the dynamic scan as they were acquired in different sessions, while the subsequent time-points of the dynamic scan were closer in time. The network faces difficulties reconstructing the initial time-point, but then recovers from it after super-resolving the first one and then maintaining its performance steadily for all subsequent time-points. This also supports the hypothesis that the DDoS-UNet learnt both spatial and temporal relationships, as shown in Eqs. 9.6 and 9.7 in Sect. 9.3.2.1.

The reconstruction (inference) time using the proposed DDoS-UNet was approximately 0.36 secs for each time-point (9 secs for 25 TPs) while reconstructing using an Nvidia Tesla V100 GPU. Fast reconstruction time, coupled with the high speed of acquisition (shown in Table 9.5), this method shows the potential to acquire-reconstruct each time-point of a 3D dynamic acquisition within 0.71 secs (for 4% of k-space with Protocol 1)—making it a potential candidate for near real-time MR acquisitions. The acquisition time can be further reduced using techniques such as parallel imaging, as shown in the earlier work [pub:13]. The focus of this research is on abdominal imaging; however, this method might also be used for other types of dynamic imaging, e.g. cardiac imaging.

9.3.2.4 Summary and Outlook

This research proposes the DDoS-UNet model to perform 3D volumetric super-resolution of low-resolution dynamic MRIs by using a subject-specific high-resolution prior planning scan as a prior information (***[RQ:1]***, ***[RQ:3]***) and exploiting the spatio-temporal relationship present in the dynamic MRI. The proposed DDoS-UNet model receives two inputs as dual-channel—the actual image that is to be super-resolved and a prior image (***[RQ:3.1]***). Initially, the available planning scan is supplied as the prior image to super-resolve the initial time-point—known as the "Antipasto" phase. Then afterwards, the earlier super-resolve image is supplied as the prior image to super-resolve the next one. The proposed network was trained using an artificially created dynamic dataset from the CHAOS abdominal benchmark dataset and then was tested using dynamic MRIs comprising 25 time-points. It was observed that even though the network was trained using a dataset with MRI acquisition parameters very different from the test set, the network was able to super-resolve the given input images with high accuracy—even for high undersampling factors. The proposed method resulted in 0.951 ± 0.017 SSIM while super-resolving the highest undersampling experimented in this research (i.e. 4% centre k-space), whereas the baseline UNet (model without supplying the super-resolved previous time-point as prior information) resulted in 0.916 ± 0.015 (***[RQ:1.1]***). The results show that the proposed network managed to mitigate the spatio-temporal problem of dynamic MRI by performing spatial super-resolution with the help of the temporal relationship present in the data without compromising the acquisition speed. Given the reconstruction speed of the proposed approach, this can be a candidate for near real-time dynamic acquisition scenarios, such as interventional MRI. The proposed approach employs a multi-channel approach to supply the prior image (initially, the high-resolution static scan, then the super-resolved volumes). However, other approaches, such as dual-branch, have also been proposed [pub:27], which might also be used to supply such prior images to the network. Such an architecture can deal with the prior image and the low-resolution image differently (i.e. different weights applied on each), whereas the current initial layer of the network treats them equally and merges them as an internal representation in the initial layer. Moreover, DDoS-UNet is interesting in interventional setup. During interventions, devices such as catheters are used, which were not present in the training set. An interesting future direction of research will be evaluate the performance of this method with such devices.

10 Working in the Hybrid Space

Even though working in the image-space with deep learning techniques to reconstruct undersampled MRIs is a common approach, several k-space (Sect. 5.2.2) and a mix of image and k-space methods (Sect. 5.2.3) have been proposed over the years. The k-space is always complex-valued, whilst the image space can either be complex-valued or can also be real-valued if only the magnitude image (or only the phase-image) is considered or if the magnitude and phase images are considered separately. Methods working in the image space have mostly focused on correcting the magnitude images. Hence, real-valued convolution operations are sufficient in this case. However, most of the approaches working with k-space [301, 316, 384] or even the methods working on a mix of the spaces [267, 297, 368] treat the complex-valued input as two separate real-valued data by splitting the complex data into real and imaginary parts and then concatenating them on the channel dimension to supply to the real-valued neural network models—discarding the richer algebraic structure of the complex data [382]. This chapter focuses on the research of complex-valued convolution neural networks and their applications in the field of undersampled MR reconstruction.

Complex-valued Convolutional Neural Network (CV-CNN) [286] has been shown to outperform ordinary convolutional networks in many scenarios [259, 376, 391, 398]. Nevertheless, it has not been commonly used so commonly, which could be attributed to the fact that the technology of complex-valued convolutions is less developed than the one of real-valued convolution [319]. With the incorporation of complex-valued convolution into the PyTorch framework [327], the possibility of developing such approaches increased. Cole et al. [382] demonstrated that the CV-CNNs could perform better than real-valued CNNs applied on complex-data split into real and imaginary parts while performing artefact reduction in the image-space of undersampled MRIs. During the course of this thesis, a python package for creating complex-valued neural networks was developed, ***pytorch-complex***, which

S. Chatterjee, *Deep Learning Innovations in MRI Reconstruction and Analysis*,
https://doi.org/10.1007/978-3-658-50741-1_10

is a wrapper over the PyTorch framework, has been made available publicly on GitHub [code:10].

10.1 Preliminary Experiments

10.1.1 CNN vs CV-CNN: Which is Better?

The preliminary experiments started by conducting a structured set of experiments to compare the real-valued CNNs (or just CNNs) and Complex-valued Convolutional Neural Network (CV-CNN). These experiments were performed by Pavan Tummala. Cole et al. [382] have demonstrated that better performance on the task of undersampled MRI reconstruction in the image-space can be achieved by preserving the complex-valued relationship present in the data by using CV-CNN, instead of splitting the data into two channels—real and imaginary, consequently destroying the algebraic structure of the complex data.

The experiments which are shown aim at comparing CNNs and CV-CNNs for the task of classification and segmentation when each pair of the networks (CNN and CV-CNN) had the same number of parameters (features). The only difference was that the CNNs and real-valued features (kernels), while the CV-CNNs had complex-valued ones. However, in terms of the exact number of trainable parameters, the CV-CNNs had double the number of trainable parameters of the CNNs, as they were complex-valued—each of them had real and imaginary parts. Hence, to be fair in terms of the number of trainable parameters, one more set of networks was added in this research, which was real-valued CNNs, but with double the number of trainable parameters as their CNN counterparts—having the same number of trainable parameters as the CV-CNN, while having twice the number of features than both CNN and CV-CNN.

10.1.1.1 Network Models

For classification, three different sizes of the ResNet [200] were used: ResNet18, ResNet34, and ResNet50. The torchvision implementations [url:6] of those were used and modified for this research. The original torchvision implementations were used as the *CNN* models and then were modified in two different ways—the number of features was increased by a factor of 2 for all the layers to create the *CNNx2* models, and all the layers of the implementation were changed into complex-valued layers with the help of pytorch-complex [code:10] package to create the *CV-CNN* models. For segmentation, three different models were used—UNet [186], Attention UNet [277], and a modified version of ReconResNet [pub:8]. Those *CNN* models were modified in a similar fashion as the classification models to create the *CNNx2* and *CV-CNN* models.

10.1.1.2 Dataset

The dataset used in this research was the BraTS 2020 dataset [169, 232, 252], containing multimodal MRIs: T1-weighted (T1w), post-contrast T1-weighted (T1ce), T2-weighted (T2), and T2 Fluid Attenuated Inversion Recovery (T2-FLAIR, or simply, FLAIR) of 346 patients with two different types of tumours: low-grade glioma (LGG, 73 patients) and high-grade glioma or glioblastoma (HGG, 273 patients). The dataset contains 3D volumes The slices without any tumour were considered as "tumour-free", and a third class was created using them, resulting in 4,731, 18,276, and 24,679 slices, for LGG, HGG, and tumour-free class, respectively. This dataset only contains real-valued magnitude images. So the input to the models did not have any phase even though they were supplied as complex-valued.

10.1.1.3 Results and Discussion

The comparative results of the CNN and CV-CNN models are shown in Tables 10.1 and 10.2, for classification and segmentation tasks, respectively. Two common observations can be made from all the results. First is that the CNNx2 models performed better than the CNN models—which can be credited to the fact that the CNNx2 models had a double number of features as the CNN models. The second observation is that the CV-CNN models performed better than both the CNN and CNNx2 models, despite having the same number of features as the CNN models (but double the number of trainable parameters), while having half the number of features as the

Table 10.1 Classification results of CNN and CV-CNN

Model	Type	F1-Score	Accuracy
ResNet18	CNN	0.819±0.101	0.809±0.056
	CNNx2	0.827±0.097	0.811±0.121
	CV-CNN	***0.829±0.119***	***0.835±0.129***
ResNet34	CNN	0.783±0.107	0.798±0.115
	CNNx2	0.798±0.085	0.805±0.121
	CV-CNN	**0.817±0.072**	**0.829±0.123**
ResNet50	CNN	0.790±0.126	0.788±0.125
	CNNx2	0.796±0.089	0.795±0.118
	CV-CNN	**0.807±0.115**	**0.819±0.097**

Table 10.2 Segmentation results of CNN and CV-CNN

Model	Type	Dice	IoU
UNet	CNN	0.752±0.011	0.679±0.121
	CNNx2	0.759±0.193	0.681±0.089
	CV-CNN	**0.789±0.106**	**0.709±0.112**
Attention UNet	CNN	0.848±0.070	0.733±0.111
	CNNx2	0.849±0.103	0.742±0.087
	CV-CNN	***0.868±0.089***	***0.771±0.105***
ReconResNet	CNN	0.829±0.139	0.743±0.098
	CNNx2	0.831±0.112	0.755±0.108
	CV-CNN	**0.853±0.540**	**0.768±0.122**

CNNx2 models (but having the same number of trainable parameters). These experiments conclusively show that the improvements are not only because the CV-CNN models have the double number of trainable parameters when the number of features is the same for both, but because of the complex-valued features (***[RQ:4.1]***).

The ResNet18 came out as the winning network architecture for the classification problem, even when it was the smallest one. This could be because, being a small less-complex model, it was less prone to overfitting, and the number of features in ResNet18 might have been enough to learn to solve this given problem. For the segmentation task, Attention UNet outperformed the other two models. Nevertheless, another interesting observation can be made regarding the ReconResNet model, which was originally proposed for the task of artefact reduction in undersampled MRIs [pub:8]. The results show that this model is also capable of performing segmentations—performing significantly better than the UNet model and obtaining scores close (even though less) to the winning Attention UNet model.

With the recent release of the pathological annotations for the fastMRI dataset [293], known as the fastMRI+ [400], this research of CNN vs CV-CNN can be significantly extended by using real complex-valued data instead of only magnitude data as it was done here. Moreover, as fastMRI contains raw data, an interesting future direction to explore will be applying the CV-CNN classification models on the k-space data and explore the possibility of classifying directly in the k-space.

10.1.2 Going Beyond the Image Space: Missing Frequency Prediction in the k-space Using a CV-CNN

Reducing artefacts or improving resolution—using models are working in the image space for both these problems, which have been shown in the earlier chapters. When the images are already heavily corrupted by artefacts, it might be difficult for a network to reconstruct fine structures which are not visible because of the obscurity caused by those artefacts. When it comes to the k-space, there is only missing data and no obscurity. With this thing in mind, this can be hypothesised that if the network directly works in the k-space, then the network will be able to reconstruct even the fine structures which are obscured in the undersampled. AUTOMAP [297] was proposed to work directly in the k-space and provide the image space as output. But the network design hinders scalability when it comes to images with larger matrix sizes. There was a few more research in this direction [262, 316], but they are based on real-valued convolutions by converting the complex-valued k-space data into two-channel (real and imaginary) real-valued data. Following the better performance observed in the earlier section (Sect. 10.1.1) and the disadvantages of splitting real and imaginary parts of the data while doing undersampled reconstruction in the image space, as shown by Cole et al. [382], a new direction of research was started employing CV-CNN to perform undersampled MRI reconstruction directly in the k-space. This work was presented at [pub:37].

10.1.2.1 Network Architecture

The network model, **KSPResNet**, was created by taking inspirations from the earlier-presented image-space model—ReconResNet [pub:8]. The proposed CV-CNN model consists of a series of residual blocks, and each comprises two complex-valued convolutions with a kernel size of three, a stride of one, and padding of one, separated by an AdaptiveCmodReLU [284] for non-linearity. The AdaptiveCmodReLU computes the complex modulus ReLU of the complex input while learning a threshold value. These complex-valued operations were performed with the help of pytorch-complex [code:10]. The undersampled k-space was supplied as input to the network, and the network provided an output k-space. The real acquired data present in the undersampled k-space were replaced in the output k-space for data consistency, creating the final output k-space. 2D inverse fast Fourier transform was performed on this output k-space to generate the reconstructed image and from the fully-sampled k-space to generate the ground-truth image. Finally, the loss was calculated using L1 loss (Eq. 4.28) by comparing the reconstructed image with the ground-truth image. The network architecture can be seen in Fig. 10.1.

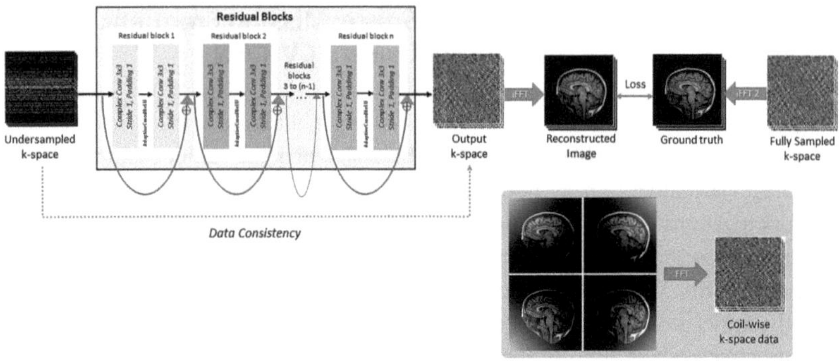

Fig. 10.1 KSPResNet: Architecture of the proposed network [pub:37]

10.1.2.2 Dataset

These preliminary experiments were performed using the fastMRI [293] T2w Axial brain dataset. The network worked directly with coil-wise images. During inference, the final coil-images were combined using sum-of-squares to obtain the final output image. Two different undersampling patterns with an acceleration factor of 4 experimented here: 1D random Variable density (Varden) sampling with a centre fraction of 0.08—resulting in ghosting artefacts in the reconstructed image and by taking only the centre of the k-space with full phase-encoding lines—causing the image to loose resolution. The network was trained using 50 volumes, validated using 25 volumes and finally tested on 50 volumes.

10.1.2.3 Results and Discussion

Figures 10.2 and 10.3 show the results for qualitative analysis, for the reconstruction of random Variable density undersampling and centre of the k-space sampling, respectively. For the centre k-space sampling, it can be observed that the network was able to reconstruct the image close to the ground-truth. For the random Varden undersampling, the network could reconstruct the image but with a few remaining undersampling artefacts. While performing qualitative analysis, the network achieved an average Structural Similarity Index (SSIM, Eq. 4.35) [74] of 0.9160.128 and 0.9080.110 for both the sampling patterns, respectively. The best resultant image produced an SSIM of 0.954 and 0.954 for the sampling patterns, respectively.

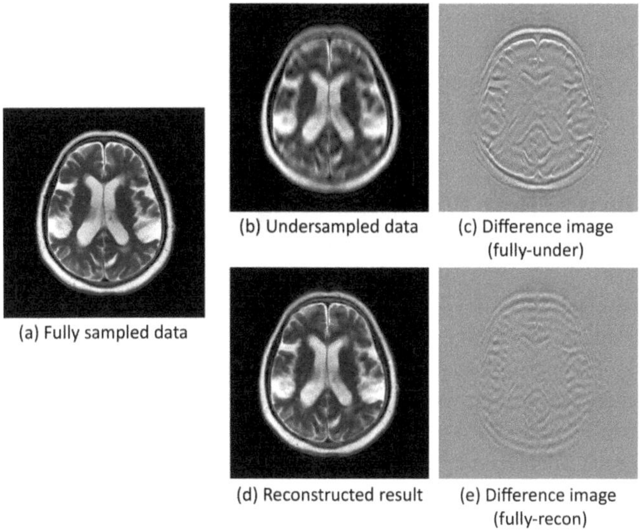

Fig. 10.2 KSPResNet results while dealing with undersampling artefacts originating from 1D Varden undersampling [pub:37]

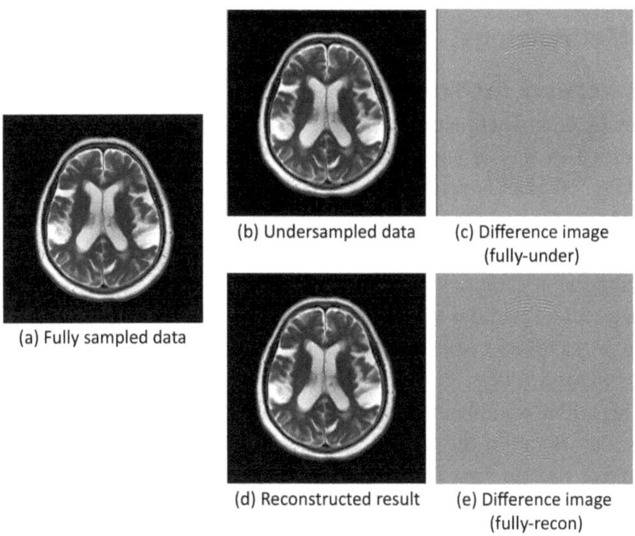

Fig. 10.3 KSPResNet results while dealing with lower-resolution due to undersampled k-space by taking only the centre [pub:37]

This work presented a novel complex residual network architecture for undersampled MR reconstruction while directly working in the k-space. The network has shown promising results in reducing artefacts and making the images sharper compared to the undersampled images.

10.2 Best of Both Worlds: Filtering in the Image Space and Predicting Missing Frequencies in the k-space

The Primal-Dual Network [249] proposed for the reconstruction of CT images has demonstrated its superiority over other deep learning based approaches for that task [389]. This approach was extended by proposing PD-UNet in Sect. 8.2 [pub:6], which demonstrated improvements in terms of speed and accuracy. Furthermore, the applicability of both PD network, as well as of PD-UNet, for the task of undersampled radial MRI reconstruction has also been demonstrated earlier (Sect. 8.2.2 [pub:37]). This chapter extends these approaches further to reconstruct undersampled Cartesian MRIs by working in the k-space—for predicting missing frequencies, and in the image space—for removing artefacts (filtering). The code of this research is available on GitHub as a part of the NCC1701 code [code:8].

10.2.1 Methodology

10.2.1.1 Network Architectures

The original Primal-Dual network (PD Network) [249], and the PD-UNet [pub:6] were modified in this research to propose Fourier-PD (Fig. 10.4) and Fourier-PDUNet (Fig. 10.5) architectures, where the primal blocks work in the image space (same as the original works), while the dual blocks work in the k-space. Because k-space is complex-valued, the dual blocks of both these networks were modified by replacing the convolution layers with complex-valued convolution layers and PReLU activation functions with Complex Rectified Linear Unit ($\mathbb{C}$ReLU) activations. The primal blocks were identical to the original ones.

Fourier-PD and Fourier-PDUNet can be construed as an unrolled iterative reconstruction algorithm optimising both the k-space, as well as the image space. During each step of the iteration, the processed k-space (output of the dual blocks) is reconstructed using Inverse Fast Fourier Transform (iFFT) and is combined with the previously processed image (output of the primal blocks) to supply to the primal blocks. The output of the primal blocks is taken back to the k-space using Fast Fourier Transform (FFT), then combined with the processed k-space and the original undersampled k-space to supply to the dual blocks. The dual blocks optimise

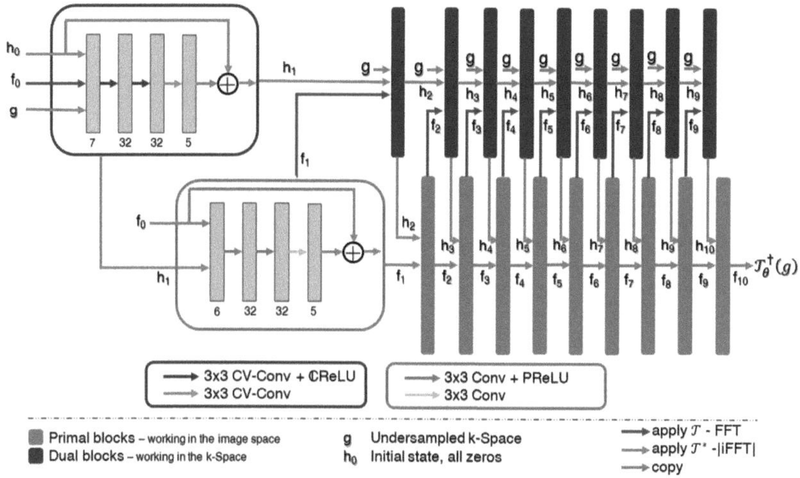

Fig. 10.4 Fourier Primal-Dual Network (Fourier-PD): Architecture of the proposed network. The yellow ochre and the violet boxes signify the primal and dual iterates. The primal block uses a fully-convolutional network as the original primal-dual network, while the proposed dual block uses a similar structure as the original primal-dual network, but with two modifications—complex-valued convolutions (CV-Conv) and Complex Rectified Linear Unit (ℂReLU) activations were employed in place of ordinary real-valued convolutions (Conv) and PReLU, respectively. It is to be noted that the second to tenth yellow ochre and violet blocks have the same architecture as the first one. The initial dual block receives all-zeros as the initial states f_0 and h_0, and the undersampled k-space g. The output of this block, along with the same f_0 is given as input to the first primal block

the k-space—by trying to predict the missing frequencies, while the primal blocks optimise the image by trying to remove the artefacts. The output of the final primal block is the final reconstruction of the model. These models improve the quality of the reconstruction over iterations while also ensuring data consistency, similar to the iterative reconstruction approaches like compressed sensing [95]. However, as these are learnt approaches with a fixed number of iterations, the speed of reconstruction is not slow and can still be used for real-time approaches. On the other hand, as the original PD network and PDUNet utilise Radon transform, while the Fourier-PD and Fourier-PDUNet use Fourier transform—which is considerably faster than the Radon transform—these models should perform faster than the earlier ones. Due to this fact, the number of iterates of the Fourier-PDUNet was increased to 10 from 2—similar to the PD and Fourier-PD networks, without compromising the reconstruction speed too much.

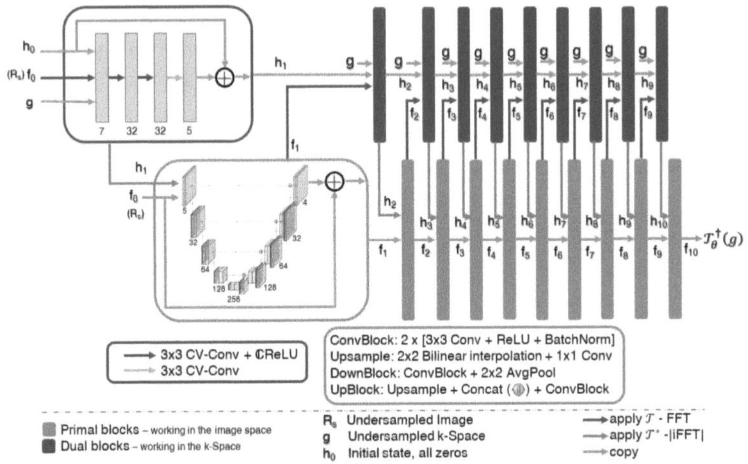

Fig. 10.5 Fourier Primal-Dual UNet (Fourier-PDUNet): Architecture of the proposed network. The yellow ochre and the violet boxes signify the primal and dual iterates, respectively. The dual block of the previously proposed (Sect. 8.2.1.2 [pub:6]) PD-UNet architecture was modified by replacing the convolution (Conv) layers and PReLU activations with complex-valued convolutions (CV-Conv) and Complex Rectified Linear Unit (ℂReLU) activations, respectively. The primal block is identical to the PD-UNet architecture. It is to be noted that the second to tenth yellow ochre and violet blocks have the same architecture as the first one. The initial dual block receives the magnitude image reconstructed by applying iFFT transform on the undersampled k-space as f_0 (unlike the original primal-dual network, which receives all zeros—a modification which was proposed earlier in Sect. 8.2.1.2 [pub:6]), all-zeros as the initial-state h_0, and the undersampled k-space g. The output of this block, along with the same f_0 is given as input to the first primal block

10.2.1.2 Implementation and Training

The models were trained by calculating the loss in the image space by comparing the final reconstruction of the model against the ground-truth fully-sampled images with the help of L1 loss (Eq. 4.28), and it was minimised using the Adam optimiser (Initial learning rate 0.001, decayed by 10 after every 25 epochs; $\beta_1 = 0.9$, $\beta_2 = 0.999$, $\epsilon = 1e{-}09$) with batch size one for 50 epochs. This network was implemented using PyTorch [327] with the help of pytorch-complex [code:10] for the complex-valued operations and was trained using Nvidia RTX 2080Ti GPUs. Once the models were trained, they were combined with the NCC1701 pipeline (Sect. 8.1.1.4) to apply the final data consistency step to obtain the final result. The code of these models is available as part of the NCC1701 code on GitHub [code:8].

10.2.1.3 Dataset and Undersampling

The raw k-space data from the fastMRI [293] dataset was used in this research. This dataset contains the brain and knee MRIs with different contrasts acquired with 1.5T and 3T MR scanners. From all the available volumes, T2w axial brain MRIs acquired at 3T with 16-channel coils with a matrix size of 768 $\times$ 396 were utilised. The dataset was undersampled using 1D random Variable density (Varden) sampling with acceleration factors of 8 and 16, and with a centre factor of 0.08 for accelerations. Currently, the proposed approach works with single-channel input. Hence, the fully-sampled, as well as the undersampled data, were combined using root-sum-of-squares after bringing them to the image space and then they were then brought back to the k-space after the coil-combination step.

10.2.2 Results

The performance of Fourier-PD and Fourier-PDUNet were compared against the undersampled reconstruction of the zero-filled k-space, as well as the ReconResNet model proposed earlier (Sect. 8.1 [pub:8]), which works only in the image space. All three models were combined with the NCC1701 pipeline (Sect. 8.1.1.4) to obtain the final results. The quantitative evaluation was performed using the Structural Similarity Index (SSIM, Eq. 4.35), Normalised Root Mean-Squared Error (NRMSE, Eq. 4.32) with the averaged Euclidean norm of the fully sampled image as the normalisation factor, Peak Signal to Noise Ratio (PSNR, Eq. 4.33), and Universal Quality Index (UQI, Eq. 4.34) [64]. The statistical significance of the differences observed in the metric values was measured using the Mann-Whitney U-test [7]

Table 10.3 Mean and standard deviation of the different metrics over the whole test set

	SSIM		NRMSE		PSNR	
	AF = 8	AF = 16	AF = 8	AF = 16	AF = 8	AF = 16
Undersampled	0.738±0.033	0.697±0.036	0.276±0.049	0.291±0.055	25.748±1.684	25.288±1.728
ReconResNet	0.926±0.017	0.913±0.019	0.161±0.019	0.193±0.018	30.342±1.255	28.791±1.077
Fourier-PD	0.927±0.018	0.912±0.019	0.158±0.020	0.196±0.017	30.514±1.340	28.627±1.080
Fourier-PDUNet	**0.931±0.018**	**0.919±0.019**	**0.152±0.022**	**0.185±0.018**	**30.914±1.451**	**29.165±1.109**

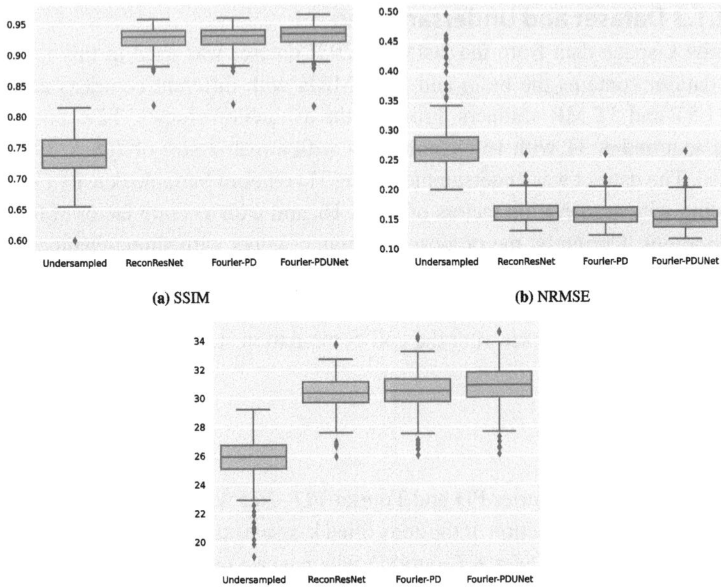

(a) SSIM

(b) NRMSE

(c) PSNR

Fig. 10.6 Quantitative comparison of the models against the undersampled MRI, for the acceleration factor of 8

The quantitative comparisons of the undersampled data and the result of the different models against the ground-truth using SSIM, NRMSE, and PSNR for the acceleration factors 8 and 16 are presented in Table 10.3 and shown with the help of box plots in Figs. 10.6 and 10.7, respectively. Fourier-PD model obtained better scores than the ReconResNet model on all three metrics for the acceleration factor 8, whilst achieving similar scores for acceleration factor 16—but all of them were statistically insignificant. The Fourier-PDUNet, on the other hand, achieved statistically significant improvements over both ReconResNet and Fourier-PD models on all three metrics—for both acceleration factors. As the results of the deep learning models were very close, they were additionally compared using UQI, shown in Fig. 10.8. Fourier-PD secured a statistically significant improvement over the ReconResNet for only acceleration factor 8, while Fourier-PDUNet obtained better scores than both the models with statistical significance. Figures 10.9 and 10.10 portray the qualitative comparison of the results obtained by the different methods. The qualitative comparison confirms the quantitative results—Fourier-PDUNet performs better than both models.

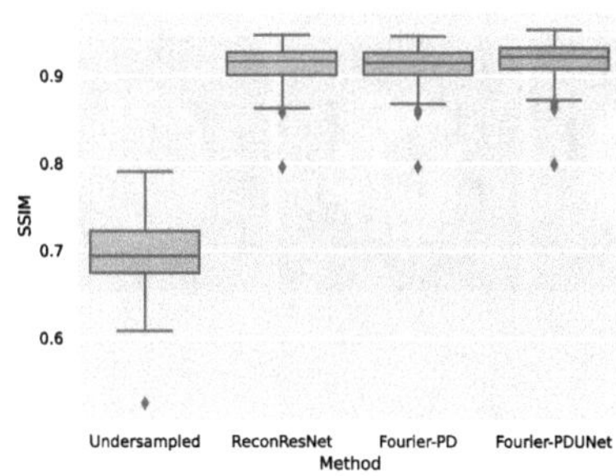

(a) SSIM

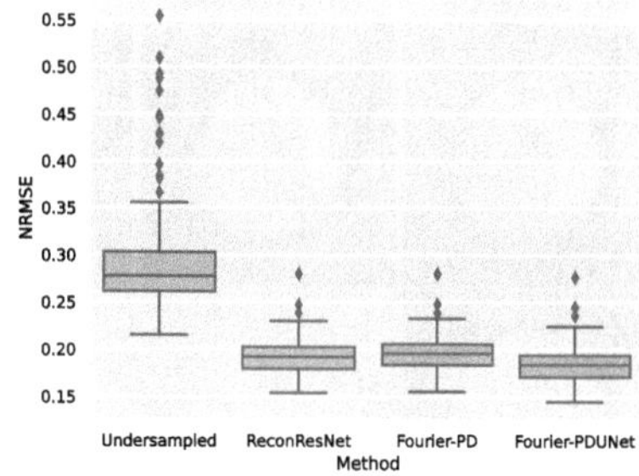

(b) NRMSE

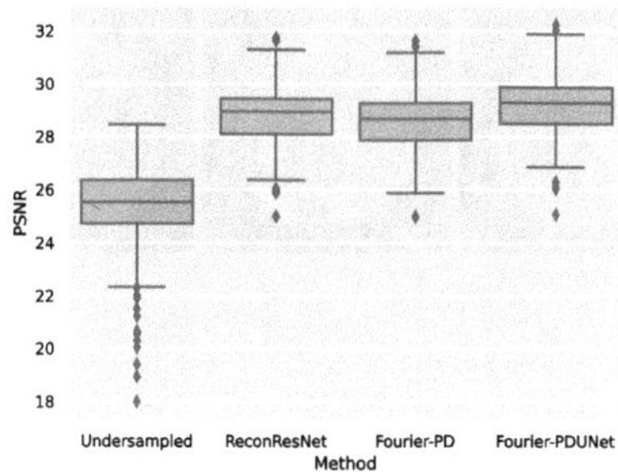

(c) PSNR

Fig. 10.7 Quantitative comparison of the models against the undersampled MRI, for the acceleration factor of 16

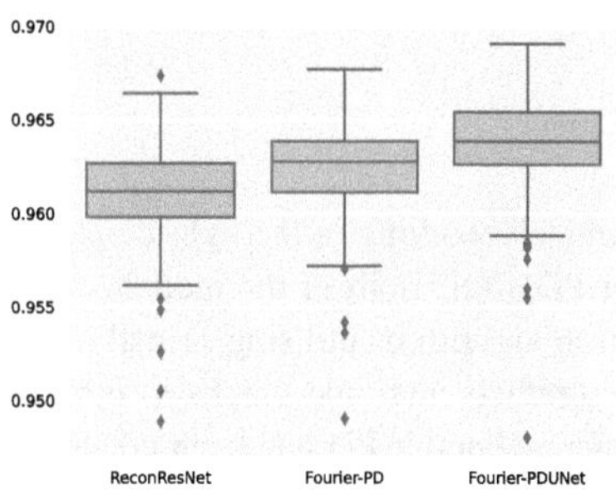

(a) AF=8

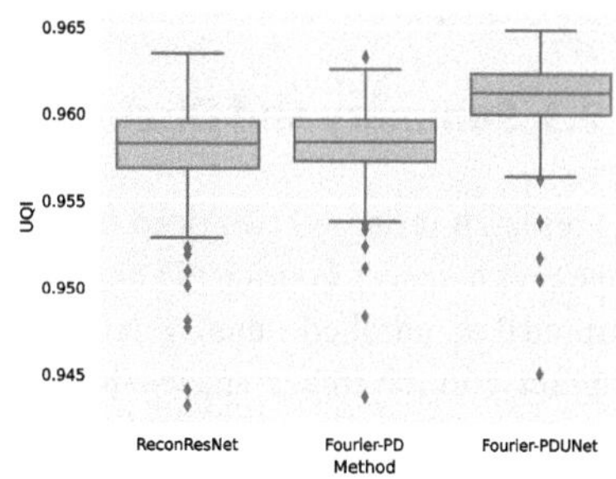

(b) AF=16

Fig. 10.8 Quantitative comparison of the models using UQI

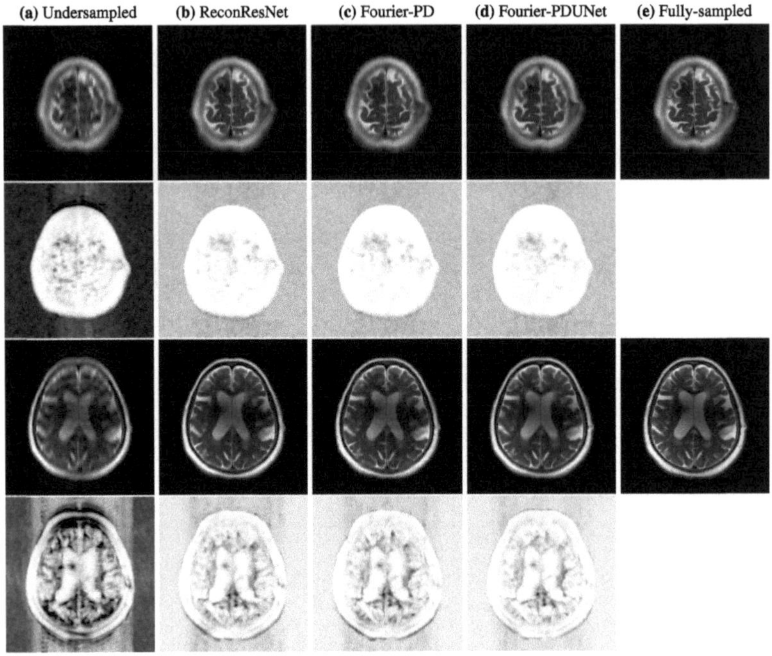

Fig. 10.9 Qualitative comparison of the models against the fully-sampled ground-truth for two example slices (Rows 1 and 3) and their corresponding SSIM maps while being compared against ground-truth (Rows 2 and 4)—for acceleration factor 8

10.2.3 Summary and Discussion

This research proposed two deep learning models working in the hybrid space of image and k-space: Fourier-PD and Fourier-PDUNet. Both of the models can be interpreted as unrolled iterative reconstruction algorithms utilising primal blocks to optimise in the image space—to reduce artefacts and dual blocks to optimise the k-space—to predict the missing frequencies. Fourier-PD has been created by extending the learnt primal-dual approach, which was originally proposed for CT reconstruction, while Fourier-PDUNet extends the PDUNet model proposed earlier in this thesis (Sect. 8.2). Both these models were extended by replacing the convolutional layers and the PReLU activation functions in the dual blocks with complex-valued convolutional layers and complex ReLU activation functions—to accommodate the complex-valued nature of the k-space. The proposed models

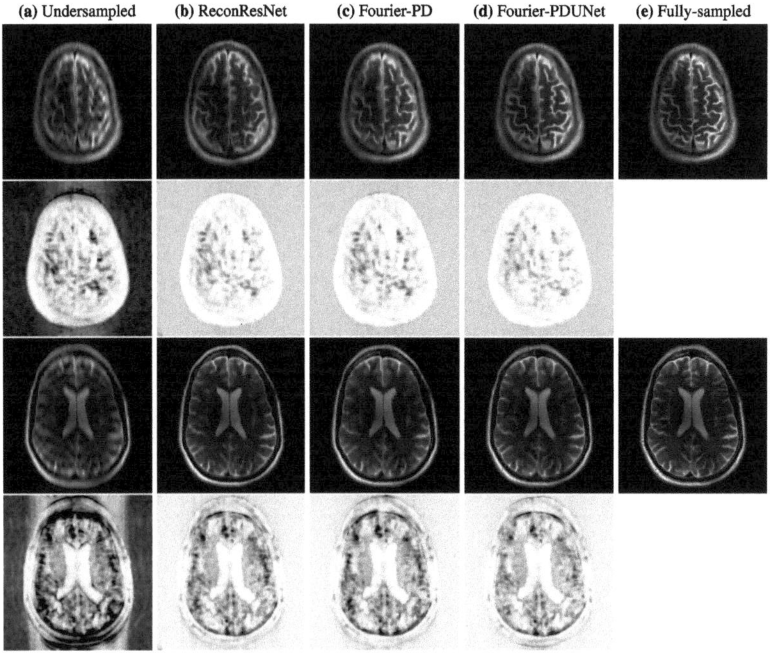

Fig. 10.10 Qualitative comparison of the models against the fully-sampled ground-truth for two example slices (Rows 1 and 3) and their corresponding SSIM maps while being compared against ground-truth (Rows 2 and 4)—for acceleration factor 16

were compared against the earlier proposed (Sect. 8.1) ReconResNet model, which works completely in the image space. All three models were incorporated into the NCC1701 framework (Sect. 8.1.1.4) to obtain the final results with data consistency. The models were applied on the artificially undersampled coil-combined dataset of T2w brain MRIs from the fastMRI dataset—for acceleration factors 8 and 16. Fourier-PDUNet achieved statistically significant improvements over both Fourier-PD and ReconResNet models (***[RQ:4]***, ***[RQ:4.2]***). However, there was no statistically significant difference between Fourier-PD and ReconResNet models. The winning model Fourier-PDUNet secured average SSIM values 0.931±0.018 and 0.919±0.019, for acceleration factors 8 and 16, respectively, while improving the undersampled MRIs which had SSIM values of 0.738±0.033 and 0.697±0.036. This research demonstrated the advantage of working in the hybrid space over pure image space (***[RQ:4]***).

Part IV
Undersampled Reconstruction as a Generalised Component of an MRI Processing Pipeline

Summary

MRIs are not only used by clinicians for performing diagnostics by visual inspections, but also used in several machine learning techniques for further processing—for automatic decision making or as part of decision support systems—to perform classification, segmentation, anomaly detection, image registration, and many other things. Undersampling can also benefit these methods because undersampling reduces scan time—making it possible to create larger datasets, and it also reduces the risk of motion artefacts due to shorter scan times—making it easier for the machine learning approaches to work with such data with fewer artefacts.

As a part of this thesis, four different machine learning tasks working with MRIs were performed. Initially, methods were developed to improve the results over the state-of-the-art methods and were published in respective manuscripts. As a part of this thesis, these methods were tested on undersampled MRIs, and by adding NCC1701 (Sect. 8.1) as a precursor—as an initial preprocessing step to those methods and consequently testing on the reconstructed MRIs. The goal of these experiments was to judge to what extent these methods can work with undersampled MRIs and how much improvement can be achieved if those undersampled MRIs are first reconstructed with NCC1701 before supplying them to the respective methods.

This part of the thesis presents five different machine learning based, to be precise, deep learning based MRI processing pipelines for four different tasks—supervised spatiospatial models for the classification of brain tumours, GP-models for supervised classification, and weakly-supervised segmentation of brain tumours using inherently-explainable models, unsupervised anomaly detection in brain MRIs using the StRegA pipeline, deformation-aware semi-supervised learning (DS6) for small vessel segmentation in 7T MRAs using noisy training data, and multi-scale inverse-consistent deformable image registration of intramodal brain MRIs using multi-scale UNet with self-constructing graph latent (MICDIR).

Each section in this chapter presents one of the above-mentioned methods—starts by briefly explaining the topic, including the aim and contributions, then explains the methodology, presents the results and discussions, and finally provides a summary and outlook of that particular research. The results subsections include the results shown in the respective publications and also include additional results obtained on the undersampled MRIs and on the NCC1701 reconstructed MRIs.

Spatiospatial Models: Brain Tumour Classification 11

The tumour classification approach described in this section was primarily performed within the context of the master thesis of Faraz Ahmed Nizamani [stud:4] and has been subsequently published in [pub:10]. The code of this project is available publicly on GitHub [code:2].

11.1 Background and Motivation

A brain tumour is the growth of abnormal cells in the brain. Brain tumours are classified based on their speed of growth and the likeness of them growing back after treatment. Detection of cancer, specifically an earlier detection, holds the potential to make a difference in treatment. An automated deep-learning method might offer a solution—predicting the grade of the tumour, consequently reducing the burden on the radiologists and improving the overall outcome of the treatment. Several deep learning models have been proposed for the task of brain tumour classification using different 2D and 3D models, some of which are discussed in Sect. 7.1.1. Spatiotemporal models [287] have shown their superiority over pure 3D models for video classification tasks by treating the time dimension differently than the spatial dimensions. This research explores the possibility of applying spatiotemporal models (ResNet(2+1)D and ResNet Mixed Convolution) as "spatiospatial" models by treating one dimension (slice dimension) differently than the other two spatial dimensions of the 3D volumetric images for the task of brain tumour classification from MRIs, while also classifying where or not the subject has a tumour. Furthermore, this research evaluates the applicability of transfer learning for this task.

S. Chatterjee, *Deep Learning Innovations in MRI Reconstruction and Analysis*,
https://doi.org/10.1007/978-3-658-50741-1_11

11.1.1 Technical Background

The spatiotemporal models introduced by Tran et al. [287] for action recognition are fundamentally 3D Convolutional Neural Networks based on ResNet. There are two spatial dimensions and one temporal dimension in video data, making the data three-dimensional. For handling such data (e.g. action recognition task), using a network with 3D convolution layers is an obvious choice. Tran et al. [287] introduced two variants of spatiotemporal models: ResNet (2+1)D and ResNet Mixed Convolution. The ResNet(2+1)D model consists of 2D and 1D convolutions, where the 2D convolutions are used spatially while the 1D convolutions are reserved for the temporal element. This gives an advantage of increased non-linearity by using non-linear rectification, which allows this kind of mixed model to be more "learnable" than conventional full 3D models. On the other hand, the ResNet Mixed Convolution model is constructed as a mixture of 2D and 3D Convolution operations. The initial layers of the model are made of 3D convolution operations, while the later layers consist of 2D convolutions. The rationale behind using this type of configuration is that the motion-modelling occurs mostly at the initial layers, and applying 3D convolution there encapsulates action better.

Apart from trying to improve the network architecture, one frequently used technique to improve the performance of the same architecture is transfer learning [118]. This is a technique for re-purposing a model for another task that is different from the task the model was originally trained for performing. Typically, the model parameters are initialised randomly before starting the training. However, in the case of transfer learning, model parameters learned from task one are used as the starting point (called pre-training), instead of random values, for training the model for task two. Pre-training has shown to be an effective method to improve the initial training process, eventually achieving better accuracy [378, pub:13].

Spatiotemporal models are typically used for video classification tasks, which are three-dimensional in nature. Their potential in classifying 3D volumetric images like MRI, considering them as "spatiospatial" models, has not been explored yet. This research explores the possibility of applying spatiotemporal models (ResNet(2+1)D and ResNet Mixed Convolution) as "spatiospatial" models by treating one dimension (slice dimension) differently than the other two spatial dimensions of the 3D volumetric images. "Spatiospatial" were employed to classify brain tumours of the different types of gliomas based on their grading as well as healthy brains from 3D volumetric MR Images using a single MR contrast, and compare their performances against a pure 3D convolutional model (ResNet3D). Furthermore, the models are to be compared with and without pre-training—to judge the usability of transfer learning for this task.

11.2 Methodology

This section explains the network models used in this research, implementation details, pre-training and training methods, data augmentation techniques, dataset information, data pre-processing steps, and finally, the evaluation metrics.

11.2.1 Network Models

Spatiotemporal models are mainly used for video-related tasks, where there are two spatial and one temporal dimension. These models deal with the spatial and temporal dimensions differently, unlike pure 3D convolution-based models. There is no temporal component in 3D volumetric image classification tasks; hence, using a 3D convolution-based model is a frequent choice. At times, they are divided into 2D slices, and 2D convolution-based models are applied to them. For the task of tumour classification, the rationale for using 3D filters is grounded in the morphological heterogeneity of gliomas [142], it is to make the convolution kernels invariant to tissue discrimination in all dimensions, learning more complex features spanning voxels, while 2D convolution filters will capture the spatial representation within the slices. Spatiotemporal models combine two different types of convolution into one model while having the possibility of reducing the complexity of the model or of incorporating more non-linearity. These advantages might be possible to exploit while working with volumetric data by considering the spatiotemporal models as "spatiospatial" models—the motivation behind using such models for a tumour classification task. In this research, the slice dimension is treated as the pseudo-temporal dimension of spatiotemporal models, and in-plane dimensions are treated as the spatial dimensions. The spatiotemporal models used here as spatiospatial models are based on the work of Tran et al. [287].

Two different spatiospatial models are explored here: ResNet (2+1)D and ResNet Mixed Convolution. Their performances are compared against ResNet3D, which is a pure 3D convolution-based model.

11.2.1.1 ResNet (2+1)D:

ResNet (2+1)D uses a combination of 2D convolution followed by 1D convolution instead of a single 3D convolution. The benefit of using this configuration is that it allows an added non-linear activation unit between the two convolutions, as in comparison to using a single 3D Convolution [287]. This then results in an overall increase of ReLU units in the network, giving the model the ability to learn even more complex functions. The ResNet(2+1)D uses a stem that contains a 2D convolution with a kernel size of seven and a stride of two, accepting one channel as an input and providing 45 channels as output; followed by a 1D convolution with a kernel size of three and a stride of one, providing 64 channels as the final output. Next, there are four convolutional blocks; each of them contains two sets of basic residual blocks. Each residual block contains one 2D convolution with a kernel size of three and a stride of one, followed by a 1D convolution with a kernel size of three and a stride of one. Each convolutional layer in the model (both 2D and 1D) is followed by a 3D batch normalisation layer and a ReLU activation function. The residual blocks inside the convolutional blocks, except for the first convolutional block, are separated by a pair of 3D convolution layers with a kernel size of one and a stride of two—to downsample the input by half. The 2D convolutions are applied in-plane, and the 1D convolutions are applied on the slice dimension. After the final convolutional block, an adaptive average pooling layer has been added, with an output size of one for all three dimensions. After the pooling layer, a dropout layer followed by a fully connected layer with n output neurons for n classes were added to obtain the final output. Fig. 11.1(a) portrays the schematic diagram of the ResNet (2+1)D architecture.

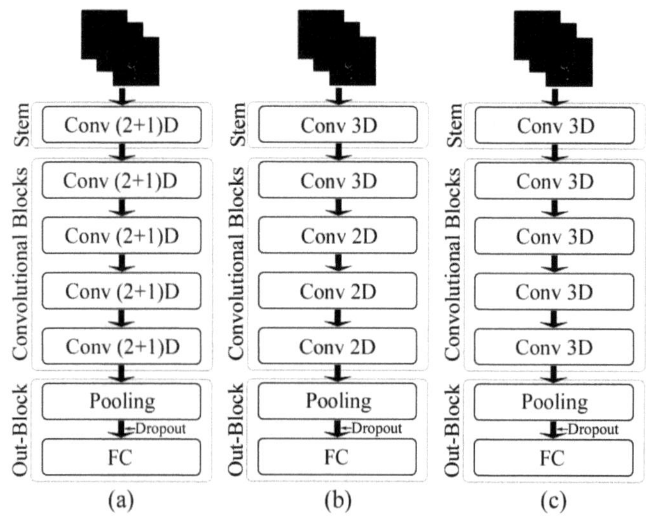

Fig. 11.1 Schematic representations of the network architectures. (a) ResNet (2+1)D, (b) ResNet Mixed Convolution, and (c) ResNet 3D [pub:10]

11.2.1.2 ResNet Mixed Convolution:

ResNet Mixed Convolution uses a combination of 2D and 3D Convolutions. The stem of this model contains a 3D convolution layer with a kernel size of (3,7,7), a stride of (1,2,2), and padding of (1,3,3)—where the first dimension is the slice dimension and the other two dimensions are the in-plane dimensions, and accepts a single channel as input while providing 64 channels as output. After the stem, there is one 3D convolution block, followed by three 2D convolution blocks. All the convolution layers (both 3D and 2D) have a kernel size of three and a stride of one, identical for all dimensions. Each of these convolution blocks contains a pair of residual blocks, each of which contains a pair of convolution layers. Similar to ResNet (2+1)D, the residual blocks inside the convolutional blocks, except for the first convolutional block, are separated by a pair of 3D convolution layers with a kernel size of one and a stride of two—to downsample the input by half. Each convolutional layer in the model (both 3D and 2D) is followed by a 3D batch normalisation layer and a ReLU activation function. The motivation behind using both modes of convolution in 2D and 3D is that the 3D filters can learn the spatial features of the tumour in 3D space while 2D can learn representation within each 2D slice. After the convolutional blocks, the final pooling, dropout, and fully connected layers are identical to the ResNet (2+1)D architecture. Fig. 11.1(b) shows the schematic representation of this model.

11.2.1.3 ResNet3D:

The performance of the spatiospatial models are compared against a pure 3D ResNet model, schematic diagram shown in Fig. 11.1(c). The architecture of the ResNet3D model used here is almost identical to the architecture of ResNet Mixed Convolution (Sect. 11.2.1.2), except for the fact that this model uses only 3D convolutions. The stem of these models are identical, the only difference being that this model uses four 3D convolution blocks, unlike ResNet Mixed Convolution, which uses one 3D convolution block, followed by three 2D convolution blocks. This configuration of ResNet3D architecture results in a 3D ResNet18 model.

11.2.1.4 Summary and Comparison:

The general structure of the network models can be divided into the following: input goes to the stem, then there are four convolutional blocks, followed by the output block—which contains an adaptive pooling layer, followed by a dropout layer, and finally a fully connected layer. ResNet Mixed Convolution and ResNet 3D have the same stem, including a 3D convolutional layer with a kernel size of (3,7,7), followed by a batch normalisation layer and a ReLU. ResNet (2+1)D uses a different stem: a 2D convolution layer with a kernel size of seven, then a 1D convolution with a kernel size of three—splitting the 3D convolution (3,7,7) used by the other models into a pair of 2D and 1D convolution: (7,7) and (3). Both 2D and 1D convolution inside this stem is followed by a batch normalisation layer and ReLU pair. The convolutional blocks in the ResNet3D and ResNet Mixed Convolution architectures follow the same architecture: two residual blocks consisting of two sub-blocks consisting of a 3D convolution with a kernel size of three, followed by a batch normalisation layer and a ReLU. On the other hand, the first convolutional block of the ResNet (2+1)D architecture uses a pair of 2D and 1D convolutions with a kernel size of three instead of the 3D convolutional layers used by the other models. The rest of the architecture is the same. It is noteworthy that this model has more non-linearity because the 3D convolutions are split into a pair of 2D and 1D convolutions; an additional pair of batch normalisation and ReLU could have been used between the 2D 1D convolution. There is one difference between the first convolutional block and the other three blocks (applicable for all three models): the second, third and fourth convolutional blocks included a downsampling pair, which consisted of a 3D convolutional layer with a kennel size of one and a stride of two, followed by a batch normalisation layer. This was not present in the first convolutional block. The convolution blocks of each of all three models double the input features by two (number of input features to the first block: 64, number of output features of the fourth (and final) block: 512). All of these models end with an adaptive average pooling layer, which forces the output to have a shape of 1x1x1, with 512 different features. A dropout with a probability of 0.3 is then applied to introduce regularisation to prevent over-fitting before supplying them to a fully connected linear layer that generates n classes as output. The width and depth of these models are comparable, but they differ in terms of the number of trainable parameters depending upon the type of convolution used, as shown in Tab. 11.1. It is noteworthy that the less the number of trainable parameters—the less the computational costs. A model with a lesser number of parameters would require lesser memory for computation (GPU and RAM), and also, the complexity of the model is lesser—reducing the overall computational costs for both training and inference. Moreover, a lesser number of trainable parameters would also reduce the risk of overfitting.

Table 11.1 Total Number of trainable parameters in each model

Model	No of Parameters
ResNet3D	33,150,522
ResNet (2+1)D	31,297,254
ResNet Mixed Convolution	11,472,963

11.2.2 Implementation and Training

The models were implemented using PyTorch [327], by modifying the Torchvision models [url:6] and were trained with a batch size of 1 using an Nvidia RTX 4000 GPU, which has a memory of 8GB. Models were compared with and without pre-training. Models with pre-training were pre-trained on Kinetics-400 [url:9], except for the stems and fully connected layers. Images from the Kinetics dataset contain three channels (RGB Images), whilst the 3D volumetric MRIs have only one channel. Therefore, the stem trained on the Kinetics dataset could not be used and was initialised randomly. Similarly, for the fully connected layer, Kinetics-400 has 400 output classes, whereas the task at hand has three classes (LGG, HGG and Healthy)—hence, this layer was also initialised with random weights.

Trainings were performed using mixed-precision [236] with the help of Nvidia's Apex library [url:1]. The loss was calculated using the weighted cross-entropy loss function (Equations 4.18 and 4.19) to minimise the under-representation of classes with fewer samples during training and was optimised using the Adam optimiser with a learning rate of $1e-5$ and weight decay coefficient $\lambda = 1e-3$. The code of this research is publicly available on GitHub [code:2].

11.2.3 Data Augmentation

Different data augmentation techniques were applied to the dataset before training the models, and for that purpose, TorchIO [394] was used. Initial experiments were performed using different amounts of augmentation and can be categorised as light and heavy augmentation, where light augmentation included only random affine (scale 0.9–1.2, degrees 10) and random flip (L-R, probability 0.25); on the other hand, heavy augmentation included the ones from light augmentation together with elastic deformation and random k-space transformations (motion, spike, and ghosting). It was observed that the training of the network with heavily augmented data not only performed poorly in terms of final accuracy, but the loss took a much longer time to converge. Therefore, only light augmentation was used throughout this research.

11.2.4 Dataset

Two different datasets were used in this work—the pathological brain images were obtained from the Brain Tumour Segmentation (BraTS) 2019 dataset, which includes images with four different MR contrasts (T1, T1 contrast-enhanced, T2 and FLAIR) [169], [217], [252]; and non-pathological images were collected from the IXI Dataset [url:12]. Among the available four types of MRIs, T1 contrast-enhanced (T1ce) is the most commonly used contrast while performing single-contrast tumour classification [292, 362]. Hence in this research, T1ce images of 332 subjects were used from the BRaTS dataset: 259 volumes of Glioblastoma Multiforme (High-Grade Glioma, HGG), and 73 volumes of Low-Grade Glioma (LGG). 259 T1 weighted volumes were chosen randomly from the IXI dataset as healthy samples to have the same number of subjects as HGG. The final combined dataset was then randomly divided into 3-folds of training and testing split with a ratio of 7:3.

11.2.5 Data Pre-Processing

The IXI images were pre-processed first by using the brain extraction tool (BET2) of FSL [73, 138]. This was done to keep the input data uniform throughout, as the BraTS images are already skull stripped. Moreover, the intensity values of all the volumes from the combined datasets were normalised by scaling intensities to [0.5,99.5] percentile, as used by Isensee et al. [270]. Finally, the volumes were re-sampled to the same voxel resolution of 2mm isotropic.

11.3 Results

The performance of the models was compared compared with and without pre-training using precision, recall, F1-score, specificity, and testing accuracy. Furthermore, a confusion matrix was used to show class-wise accuracy. Figures 11.2, 11.3, and 11.4 show the average accuracy over 3-fold cross validation using confusion metrics, for ResNet (2+1)D, ResNet Mixed Convolution, and ResNet 3D, respectively.

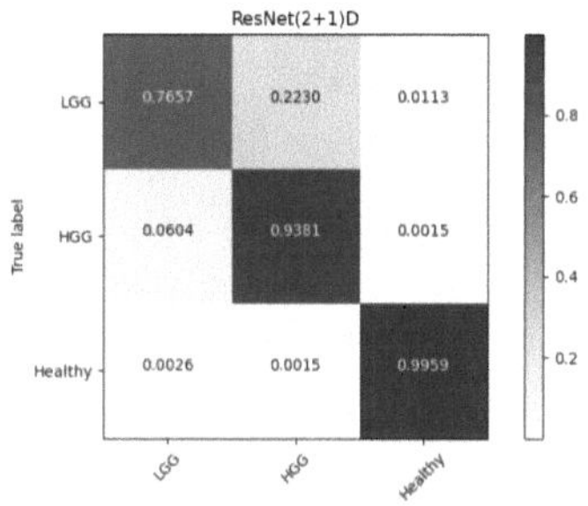

(a) Without pre-training

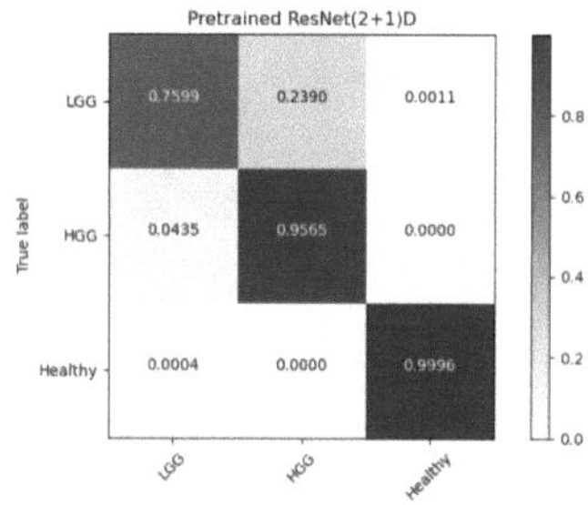

(b) With pre-training

Fig. 11.2 Confusion Matrix for 3-fold cross-validation on ResNet(2+1)D [pub:10]

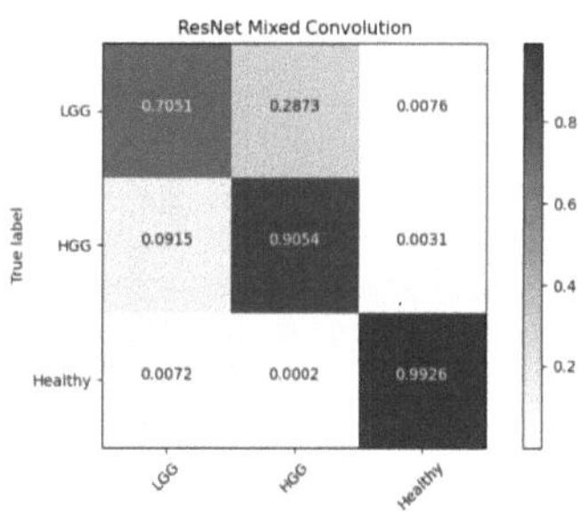

(a) Without pre-training

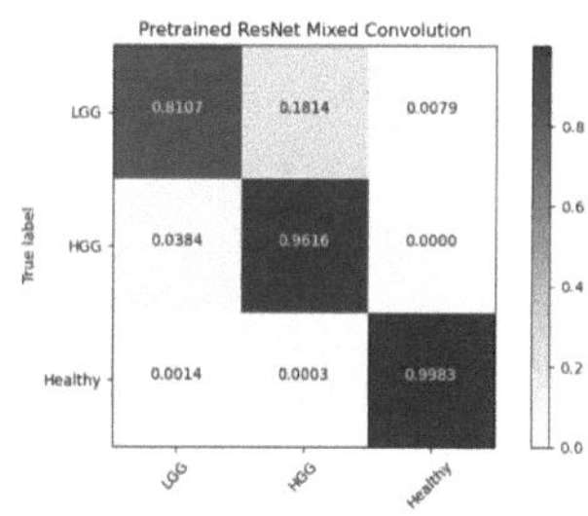

(b) With pre-training

Fig. 11.3 Confusion Matrix for 3-fold cross-validation on ResNet Mixed Convolution [pub:10]

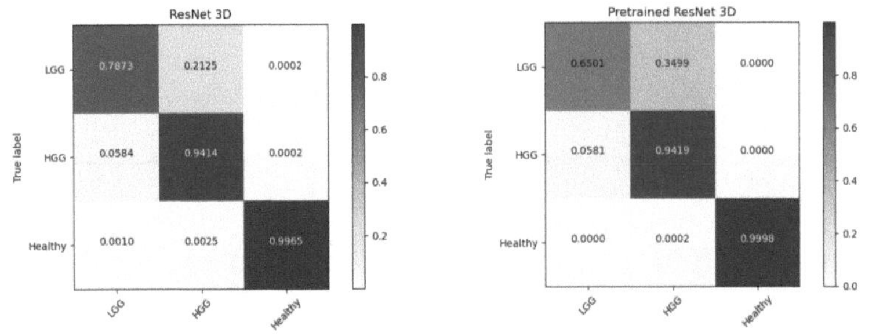

(a) Without pre-training **(b)** With pre-training

Fig. 11.4 Confusion Matrix for 3-fold cross-validation on ResNet3D18 [pub:10]

Figure 11.5 shows the class-wise performance of the different models, both with and without pre-training, using precision, recall, specificity, and F1-score.

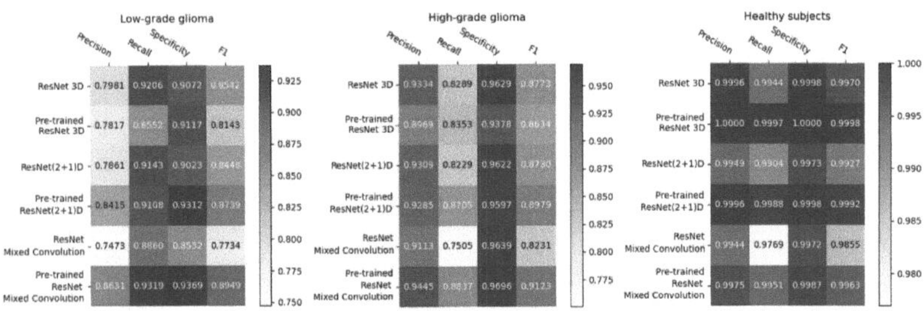

(a) (b) (c)

Fig. 11.5 Heatmaps showing the class-wise performance of the classifiers, compared using Precision, Recall, Specificity, and F1-score: (a) LGG, (b) HGG, and (c) Healthy [pub:10]

11.3.1 Comparison of the models

The mean F1-score over 3-fold cross-validation was used as the metric to compare the performance of the different models. Tables 11.2, 11.3, and 11.4 show the results of the different models for the classes LGG, HGG, and Heathy, respectively; and finally Table 11.5 shows the consolidated scores.

Table 11.2 Low-Grade Glioma—performance comparison of the models (* denotes the overall winning model)

Low-Grade Glioma	
Model	mean F1-score
ResNet 3D	0.8542±0.049
Pre-trained ResNet 3D	0.8143±0.048
ResNet(2+1)D	0.8448±0.019
Pre-trained ResNet(2+1)D	0.8739±0.041
ResNet Mixed Convolution	0.7734±0.031
Pre-trained ResNet Mixed Convolution*	**0.8949±0.033**

Table 11.3 High-Grade Glioma—performance comparison of the models (* denotes the overall winning model)

High-Grade Glioma	
Model	mean F1-score
ResNet 3D	0.8773±0.034
Pre-trained ResNet 3D	0.8634±0.042
ResNet(2+1)D	0.8730±0.022
Pre-trained ResNet(2+1)D	0.8979±0.032
ResNet Mixed Convolution	0.8231±0.027
Pre-trained ResNet Mixed Convolution*	**0.9123±0.029**

Table 11.4 Healthy brain—performance comparison of the models (* denotes the overall winning model)

Healthy brain	
Model	mean F1-score
ResNet 3D	0.9970±0.005
Pre-trained ResNet 3D	**0.9998±0.0002**
ResNet(2+1)D	0.9927±0.009
Pre-trained ResNet(2+1)D	0.9992±0.001
ResNet Mixed Convolution	0.9855±0.004
Pre-trained ResNet Mixed Convolution*	0.9963±0.002

For Low-Grade Glioma (LGG), ResNet Mixed Convolution with pre-training achieved the highest F1-score of 0.8949 with a standard deviation of 0.033. The pre-trained ResNet(2+1)D is not far behind, with 0.8739±0.033.

For the High-Grade Glioma (HGG) class, the highest F1 was achieved by the pre-trained ResNet Mixed Convolution model, with an F1-score of 0.9123±0.029. This is higher than the best model's F1-score for the class LGG. This can be expected because of the class imbalance between LGG and HGG. As with LGG, the second-best model for HGG is also the Pre-trained ResNet(2+1)D with the F1-score of 0.8979±0.032.

The healthy brain class achieved the highest F1-score of 0.9998±0.0002, with the pre-trained ResNet 3D model, which can be expected because of the complete absence of any lesion in the MR images making it far less challenging for the model to learn and distinguish it from the brain MRIs with pathology. Even though the pre-trained ResNet 3D model achieved the highest mean F1-score, all pre-trained models achieved similar F1-scores, i.e. all the mean scores are more than 0.9960—making it difficult to choose a clear winner.

Table 11.5 Consolidated comparison of the models (* denotes the overall winning model)

	Consolidated Scores	
Model	macro F1-score	weighted F1-score
ResNet 3D	0.9095	0.9269
Pre-trained ResNet 3D	0.8925	0.9171
ResNet(2+1)D	0.9035	0.9220
Pre-trained ResNet(2+1)D	0.9237	0.9393
ResNet Mixed Convolution	0.8607	0.8881
Pre-trained ResNet Mixed Convolution*	**0.9345**	**0.9470**

ResNet Mixed Convolution with pre-training came up as the best model for both classes with pathology (LGG and HGG) and achieved a similar score as the other models while classifying healthy brain MRIs, as well as based on macro and weighted F1-scores—making this model as the clear overall winner. It can also be observed that the spatiospatial models performed better with pre-training, but ResNet 3D performed better without pre-training.

11.3.2 Comparison against literature

This sub-section compares the best model from the previous sub-section (i.e. ResNet Mixed Convolution with pre-training) against seven other research papers (in no specific order), where they classified LGG and HGG tumours. Mean test accuracy was used as the metric to compare the results as that was the common metric used in those papers.

Starting from Shahzadi et al. [285], where they used LSTM-CNN to classify between HGG and LGG, using T2-FLAIR images from the BraTS 2015 dataset. Their work focuses on using smaller sample size, and they were able to achieve 84.00% accuracy [285]. Pei et al. [329] achieved even less classification accuracy of 74.9% although they did use all of the available contrasts of the BraTS dataset, and their method performed segmentation using a UNet-like model before performing classification. Ge et al. [264] use a novel method of fusing the contrasts into multiple streams to be trained simultaneously. Their model achieved an accuracy of 90.87% overall on all the contrasts, and they achieved 83.73% on T1ce. Mzoughi et al. [362] achieved 96.59% using deep convolutional neural networks on T1ce images. Their work does not present any other metric for their results, except for the overall accuracy of their model, which makes it difficult to compare against their results. Next, Yang et al. [292] did similar work; they used pre-trained GoogLeNet on 2D images, achieving an overall accuracy of 94.5%. They did not use the BraTS dataset, but the purpose of their work was similar—to classify glioma tumours based on LGG and HGG grading. Their dataset had fewer samples of LGG and HGG class in comparison to this research, with the former having 52 samples, and later 61 samples [292]. Ouerghi et al. [393] used different machine learning methods in their paper to train on the fusion images, one of which is the random forest, on which they achieved 96.5% for classification between High-Grade Glioma and Low-Grade Glioma. Finally, the Zhuge et al. [379] achieved an impressive 97.1% using Deep CNN for classification of glioma based on LGG and HGG grading, beating the proposed model by 0.12%. This difference can be explained by two factors, 1) their use of an additional dataset from The Cancer Imaging Archive (TCIA) in combination with BraTS 2018 2) and their use of four different contrasts—both these factors increase the size of the training set significantly. Furthermore, no cross-validation has been reported in their paper. Table 11.6 shows the complete comparative results.

Table 11.6 Comparisons against other published works *s=specificity †cross-validated ‡State of the art

Study	Method	Contrast	Dimension	Test Accuracy
Shahzadi et al. [285]	CNN with LSTM	T2-FLAIR	3D	84.00 %
Pei et al. [329]	Similar to UNet for Segmentation, Regular CNN for classification	T1, T1ce, T2, T2-FLAIR	3D	74.9%
Ge et al. [264]	Deep CNN	T1, T2 , T2-FLAIR	2D	90.87%
Yang et al. [292]	Pre-trained GoogLeNet	T1ce	2D	94.5%†
Mzoughi et al. [362]	Deep CNN	T1ce	3D	96.49%
Zhuge et al. [379]	Deep CNN	T1, T1ce, T2, T2-FLAIR	3D	97.1% s*=0.968
Ouerghi et al. [393]‡	Random forest	T1, T2, T2-FLAIR	2D	96.5%
This research	**Pre-trained ResNet Mixed Convolution spatiospatial model**	**T1ce**	**3D**	**96.98%† s=0.9684**

11.3.3 Working with Undersampled MRIs

The initial fold of the cross-validation was used in this analysis. The test volumes were undersampled using MRUnder [url:8, code:7] with five different sampling patterns: four Cartesian sampling patterns—1D Varden 30%, 1D Varden 10%, 2D Varden 10%, 2D Varden 5% and one radial sampling pattern—golden angle acquisition with 30 spokes, resulting in MR acceleration factors of 3, 10, 10, 20, and 17, respectively. The undersampled MRIs were then reconstructed slice-wise in 2D using the NCC1701 (Sect. 8.1) pipeline with the help of the already trained ReconResNet models for the respective sampling patterns—trained on brain-extracted T1w MRIs of the IXI dataset [url:12]. Then the reconstructed 2D slices were stacked together to obtain the final 3D volumes. The undersampled and the reconstructed volumes were then supplied to the trained spatiospatial models (ResNet Mixed Convolution and ResNet (2+1)D), as well as the fully-convolution ResNet 3D model.

Table 11.7 presents the quantitative results of this evaluation. It can be observed that the models were robust against the undersampling artefacts, resulting in the lowest F1-score of 76.02 for an acceleration factor of 20, while the NCC1701 reconstruction improved that result to 79.01. The NCC1701 improved the results over the undersampled MRIs for all undersampling patterns. For 1D Varden 30% (AF: 3) and 2D Varden 10% (AF: 10) resulted in similar results as the fully-sampled MRIs. One interesting observation can be made from the results—the fully-convolution ResNet3D model is more robust than the spatiospatial models. On the fully-sampled dataset, the ranking of the models was ResNet Mixed Convolution, ResNet (2+1)D, and ResNet3D; while working with undersampled and reconstructed data, this trend has been flipped.

Table 11.7 Resulting weighted F1-scores for the different models while working with the fully-sampled, undersampled, and NCC1701 reconstructed MRIs

Undersampling Pattern	Acc. Factor	Type	ResNet3D	ResNet (2+1)D	ResNet MixedConv
–	1	Fully	89.06	89.90	90.16
1D Varden 30%	3	Under	85.16	82.45	81.16
		Recon	90.11	86.01	87.62
1D Varden 10%	10	Under	80.27	79.87	77.27
		Recon	83.15	81.22	80.87
2D Varden 10%	10	Under	88.12	86.71	80.12
		Recon	90.07	87.70	87.65
GA Radial 30sp	17	Under	79.08	78.21	73.11
		Recon	83.51	83.28	78.12
2D Varden 5%	20	Under	76.02	74.82	73.52
		Recon	79.01	76.99	76.09

11.4 Discussion

The F1-scores of all the models in classifying healthy brains were very close to one, as segregating healthy brains from brains with pathology is a comparatively simpler task than classifying the grade of the tumour. Furthermore, using two different datasets for the healthy and pathological brain, MRIs could have also introduced a dataset bias. In classifying the grade of the tumour, the pre-trained ResNet Mixed Convolution model performed best, while in classifying healthy brains, all the three pre-trained models performed similarly. For comparing the models based on consolidated scores, macro and weighted F1-scores were used. However, the macro F1-score is to be given more importance as the dataset was imbalanced. Both of the metrics declared the pre-trained ResNet Mixed Convolution as the clear winner.

One interesting observation that can be made from the confusion matrices is that the classification performance of the models for the LGG class has been lower than the other two classes. Even the best performing model managed to get an accuracy of 81% for LGG while achieving 96% for HGG and nearly perfect results for healthy. This might be attributed to the fact that the dataset was highly imbalanced (Sec. 11.2.4), i.e. 259 volumes each for HGG and healthy, while having 73 volumes for LGG. Even though weighted cross-entropy loss (Eq. 4.18) was used in this research to deal with the problem of class imbalance, increasing the number of LGG samples or employing further techniques to deal with this problem further and might improve the performance of the models for LGG [318].

It is noteworthy that the pre-trained ResNet Mixed Convolution resulted in the best classification performance, even though it is the model with the fewest number of trainable parameters (see Table 11.1). Moreover, it is to be noted that both spatiospatial models performed better than the pure 3D ResNet18 model, even though they had fewer trainable parameters than the 3D ResNet18. Fewer of trainable parameters can reduce the computational costs, as well as the chance of overfitting. It was hypothesised that the increased non-linearity due to the additional activation functions between the 2D and 1D convolutions in (2+1)D convolutional layers helped the ResNet (2+1)D model to achieve better results than ResNet3D, and the reduction of trainable parameters while having a similar number of layers, in turn preserving the level of non-linearity, contributed to the success of ResNet Mixed Convolution. Even though it has been seen that the spatiospatial models performed better, it is worth mentioning that the spatiospatial models do not adequately maintain the 3D nature of the data—the spatial relationship between the three dimensions is not preserved within the network like a fully 3D network as ResNet3D—which is a limitation of this architecture, which might have some unforeseen adverse effects. One of the possible adverse effects can be seen while working with undersampled and reconstructed MRIs. The presence of artefacts and change in the noise pattern of the images adversely impacted the spatiospatial models—causing ResNet3D to outperform them, while ResNet (2+1)D performed better than the ResNet Mixed Convolution model, which was the winning model in the case of the fully-sampled dataset. It can be hypothesised that the inter-slice relationship, which is not appropriately preserved by the spatiospatial models, assist the ResNet3D model in being more robust against the artefacts. Between the two spatiospatial models, ResNet (2+1)D performed better than ResNet Mixed Convolution on the undersampled and reconstructed datasets also supports the hypothesis that ResNet (2+1)D maintains the inter-slice relationship better because it uses (2+1)D convolutions throughout the network, while the other uses 3D convolutions followed by 2D ones. Another interesting observation can be made while working with the undersampled MRIs—for 1D Varden 30% and 2D Varden 10%, the ResNet3D model resulted in better F1-scores than the fully-sampled dataset. The author hypothesises that the reduction of the Gaussian noise present in the original MRI dataset by the smoothing observed in the reconstruction might have aided the model in performing better. It is worth mentioning that the NCC1701 models were trained on T1w MRIs, but the tested MRIs were T1ce—which might have some negative impact or bias in the results.

In this research, the slice dimension in the axial orientation was considered as the "specially-treated" spatial dimension of the spatiospatial models, which can also be seen as the pseudo-temporal dimension of the spatiotemporal models. It can be hypothesised that using the data in sagittal or coronal orientation in a similar way might also be possible to exploit the advantages of such models, which it is yet to be tested.

It can also be observed that the pre-trained models were the winners for all three different classes. However, the effect of pre-training was not the same in all three models. For both the spatiospatial models, pre-training improved the model's performance, but in different amounts: 2.24% improvement for ResNet (2+1)D and 8.57% for ResNet Mixed Convolution (based on macro F1-scores). However, pre-training had a negative impact on the 3D ResNet18 model (for two out of three classes), causing it to decrease the macro F1-score by 1.87%. Average macro F1-scores for all the models with and without pre-training (0.9169 with pre-training, 0.8912 without pre-training) show that the pre-training resulted in an overall improvement of 2.88% across models. It is noteworthy that the pre-trained networks were initially trained on RGB videos. Pre-training them on MRI volumes or MR videos (dynamic MRIs) might further improve the performance of the models.

Regarding the comparisons to other published works, an interesting point to note is that the previous papers only classified different grades of brain tumours (LGG and HGG), whereas this research also classified healthy brains as an additional class. Thus, the results are not fully comparable as more classes increase the difficulty of the task. Even then, the results obtained by the winning model are better than all previously published methods, except for one, which reported comparable results to the ResNet Mixed Convolution (that paper reported 0.12% better accuracy, and 0.41% less specificity). However, that paper used four different contrasts and an additional dataset apart from BraTS, making them have a larger dataset for training.

11.5 Summary and Outlook

This research shows that the spatiotemporal models, ResNet(2+1)D and ResNet Mixed Convolution, working as spatiospatial models, could improve the classification of grades of brain tumours (i.e. low-grade and high-grade glioma), as well as classifying brain images with and without tumours, while reducing the computational costs. A 3D ResNet18 model was used to compare the performance of the spatiospatial models against a pure 3D convolution model. Each of the three models was trained from scratch and also trained using weights from pre-trained models that were trained on an action recognition dataset—to compare the effectiveness of pre-training in this setup. The final results were generated using cross-validation with three folds. It was observed that the spatiospatial models performed better than a pure 3D convolutional ResNet18 model, even though having fewer trainable parameters. It can be observed further that pre-training improved the performance of the models. Overall, the pre-trained ResNet Mixed Convolution model was observed to be the best model in terms of F1-score, obtaining a macro F1-score of 0.9345 and a mean test accuracy of 96.98%, while achieving 0.8949 and 0.9123 F1-scores for Low-Grade Glioma and High-Grade Glioma, respectively. This study shows that the spatiospatial models have the potential to outperform a fully 3D convolutional model. All the three models were seen to be robust against undersampling artefacts, while the NCC1701 reconstruction improved the results over the undersampled MRIs, and resulted in comparable scores up to an acceleration factor of 10—the highest decrease in weighted F1-score was from 90.16 to 87.65, while also securing the best F1-scores of 83.51 and 79.01 for the acceleration factors 17 (radial) and 20 (Cartesian), respectively. However, it is noteworthy that the ResNet3D model performed better than the spatiospatial models while working with the undersampled and reconstructed MRIs—which can be seen as an adverse effect of not preserving the 3D volumetric spatial relationships by the spatiospatial models. Also, it is worth mentioning that the experiments were only shown for a specific task here—brain tumour classification, using one dataset—BraTS. These models should be compared for other tasks in the future to build a common consensus regarding the spatiospatial models. One limitation of this study is that it only used T1 contrast-enhanced images for classifying the tumours, which already resulted in good accuracy. Incorporating all four available types of images (T1, T1ce, T2, T2-FLAIR) or any combination of them might improve the performance of the model even further.

GP-models: Brain Tumour Classification 12

The tumour classification approach described in this section was primarily performed within the context of the master thesis of Hadya Yassin [stud:2] and has been subsequently published in [pub:1], while the initial findings were presented in [pub:32]. The code of this project is available publicly on GitHub [code:1].

12.1 Background and Motivation

Several deep learning based models have been proposed in the recent years for the task of brain tumour classification (Sect. 7.1.1) and segmentation (Sect. 7.1.2), including the method presented in the earlier chapter (see Chap. 11). However, some critical issues must be addressed before deep learning methods can be widely accepted and adopted in the medical field. Because of the complexity of the CNNs reasoning process, they tend to be obscure and are frequently referred to as black boxes, as discussed in Sect. 4.7. While some researchers came up with post-hoc methods to explain and interpret such models [pub:9], others try to create inherently explainable-interpretable models. Additionally, most deep learning based segmentation models are based on supervised learning and require manually annotated data, which is time-consuming and imposes a big challenge. Nevertheless, labelling the data with the type of tumour or whether it is present or not is much easier for radiologists and, hence, less time-consuming. [224] has shown the potential of training a segmentation model in a weakly-supervised manner using regression labels for lesion counts. This research aims to extend that research and develop fully automated, inherently-explainable CNN models for multi-class classification of brain tumours, which are also capable of tumour segmentation without the need for segmentation labels.

S. Chatterjee, *Deep Learning Innovations in MRI Reconstruction and Analysis*, https://doi.org/10.1007/978-3-658-50741-1_12

12.2 Methodology

12.2.1 Network Models

The proposed CNNs are based on modified segmentation or reconstruction models—modified to work as a multi-class classifier with only one global class label per image. The original models output *n* equally sized output images same as the size of the input, where *n* is the number of segmentation classes that were required. In this proposed approach, prior to applying the final fully-connected convolution layer, the modified CNN model is coupled with a global pooling layer in the training stage to aggregate the 3D output into a single-pixel (neuron) for each class. As a result, classification image-level labels work for training the model in a supervised manner, as an alternative to the traditional cumbersome segmentation labels during the network's training—making the segmentation training of the network weakly-supervised. During the testing stage, the global pooling layer can be removed to obtain the heatmaps with the same size as the input, or it can be kept as it is (same as the training stage) to obtain the class. The obtained heatmaps show the region-of-interest of the network—portraying the network's focus area—making the network inherently explainable. The resulting heatmaps are then suppressed by setting the negative value to zero. Without using any post-hoc interpretability methods, the model's focus areas can be understood simply by glancing at the heatmaps. Afterwards, the segmentation mask is created using Otsu thresholding. The general overview of the models is shown in Fig. 12.1.

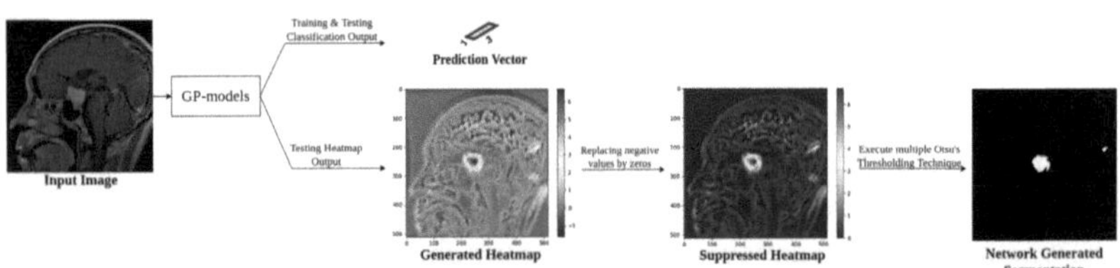

Fig. 12.1 Workflow of the GP-models [pub:1]

Three different segmentation/reconstruction models were explored in this research: UNet [186]—which was also used by [224], ReconResNet [pub:8], and ShuffleUNet [pub:18], to construct GP-UNet, GP-ReconResNet, and GP-ShuffleUNet, respectively. The resultant network architectures are shown in Figs. 12.2, 12.3, and 12.4, respectively. Original UNet and ShuffleUNet models were modified by adding a dropout layer with a probability of 0.5 in the GP-models to avoid overfitting, whereas probability of the dropout layer which is already implemented in ReconResNet was increased to 0.5. The proposed GP-models were compared against two non-GP models: InceptionV3 [211] and ResNeXt50 [245], which were chosen by comparing the performance of various other models. Table 12.1 shows the number of parameters and the approximate training time.

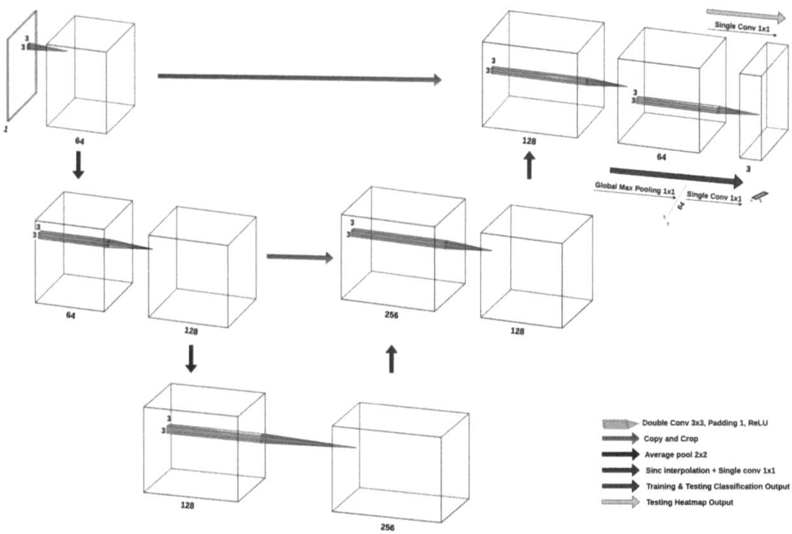

Fig. 12.2 The original Network Architecture of the baseline GP-UNet model [224] was modified by changing the up-pooling mechanism from transposed-convolution to interpolation+convolution (Sinc Up-sample method) and changing the output convolution filter number from 2^{5+i} to 2^{6+i}, $i = 0, 1, 2$ corresponding to the original depth of the model (depth=3). A dropout layer with a probability value of 0.5 was added to the model during training [pub:1]

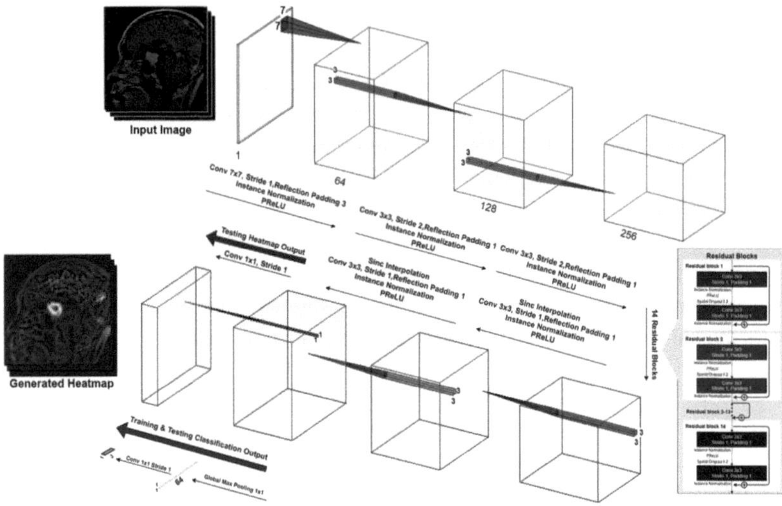

Fig. 12.3 The Network Architecture of the proposed GP-ReconResNet model. When training and testing on the BraTS 2020 dataset, a dropout layer with a probability value of 0.5 instead of 0.2 was used in the model [pub:1]

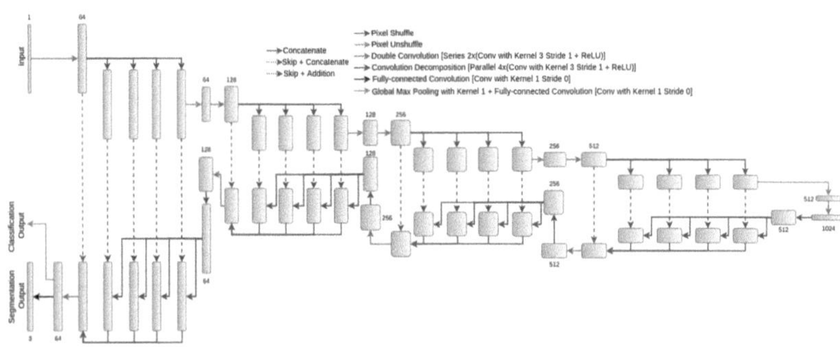

Fig. 12.4 The Network Architecture of the proposed GP-ShuffleUNet model, created by modifying the original ShulffleUNet model [pub:18] © 2021IEEE. The original model was altered by adding a global pooling layer before the fully connected convolution in the training stage transforming the model to a GP-Model, while also adding a dropout layer with a probability value of 0.5 [pub:1]

Table 12.1 Number of parameters and training time for each model

Model Name	No. of Parameters	Training Time
InceptionV3	21.8 M	$\sim$ 2 day
ResNeXt50	23.0 M	$\sim$ 2 day
GP-UNet	1.90 M	$\sim$ 4 days
GP-ShuffleUNet	26.4 M	$\sim$ 13 days
GP-ReconResNet	17.3 M	$\sim$ 9 days

12.2.2 Implementation

The implementation was done using PyTorch and PyTorch-lightning, and the experiments were performed using an Nvidia GeForce RTX 2080 TI GPU. The data was partitioned into three sets: training, validation, and testing, with real-time spatial augmentation during the training phase. The networks were trained with the help of automatic mixed-precision for 300 epochs for the second dataset, with various batch sizes, depending on when each model converged and how large it was memory-wise, keeping the effective batch size as 128 by accumulating gradients of multiple batches before backpropagating. The loss was calculated using the weighted cross-entropy loss function (Equations 4.18 and 4.19) and was optimised using the Adam optimiser with a learning rate of 0.001. To avoid overfitting, along with adding dropout layers inside the network models, L2 regularisation weights decay factor of 5e-4 was also employed. To avoid exploding gradients, a gradient clip of value one was applied. Finally, post-hoc interpretability methods [pub:9] were applied to all GP and non-GP models—to be able to compare their interpretability.

12.2.3 Dataset

The network models were employed for the task of brain tumour classification using two different datasets. The first dataset [222] contains 3, 064 2D T1-weighted contrast-enhanced MRI slices from 233 patients with three types of brain tumours: meningioma (708 slices), glioma (1426 slices), and pituitary tumour (930 slices), in three different orientations: axial (993 slices), coronal (1046 slices), and sagittal (1025 slices). The second dataset used in this research is the BraTS 2020 dataset [169, 232, 252]. This dataset contains multimodal scans acquired with different clinical protocols and scanners from 19 institutions of 346 patients with two different types of tumours: Low-Grade Glioma (LGG, 73 patients) and High-Grade Glioma or glioblastoma (HGG, 273 patients), each having 3D volumes of T1-weighted (T1w), post-contrast T1-weighted (T1ce), T2-weighted (T2), and T2 FLuid Attenuated Inversion Recovery (T2-FLAIR, or simply, FLAIR) MRIs in the axial orientation—a total of $4 \times 47,686$ slices after ignoring the complete blank slices. The slices without any tumour were considered as "tumour-free" and a third class was created using them, resulting in $4,731$, $18,276$, and $24,679$ slices, for LGG, HGG, and tumour-free class, respectively. The images were manually segmented by one to four raters, and the annotations were agreed upon by neuroradiology experts. All four types of MRIs were concatenated on the channel dimension and were supplied to the networks in a 2D, slice-wise manner. The four-channel approach was inspired by the belief that different contrasts provide varying degrees of diagnosis strength and assurance. In a sense, the entire tumour or parts of it might appear clearer to the human eye and achieve greater evaluation results using one contrast versus the other[167]. Three-fold cross-validation was performed by randomly creating subsets with the ratio of 60:15:25 using stratified sampling for training, validation, and testing, respectively.

12.2.3.1 Augmentation

Table 12.2 illustrates the real-time augmentation techniques performed only on the training set, in which each image in the set is randomly rotated in the range of -330 to $+330$, horizontally or vertically flipped, each with a 50% chance of occurring independently from one another. Augmentation can be proven helpful to prevent overfitting and enhance model robustness [140, 213].

Table 12.2 The applied spatial augmentation techniques and their respective parameters

Methods	Parameter
Flip horizontally	50% probability
Flip vertically	50% probability
Rotation	$\pm 330°$

12.3 Results

The F1-score was used as the primary quantitative metric for evaluating the models' classification performance. When having similar F1-scores, other measures such as confusion matrix, accuracy, recall, and precision were employed to aid in the decision-making process. To account for data imbalance in the BraTS dataset, a weighted average of the matrix values was used. The Sßrensen-Dice coefficient (Eq. 4.38) was used as the quantitative metric assessing the GP-models' segmentation performance. As for the last aspect, the model's inherent explainability and transparency were assessed by visually inspecting the generated localisation heatmaps, revealing the models' classification decision basis. Furthermore, post-hoc interpretability methods were applied to the best performing GP- and non-GP-models to be able to compare interpretability aspects in both.

For the first set of experiments with the first dataset [222], trainings were performed by supplying individual 2D slices of T1ce MRIs as single-channel input to the network models and were trained to perform 3-class classification for the type of brain tumour: meningioma, glioma, and pituitary tumour. For the second set of experiments with the BraTS dataset [169, 232, 252], trainings were performed by concatenating the four multi-modal MRIs: T1, T1ce, T2, and FLAIR on the channel dimension and were trained for 3-class classification for tumour-free and two different types of brain tumours—LGG and HGG. It is worth mentioning that both the datasets are imbalanced, whereas the second dataset is much more imbalanced compared to the first one. The number of slices in the first dataset was with the ratio 23:30:47 for meningioma:pituitary:glioma classes, while the ratio in the second dataset was 10:38:52 for LGG:HGG:tumour-free classes. Even though the first dataset poses less challenge in terms of class imbalance, it provides two different challenges: it is considerably smaller and contains a mix of three orientations, while the second dataset is much larger and contains only MRIs acquired in the axial orientation. The models were initially evaluated for their classification performance, followed by their segmentation performance. Finally, additional evaluations were performed with the interpretability techniques.

12.3.1 Experiments with Dataset #1

Table 12.3 shows the quantitative results for the different GP-models and the non-GP baseline models. The GP-ReconResNet performed the best among the GP-models, scoring an F1-score of 0.95, while the GP-ShuffleUNet and GP-UNet secured the second and third positions, respectively, with 0.93 and 0.85 F1-scores. The best non-GP model got 0.97 F1-score, but those models are neither explainable nor can be used for weakly-supervised segmentation. However, the segmentation quality of the GP-models was not that good. The best performing GP-ReconResNet resulted in a median Dice score of only 0.10±0.0365. Fig. 12.5 shows a few good segmentation examples of the segmentation in all three orientations for the GP-ReconResNet. It can be observed that the heatmaps show the tumour appropriately localised, and the post-processing helped in obtaining the segmentations. It is worth mentioning that these images were "hand-picked" and the segmentation quality was not that great in general.

Table 12.3 Resulting classification metrics (Precision, Recall, F1-score, Accuracy) for all the models, and the segmentation metrics (median Dice score) for the GP-models, on the first dataset [222]

Model Name	Precision	Recall	F1-Score	Accuracy	Med (DSC)
InceptionV3	0.96	0.97	0.97	0.97	–
ResNeXt50	0.96	0.96	0.96	0.97	–
GP-ReconResNet	0.96	0.95	0.95	0.95	0.10±0.0365
GP-ShuffleUNet	0.94	0.92	0.93	0.93	0.04±0.0069
GP-UNet	0.86	0.85	0.85	0.85	0.05±0.0115

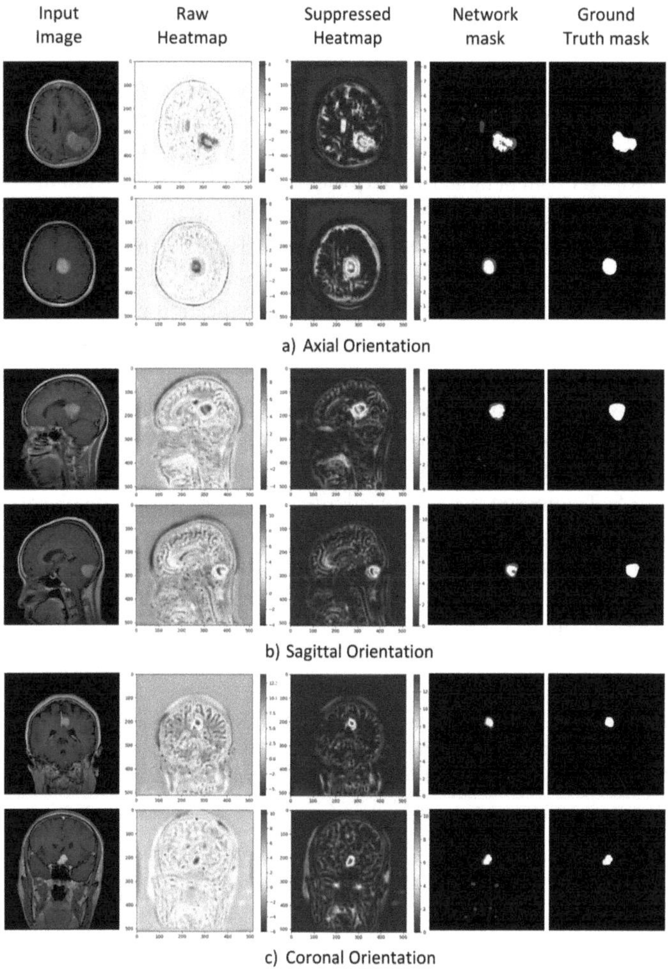

Fig. 12.5 Example results of the best-performing GP-ReconResNet from the 1st Dataset [222], in a) Axial, b) Sagittal and c) Coronal orientations. 1st column contains the input slices. 2nd column contains the model's prediction known as the raw heatmaps, where the red areas influenced the classification outcome negatively, and the blue areas influenced the classification outcome favourably. 3rd column contains the suppressed heatmaps, where negative values are suppressed to obtain positive attributions only. 4th column contains the network-generated final masks = the suppressed heatmap + Otsu thresholding. The mask is then compared to the ground-truth mask, where white indicates true segmentation, blue indicates over-segmentation, and red indicates under-segmentation. 5th column contains the ground truth mask [pub:1]

12.3.2 Experiments with Dataset #2: BraTS 2020 Dataset

On the BraTS dataset, the models show a different trend compared to the first dataset. GP-UNet and GP-ShuffleUNet resulted in 0.93 F1-score, while GP-ReconResNet resulted in 0.92. In terms of accuracy, GP-ShuffleUNet got 0.94, while GP-UNet and GP-ReconResNet both obtained 0.93. However, both the non-GP baselines resulted in 0.95 F1-score and accuracy. Even though the non-GP models resulted in better scores, it is worth mentioning that the non-GP models are not inherently explainable and do not assist in producing segmentations like the GP-models. Table 12.4 shows the complete results on different metrics. The classification performance of these models was further shown using confusion matrices in Fig. 12.7. The accuracy on a subset of the test set considering only the slices for the two tumour classes for the GP-models was compared against other state-of-the-art methods which also have used the BraTS dataset (but different versions), and it can be seen that the GP-models outperform all the other methods, while also being inherently explainable—shown in Table 12.5. The reason behind considering only LGG and HGG classes in this analysis was that the compared methods only have considered these classes in their analyses. It is to be noted, though, that the methods that worked with 3D volumes included the tumour-free slices—part of the volumes but classified the whole volume as LGG or HGG class.

Table 12.4 Resulting classification metrics (Precision, Recall, F1-score, Accuracy) for all the models, and the segmentation metrics (median Dice score) for the GP-models, on the BraTS 2020 dataset

Model Name	Precision	Recall	F1-Score	Accuracy	Dice score
InceptionV3	0.95	0.95	0.95	0.95	–
ResNeXt50	0.95	0.95	0.95	0.95	–
GP-UNet	0.93	0.93	0.93	0.93	0.67±0.08
GP-ShuffleUNet	0.94	0.94	0.93	0.94	0.53±0.08
GP-ReconResNet$_1$	0.93	0.93	0.92	0.93	0.50±0.07
GP-ReconResNet$_2$	0.93	0.93	0.92	0.93	0.60±0.06

Table 12.5 Comparison of the proposed GP-models with the previously published non-deep learning (non-DL) and deep learning (DL) works († cross-validated) for the BraTS dataset

Study / Model	Contrast	Model Type	Explainable	Test Accuracy
[230]	T1, T1ce, T2, T2-FLAIR	non-DL 3D	N	88.31%
[223]	T1, T1ce, T2, T2-FLAIR	non-DL 3D	N	89.81%†
[285]	T2-FLAIR	DL 3D	N	84.00%
[264]	T1, T2, T2-FLAIR	DL 2D	N	90.87%
[292]	T1ce	DL 2D	N	94.50%†
[362]	T1ce	DL 3D	N	96.49%
[379]	T1, T1ce, T2, T2-FLAIR	DL 3D	N	97.10%
[pub:10]	T1ce	DL 3D	N	96.98%†
GP-UNet	**T1, T1ce, T2, T2-FLAIR**	**DL 2D**	**Y**	**98.27%**†
GP-ShuffleUNet	**T1, T1ce, T2, T2-FLAIR**	**DL 2D**	**Y**	**98.74%**†
GP-ReconResNet	**T1, T1ce, T2, T2-FLAIR**	**DL 2D**	**Y**	**97.99%**†

Coming to the segmentation performance, the Dice scores show a little different trend than the classification scores, shown in Table 12.4. Two different methods were used for generating the final segmentation masks from the heatmaps: multi-Otsu thresholding with an offset and a fixed threshold of 0.5. While for the GP-UNet and GP-ShuffleUNet the first approach resulted in better segmentation, for GP-ReconResNet the second approach worked better. Hence for GP-ReconResNet, both results are furnished, denoted as GP-ReconResNet$_1$ and GP-ReconResNet$_2$, respectively for the two approaches. In terms of the segmentation performance, GP-UNet secured the first position among the GP-models, scoring a median Dice of 0.67 ± 0.08, while second version of GP-ReconResNet segmentation (GP-ReconResNet$_2$) obtained 0.60 ± 0.06 and GP-ShuffleUNet obtained 0.53 ± 0.08. The resulting Dice scores have been visualised using histograms in Fig. 12.6. Figures 12.8 and 12.9 show the generated heatmaps and the segmentation results of the GP-models, for LGG and HGG classes, respectively.

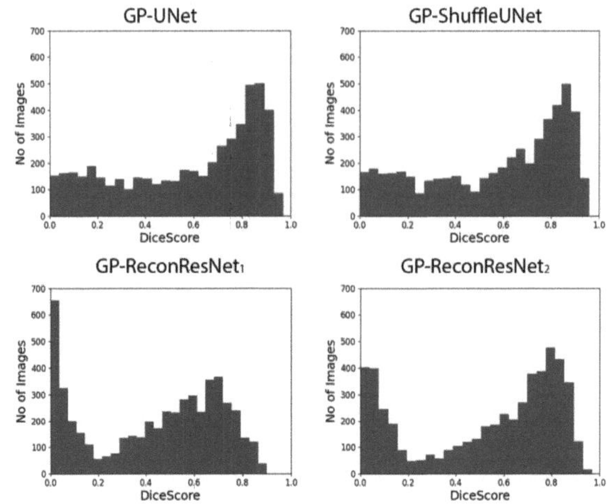

Fig. 12.6 Histogram plots displaying the distribution of the Dice scores for the segmentation results of each of the GP-models on the BraTS 2020 dataset, namely GP-UNet after an Otsu thresholding with an offset of 0.3, GP-ShuffleUNet after an Otsu thresholding with an offset of 0.1, GP-ReconResNet$_1$ after an Otsu thresholding with an offset of 1.6 and GP-ReconResNet$_2$ with a fixed threshold of 0.5 [pub:1]

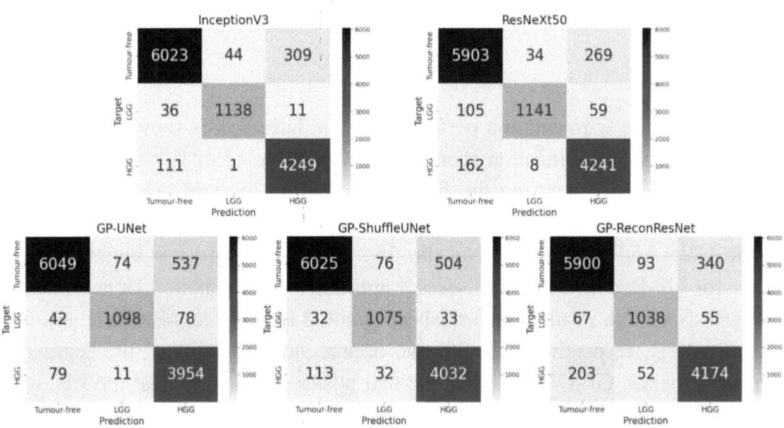

Fig. 12.7 Confusion matrices of the GP- and non-GP- models, where the three classification labels depicted as HGG for High-Grade Glioma (glioblastoma), LGG for Low-Grade Glioma, and tumour-free [pub:1]

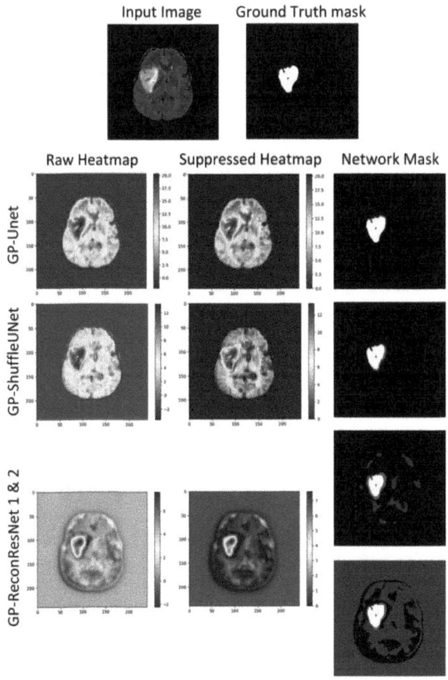

Fig. 12.8 Example results of the GP-models' correctly classified LGG tumour (class 1) from the BraTS 2020 dataset. 1st row is the input slice, the ground truth mask is used for comparison, and the rest of the rows are the models' outputs. For the rest of the rows: 1st column contains the models' predictions known as the raw heatmaps, where the red areas influenced the classification outcome negatively, and the blue areas influenced the classification outcome favourably; 2nd column contains the suppressed heatmaps, where negative values are suppressed to obtain positive attributions only; 3rd column contains the networks' generated final masks = the suppressed heatmap + thresholding [pub:1]

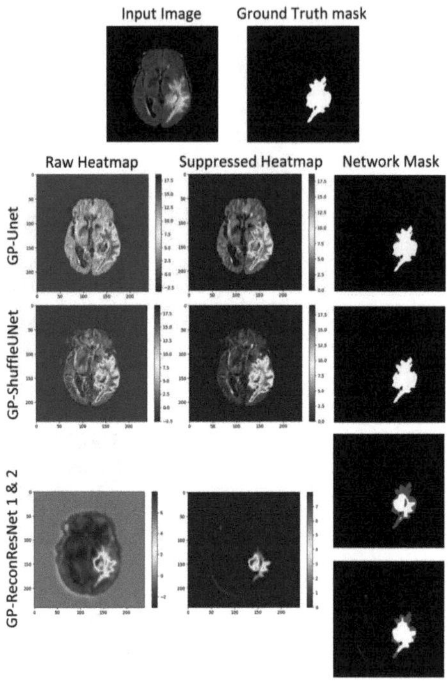

Fig. 12.9 Example results of the GP-models' correctly classified HGG tumour (class 1) from the BraTS 2020 dataset. 1st row is the input slice, the ground truth mask is used for comparison, and the rest of the rows are the models' outputs. For the rest of the rows: 1st column contains the models' predictions known as the raw heatmaps, where the red areas influenced the classification outcome negatively, and the blue areas influenced the classification outcome favourably; 2nd column contains the suppressed heatmaps, where negative values are suppressed to obtain positive attributions only; 3rd column contains the networks' generated final masks = the suppressed heatmap + thresholding [pub:1]

12.3.3 Post-hoc Interpretability

The GP-models presented here are explainable by nature, giving them an advantage over the competition in terms of model transparency and explainability portrayed. The focus area of the network can be understood from the generated heatmaps by those GP-models. The non-GP baselines are not explainable or interpretable, and typically post-hoc interpretability methods can be applied to understand the focus area of those networks [pub:3]. Therefore, two interpretability methods, one feature-based model attribution technique—occlusion, and one gradient-based model attribution technique—Guided Backpropagation (GBP), were applied to all the models to generate the interpretability results using the TorchEsegeta pipeline [pub:9]. Figures 12.10 and 12.11 show the interpretability results for all the five models, for LGG and HGG, respectively. As the GP-models are inherently explainable, the resultant heatmaps are also compared against the interpretability results.

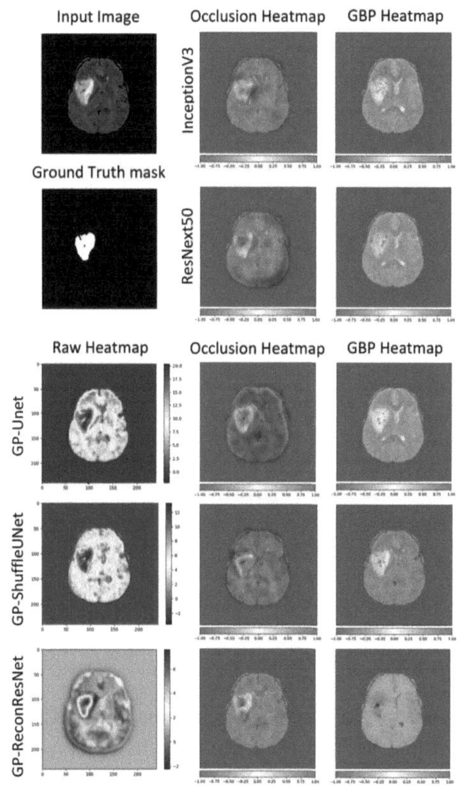

Fig. 12.10 Example interpretability results of the GP- and non-GP models, correctly classified LGG (class 1). 1st column contains the input image, ground-truth mask, and the raw heatmaps of the GP-models. 2nd and 3rd columns contain the post-hoc interpretability attributions using the occlusion and the Guided Backpropagation (GBP) methods, respectively—overlaid as heatmaps on the input slices [pub:1]

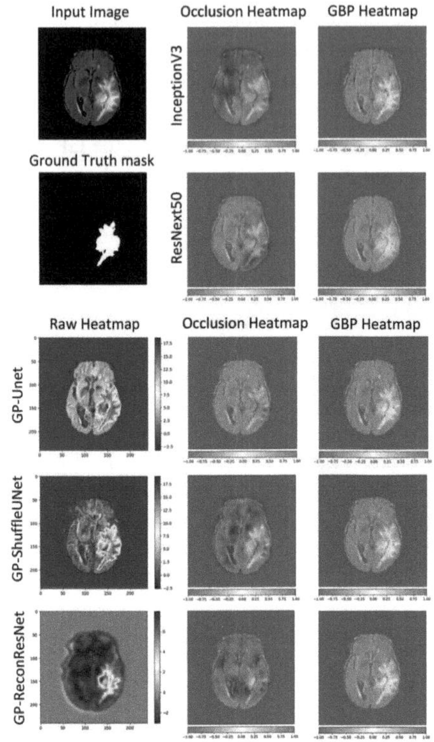

Fig. 12.11 Example interpretability results of the GP- and non-GP models, correctly classified HGG (class 2). 1st column contains the input image, ground-truth mask, and the raw heatmaps of the GP-models. 2nd and 3rd columns contain the post-hoc interpretability attributions using the occlusion and the Guided Backpropagation (GBP) methods, respectively—overlaid as heatmaps on the input slices [pub:1]

For the GP-models, by comparing the heatmaps against the interpretability results, it can be said that the interpretability results do agree with the heatmaps. The occlusion results show that the GP-models focused on the tumour. For GP-ReconResNet however, some questionable results can be seen. For LGG class, some negative attributions can be seen in the bottom side of the tumour, while for HGG class, the focus is scattered in different parts of the brain—but missing the parts of the tumour, which is shown as the most important focus region according to the heatmap. GBP has also shown the centre of the focus area(s) of the network, and they agree with the heatmaps. However, the area covering the attribution is considerably smaller than occlusion, as well as the heatmaps. This analysis shows that the interpretability results can be trusted, but might they should be considered with caution.

The interpretability results of the non-GP baselines show that the reasoning done by the ResNeXt50 model is better than the InceptionV3, even though they both have secured the same classification scores. The positive, as well as the negative attributions for the InceptionV3, are scattered nearly all over the brain.

12.3.4 Working with Undersampled MRIs

The initial fold of the cross-validation of the BraTS dataset was used in this analysis. The test volumes were undersampled using MRUnder [url:8, code:7] with five different sampling patterns: four Cartesian sampling patterns—1D Varden 30%, 1D Varden 10%, 2D Varden 10%, 2D Varden 5% and one radial sampling pattern—golden angle acquisition with 30 spokes, resulting in MR acceleration factors of 3, 10, 10, 20, and 17, respectively. The undersampled MRIs were then reconstructed slice-wise in 2D using the NCC1701 (Sect. 8.1) pipeline with the help of the already trained ReconResNet models for the respective sampling patterns—trained on brain-extracted T1w MRIs of the IXI dataset [url:12]. As the focus of NCC1701 was on the T1w imaging, the GP-models were trained using four different contrasts— GP-models were again freshly trained on T1ce images for these experiments. Then the undersampled and the reconstructed 2D slices were supplied to the GP-UNet and GP-ReconResNet models to evaluate the effect of undersampling and NCC1701 reconstruction on their classification performance.

Table 12.6 presents the quantitative results of this evaluation. It can be observed that the models were robust against the undersampling artefacts, resulting in the lowest F1-score of 0.77 for an acceleration factor of 20, while the NCC1701 reconstruction improved that result to 0.82—both of which are higher than the spatiospatial models discussed in the earlier section (Sect. 11, quantitative results in Table 11.7)—showing that these models are more robust against artefacts than the former ones. The NCC1701 improved the results over the undersampled MRIs for all undersampling patterns. Moreover, in every scenario of undersampled data, GP-ReconResNet performed better than GP-UNet, while working with reconstructed data—both models resulted in the same F1-score for three out of five undersampling patterns, and GP-ReconResNet performed better for the other two.

Table 12.6 Resulting weighted F1-scores for the different models while working with the fully-sampled, undersampled, and NCC1701 reconstructed MRIs

Undersampling Pattern	Acc. Factor	Type	GP-UNet	GP-ReconResNet
–	1	Fully	0.91	0.93
1D Varden 30%	3	Under	0.80	0.81
		Recon	0.86	0.86
1D Varden 10%	10	Under	0.77	0.78
		Recon	0.82	0.82
2D Varden 10%	10	Under	0.80	0.80
		Recon	0.84	0.85
GA Radial 30sp	17	Under	0.70	0.73
		Recon	0.78	0.79
2D Varden 5%	20	Under	0.79	0.80
		Recon	0.84	0.84

12.4 Discussion

The proposed GP-models have demonstrated comparable classification performance with the non-GP baseline models. However, they are more advantageous than the baseline non-GP models, as they are inherently-explainable, which can be more trustworthy for the decision-makers like clinicians than the other opaque models. But most importantly, they have shown the capabilities for the task of performing brain tumour segmentation without training them using any segmentation labels.

GP-models require more time in comparison to non-GP-models and vary from one another depending on the number of parameters, as shown in Table 12.1. More time is needed for GP-models because of the expanding part of the networks. Moreover, GP-ShuffleUNet, which has the highest number of parameters (26.4 M), consequently takes the longest time to train—($\sim$ 4 days). Even though for the classification performance, GP-UNet and GP-ShufflUNet have achieved the same F1-score (0.93) on the second (BraTS) dataset, GP-ShuffleUNet got 0.94 accuracy while GP-UNet got 0.93. From the perspective of segmentation performance, GP-UNet got 0.67±0.08 Dice score while GP-ShufflUNet got 0.53±0.08. On the first dataset, GP-ShuffleUNet got lesser scores compared to GP-UNet. The results show that the GP-ShuffleUNet performs better when it is trained on a larger dataset and does not encounter additional difficulties like different orientations mixed in the training set.

Considering all these factors, GP-UNet should be preferred over GP-ShuffleUNet. On the second dataset, GP-ReconResNet has achieved the same accuracy as the GP-UNet (0.93) but got 0.01 less F1-score (0.92); however, the segmentation performance of GP-ReconResNet has been seen to be lower than the GP-UNet—a median Dice score of 0.60±0.06. Since GP-UNet has the least trainable parameters, reducing the chance of overfitting and superior overall performance, this can be considered the best performing model on the second dataset. When the models were trained only on T1ce images, instead of all four contrasts, GP-ReconResNet resulted in a better F1-score than before (0.93), while GP-UNet resulted in a lower score (0.91). Moreover, on the first dataset, GP-ReconResNet obtained 0.10 more F1-score and accuracy than the GP-UNet model (0.95 and 0.85, respectively). From these, it can be concluded that the GP-ReconResNet is more robust against variations in the dataset (different orientations mixed together) and requires less training data (performed better when the training set was considerably smaller, or when it had only one contrast instead of four). However, the results also indicate that the GP-ReconResNet model performs better when the dataset is less imbalanced. When the training dataset is larger, and all MRIs are in one orientation, GP-ReconResNet and GP-UNet result in comparable performance. Although from the perspective of segmentation quality—all the models performed poorly on the first dataset, resulting in a best median Dice score of 0.10±0.0365. This could be attributed to the fact that this dataset has a mix of different orientations, and also, the amount of training data is very less. The models could handle those problems and perform classification correctly, but could not generalise enough to have good segmentation performance.

While working with undersampled and NCC1701 reconstructed data, GP-ReconResNet outperformed GP-UNet in most scenarios—in all undersampled experiments and two out of five experiments working with reconstructed data (while for the other three, both models resulted in the same score). This shows that GP-ReconResNet not only demands less training data, but is also more robust against artefacts. While working with fully-sampled data, GP-models performed better than the spatiospatial models discussed in the earlier section (Sect. 11). But, while facing undersampled and reconstructed data, the results are mixed. While working with the most challenging undersampling patterns (1D Varden 10% and 2D Varden 5%), the GP-models performed better than the spatiospatial models. But for the other three undersampling patterns, the spatiospatial models performed better. It is worth mentioning though that the fully 3D model (ResNet3D) performed better than both the spatiospatial (Table 11.7), as well as GP-models (Table 12.6) for most of the undersampling. But for the most challenging undersampling (2D Varden 5%), both the GP-models outperformed the ResNet3D model.

The generated heatmaps bring two major advantages in favour of the GP-models—they make the models explainable, which can help decision-makers like radiologists to have more trust in this kind of non-opaque model, and they make these classification models to be used as a decision support system to localise regions of interest, like in this case, tumours. By applying post-processing techniques to the heatmaps, the final segmentations are obtained, which are imperfect and have scope for improvements. However, given that the no segmentations labels were used for training, and the segmentation was only trained in a weakly-supervised manner, a median Dice of 0.67 ± 0.08 is promising. By modifying the post-processing techniques, the segmentation performance might be improved further. The segmentation quality for all the methods was better for HGG than LGG (especially for GP-ReconResNet)— which can be attributed to the fact that the training data was highly imbalanced and was leaning more towards the HGG class. Hence, the results can be improved by reducing the imbalance in the dataset or employing further techniques to handle imbalanced datasets.

Table 12.5 compares the proposed GP-models to previously published glioma tumour grading models, which use the BraTS dataset (but different versions). Since the models of this research are multi-class classifiers, test accuracy was calculated for the tumour-only subset of the data. The proposed models outperform the other studies, with GP-ShuffleUNet receiving the highest score of 98.74%. On the plus side, the 2D proposed models, which are more resource-efficient, perform better even than the 3D models in the table, while the models here are also explainable by nature—giving them an advantage over the competition in terms of model transparency and explainability.

Finally, coming to the point of post-hoc interpretability techniques. The GP-models are inherently-explainable, but to understand the non-GP baselines, post-hoc interpretability techniques are required. Hence, a gradient-based technique—GBP and a feature-based technique—occlusion, were applied to all the models. For the GP-models, the interpretability results (also known as attention maps) were compared against the heatmaps of the networks—as they can be considered as a kind of "ground-truth". The models generally had logical attention maps, but there were some discrepancies and conflicting results between methods. However, assuming that the post-hoc methods are reliable (most of them agree with the networks' heatmaps), GP-UNet, GP-ShuffleUNet, and ResNeXt50 share similar outcomes, demonstrating proper tumour focal regions in both classes, followed by GP-ReconResNet, which has superior outcomes than InceptionV3. Even though InceptionV3 has the highest classification results, the attention maps show otherwise; the positive and negative attributions are scattered compared to other models and do not offer the proper focus on the tumour for both tumour classes. A similar outcome can also be observed in a study of deep learning methods for COVID-19 classification using Chest-Xrays, where for certain models, the classification results are high, but post-hoc methods prove the models focus on different regions than the desired target[pub:3].

12.5 Summary and Outlook

This research presented three inherently-explainable deep learning models—GP-UNet, GP-ReconResNet, and GP-ShuffleUNet, and showed their applicability for the task of multi-class brain tumour classification. Segmentation and reconstruction models were converted into the GP-models by applying a global pooling layer before the final convolution layer of each of the models, and they were trained using classification labels. During inference, the global pooling layer was disabled to obtain the models' localisation maps (raw heatmaps), while the pooling layer was kept as it was during training—to obtain the classification results. The segmentation results were obtained by post-processing the raw heatmaps without training the models using any segmentation labels. The models have achieved comparable performance with the non-explainable baseline classifiers (best accuracy for GP-models: 0.94, best accuracy for non-GP baselines: 0.95) while demonstrating their advantage in terms of explainability. All the GP-models resulted in better classification accuracy for the task of tumour classification (ignoring the tumour-free class) than the existing studies, while the best model secured 98.74% test accuracy. Moreover, this research has shown the potential of using the GP-models to perform segmentation while being trained in a weakly-supervised manner using only the classification labels, resulting in the best median Dice score of 0.67 ± 0.08. The high accuracy, coupled with the heatmaps and the segmentations, have shown the potential of using these models as decision support systems.

StRegA: Anomaly Detection 13

The original work of this anomaly detection approach was performed in collaboration with Alessandro Sciarra and has been published in [pub:12], while some of the initial results were presented in [pub:38] and [pub:31]. The code of this project is available publicly on GitHub [code:12].

13.1 Background and Motivation

Anomaly detection is an approach that distinguishes anomalies completely based on characteristics that describe regular data. The model initially learns on a dataset without anomalies—to learn the distribution of the features representing anomaly-free images. Since the feature of possible irregularities is not learned during training, they stand from the normal distribution and can be consequently detected during inference. Anomaly detection solutions are useful in cases where anomalies are to be detected, but their manifestation is not known, and they even might occur very rarely—making it difficult to create a large dataset for supervised training [170]. Anomaly detection in medical imaging is defined as identifying relevant indicators of diseases and differentiating them from those of typical healthy tissue characteristics. Unsupervised Anomaly Detection (UAD) systems can directly learn the data distribution from a large cohort of un-annotated healthy subjects to detect out-of-distribution or anomalous samples to identify cases for further inspection. As anomaly detection does not depend on reference annotations, this approach does not require any human input and can therefore be used to identify any medical condition. Unsupervised segmentation of brain regions for anomaly detection is

S. Chatterjee, *Deep Learning Innovations in MRI Reconstruction and Analysis*,
https://doi.org/10.1007/978-3-658-50741-1_13

clinically relevant as it can assist radiologists. Several methods have been proposed over the years for the task of unsupervised anomaly detection, and some of them are discussed in Sect. 7.2. However, many of such techniques fail to perform well when it comes to real clinical data, resulting in Dice scores less than or close to 50% [298, 347]. This research aims to extend the research field of anomaly detection by creating a UAD pipeline capable of working with clinical data for the detection of clinically-relevant anomalies (e.g. lesions) and to evaluate its performance for the task of brain tumour detection. One possible strategy could be to simplify the input to the model—making the model less dependent on the image properties and more on the structures themselves—that has not been explored so far. This research explores the possibility of improving the performance of the model on clinical data by simplifying the input.

13.2 Methodology

Autoencoders and their flavours, specially VAEs, are commonly used for unsupervised anomaly detection, as discussed in 7.2. The models discussed in that section were considered while selecting the baselines, and finally, the best performing baseline was selected and modified to develop the proposed approach.

13.2.1 StRegA: Segmentation Regularised Anomaly

The proposed unsupervised anomaly detection pipeline, StRegA: Segmentation Regularised Anomaly—shown in Fig. 13.1, includes a modified compact version of the Context-Encoding Variational Autoencoder (ceVAE) [298]—Compact Context-Encoding Variational Autoencoder (cceVAE), combined with pre and post-processing steps. Anomaly detection is performed in 2D, per-slice basis. The input volumes are preprocessed in 3D, then slices are supplied to the model, followed by that, post-processing steps are applied, and finally, the slices are stacked together to obtain the final result in 3D. It is worth mentioning by splitting the 3D volumes into 2D slices, the spatial context within the volumes is lost. But this was done to make the training process feasible in terms of GPU requirements.

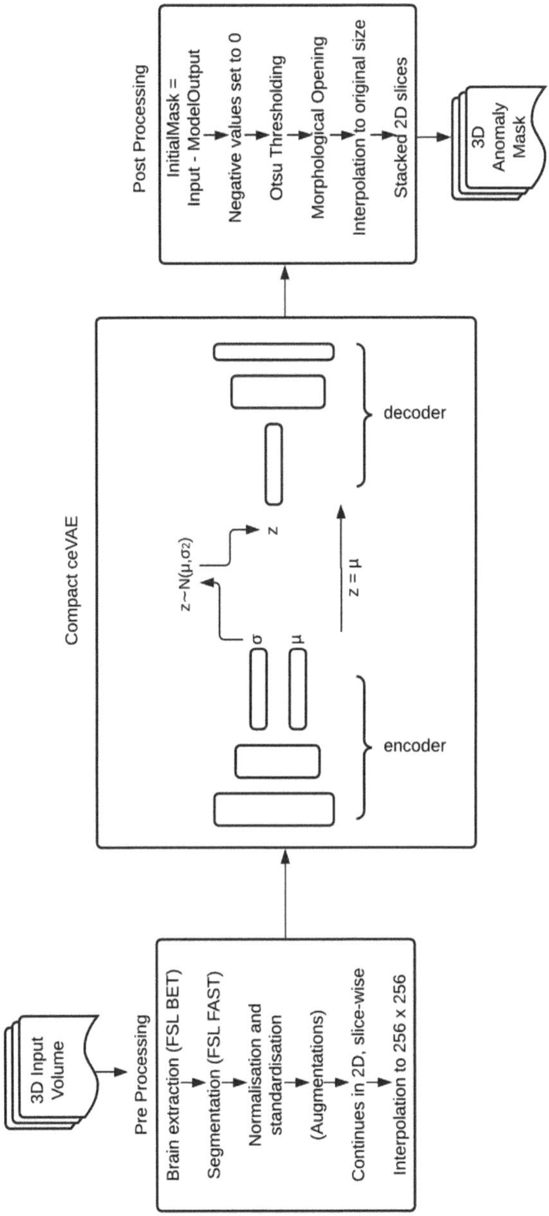

Fig 13.1 The proposed StRegA pipeline [pub:12]

13.2.1.1 Preprocessing

The first preprocessing step was to segment the input image into different tissue types and obtain the corresponding segmentation mask. Segmentation means partitioning the input images into various segments. Segmentation can simplify the input given to the model, consequently improving the performance of the anomaly detection system. This was performed by supplying the 3D brain images to the automatic segmentation tool of FSL ([138])—FAST, which segmented the brain into three different tissue types: grey matter, white matter, cerebrospinal fluid (CSF); and background—a total of four classes. The underlying principle of FAST is formulated on a hidden Markov random field model and an associated expectation-maximisation algorithm. The entire process is automated and is capable of producing a probabilistic/partial volume tissue segmentation. Apart from segmenting the input volume, this tool also corrects spatial intensity variations (also known as bias field or RF inhomogeneities). The segmentation mask was normalised with the intensity distribution of the dataset—normalised with zero mean and unit variance. The volumes were divided into 2D slices in the axial orientation, and then they were resized to 256Œ256 in-plane using bilinear interpolation before feeding them into the network. During the training phase, in addition to the above-mentioned preprocessing steps, data augmentation was also performed. The augmentation techniques used can be split into two groups: intensity-based and spatial augmentation. Intensity augmentation included bias field artefacts, Gaussian noise, random gamma, and ghosting artefacts. At the same time, spatial augmentation consisted of horizontal flipping, vertical flipping, affine transformations, and rotation with a degree between –15 and +15.

13.2.1.2 Network Model: cceVAE

This work proposes a compact modified version of the ceVAE model [298], explained in Sect. 7.2.7: Compact ceVAE (ccVAE). A representational diagram of the network has been shown in Fig. 13.2. This is a compact version as it has a smaller encoder and decoder than the original ceVAE—having a symmetric encoder-decoder with 64, 128, 256 feature maps and a latent variable size of 256. It is hypothesised that a more compact version of the network will reduce any possibilities of overfitting, especially for data that is more simplified than un-processed slices—where many learnable parameters might not be required. This proposed model also uses residual connections and batch normalisation layers—which were not present in ceVAE. The preprocessed images are fed into this network model, and the network tries to reconstruct the input. During training, anomaly-free (in the experiments of this research: healthy brain MRIs) images are provided, and the network attempts to learn a Gaussian distribution of the given anomaly-free dataset. While

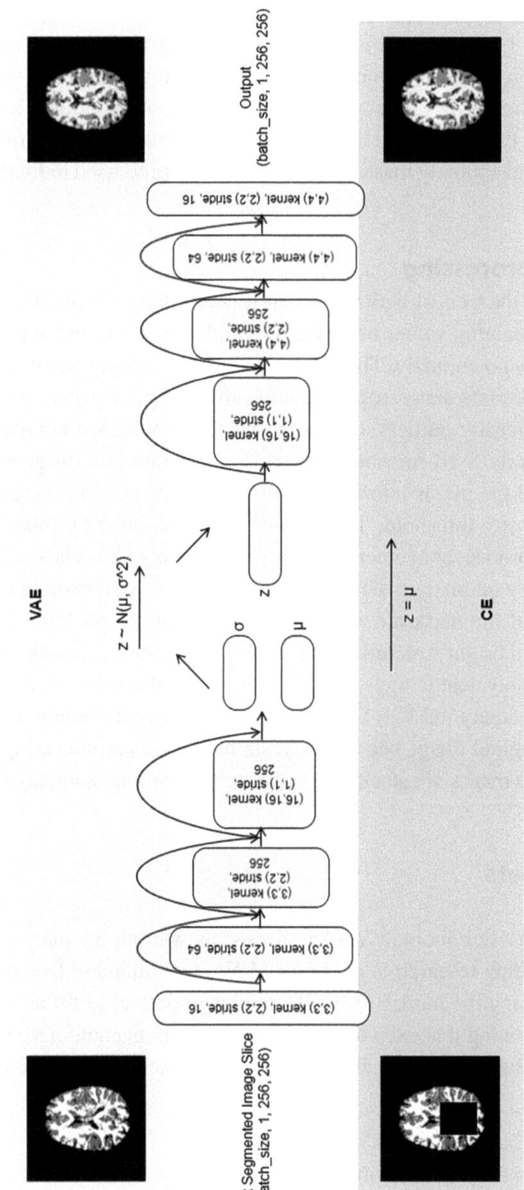

Fig 13.2 The proposed Compact ceVAE architecture: cceVAE. Each of the convolution blocks contains a convolution layer with the mentioned configuration, followed by a batch normalisation layer [pub:12]

inference, the network attempts to reconstruct the input image from the learnt Gaussian distribution. If the network encounters images with anomalies—they are then out-of-distribution samples. Hence, the network should "in-theory" fail to reconstruct the anomalies as they do not belong to the learnt distribution. The reconstruction difference is then computed by subtracting the model's output from its input to generate the initial anomaly mask-which is then post-processed to finally detect the anomaly.

13.2.1.3 Postprocessing

After subtracting the reconstruction from the model's input, the initial anomaly mask is obtained—containing values between zero and one, where one implies anomaly while zero means no anomaly. Then this mask is first thresholded to eliminate any pixel having a value less than zero by replacing all the negative values with zero. Then to remove the intensity outliers, the mask was thresholded using Otsu's method[1]. This is a frequently used method for performing automatic image thresholding, which divides all the pixels into two classes—background and foreground, based on a single intensity threshold. The threshold is calculated by minimising intra-class intensity variance, or in other words, by maximising inter-class variance[2]. This results in a binary anomaly mask. After performing Otsu thresholding, the small irrelevant parts of the mask are removed by applying morphological opening[3]—which removes all bright structures of an image with a surface smaller than the area threshold. It is equivalent to applying dilation on top of erosion on the image. This gives the final anomaly mask. It is then resized through interpolation to whichever input size the original image had before resizing during preprocessing. Finally, all the slice-wise 2D masks are stacked together to get the final anomaly mask in 3D.

13.2.2 Datasets

Training unsupervised anomaly detection models warrant anomaly-free datasets. As the focus of this research is on brain MRIs, two anomaly-free datasets were selected for training the models: MOOD challenge dataset [399] and IXI dataset[4] It is worth mentioning that even though both the datasets contain MRIs only from normal, healthy subjects—there might still be some anomalies present in some of

[1] https://en.wikipedia.org/wiki/Otsu' s_method

[2] http://www.labbookpages.co.uk/software/imgProc/otsuThreshold.html

[3] https://en.wikipedia.org/wiki/Opening_(morphology)

[4] IXI Dataset: https://brain-development.org/ixi-dataset/.

the MRIs that the experts missed. However, in this research, following the authors of the datasets—these datasets were considered to be comprised of only anomaly-free healthy brain MRIs. MOOD dataset comprises 800 T1w brain MRIs with a matrix size of 256 × 256 × 256—700 of which were used for training, and 100 were held out for testing, while the IXI dataset contains MRIs of nearly 600 subjects—all of which were used in training, acquired using 1.5T and 3T MRIs in three different hospitals, with different MRI sequences: T1w, T2w, PDw, MRA, and DTI. In this research, T1w and T2w images from the IXI dataset were used separately—for training separate T1 and T2 models. Trainings were performed by combining MOOD with one of the IXI contrast: MOOD + IXIT1 and MOOD + IXIT2. The motivation behind combining datasets (and contrasts) was to make the model more robust against differences in contrasts in the test set than in the training set.

For evaluating the approach with a dataset with clinical anomalies, 80 randomly selected volumes from BraTS 2017[5] dataset was used, which contains four types of MRIs with high and low-grade brain tumours: T1-weighted, and contrast-enhanced T1-weighted (with contrast agent), T2-FLAIR, and T2-weighted. This research used contrast-enhanced T1-weighted (T1ce) and T2-weighted MRIs. Additionally, the model was also tested on the "toy test set" provided in the MOOD dataset, comprising one anomaly-free and three anomalous MRIs containing artificial anomalies. It is noteworthy that the image contrasts, acquisition parameters, and other image properties of the BraTS dataset are very different from the training sets. Hence, a controlled synthetic anomalous dataset was created as a part of this research—to evaluate the approach on a dataset that contains anomalies, but the contrast and other image characteristics are similar to the training set.

13.2.2.1 Generating Synthetic Anomalous Dataset

An artificial controlled test set was created by superimposing anomalies from the BraTS dataset onto the held-out test-set containing anomaly-free brain MRIs. This was performed to evaluate the performance of the methods when the non-anomalous parts of the images have similar contrast and other characteristics as the training dataset—to solely evaluate the performance of the model when the model doesn't encounter any other difficulties, while it also ensures that the artificial "tumours" are realistic (Fig. 13.3).

[5] BraTs: https://www.med.upenn.edu/sbia/brats2017/data.html

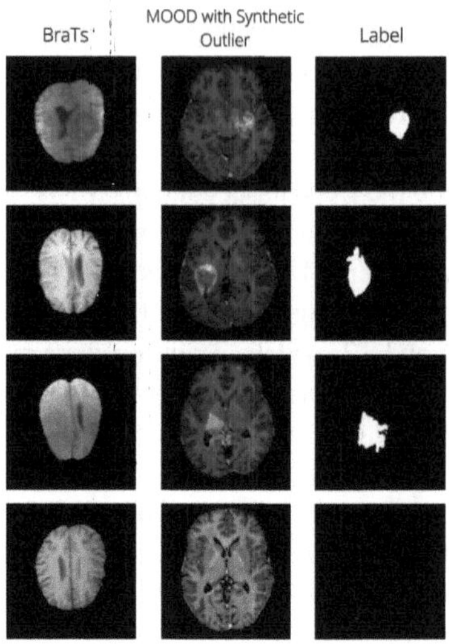

Fig 13.3 Synthetic anomalous images, generated using the MOOD dataset by taking anomalies from the BRATS data set, including one anomaly-free example in the last row [pub:12]

To generate the synthetic anomalous dataset, the tumour labels from the BraTS dataset were initially multiplied with their respective BraTS T1ce images to segment only the tumours and extract only the anomalies. Then the intensity values of them were normalised between 0 and 1, and they were reshaped to 256 × 256 using interpolation. These anomalies were now superimposed on top of the anomaly-free held-out test-set in a similar position as the original image to generate the synthetic anomalous dataset. Consider a BraTS image sample B, and its corresponding segmented mask S, then the superimposed MOOD image (from the held-out test-set) sample A of patch h can be calculated as:

$$\begin{aligned} A'_i &= A_i + M_i; \forall i \in h \text{ ,where} \\ M_j &= B_j - S_j; \forall j \in h \end{aligned} \tag{13.1}$$

13.3 Experiments and Evaluations

The methods were compared based on their accuracy in segmenting the anomalies, calculated with Sßrensen-Dice coefficient (Eq. 4.38). Initial experiments were performed using a vanilla VAE [pub:38], following that, different State Of The Art models (the ones mentioned in Sect. 7.2) were compared. Finally, the three best-performing models in this current experimental setup were chosen as baselines, and final comparisons were performed against the proposed method.

13.3.1 Initial Selection of Baseline Models

The initial approach of using the vanilla VAE [pub:38] was able to detect simple contrast anomalies—anomalies having very different contrast than the rest (e.g. MOOD toy dataset), but in most cases failed to generate an acceptable reconstruction as it was unable to provide proper latent space representation. SkipAE [341] resulted in better detection of anomalies. While the results on the synthetic data and the MOOD toy test data were better, the more complicated anomalies were harder to detect in the reconstructions. Others models, which are explained in Sect. 7.2 were also experimented with, and it was found that the ceVAE [298] (Sect. 7.2.7) performed the best when it comes to segmenting pathological anomalies. Hence, this was chosen as the primary baseline of this research. The second and third best performing models, GMVAE [347] (Sect. 7.2.6) and SkipAE [341] (Sect. 7.2.3) respectively, were also chosen as additional baselines.

13.3.2 Method Development

This research focused on simplifying the input to the backbone model—by introducing different preprocessing techniques. It was observed that using segmentation as a preprocessing step assisted the models most and improved the model's performance which was tested with all the baseline models. Different other preprocessing techniques were also experimented with, and the final set of techniques was made part of the StRegA pipeline—explained in Sect. 13.2.1.1. When experimenting with the number of layers in the encoder and decoder of the ceVAE model with these preprocessing techniques, it was observed that the results on the test datasets improved significantly when a more compact model was used. This could be due to the fact that a more compact model reduces the risk of over-fitting on the relatively simpler segmented data. Thus after several experiments, having symmetric 64, 128,

256 feature maps and a latent variable size of 256 yielded the best generalisation in reconstruction quality. Implementing skip connections as was used in SkipAE provided no meaningful improvements, but residual connections with batch normalisation provided better results and were thus included in the model—results in the final cceVAE model (Sect. 13.2.1.2).

13.3.2.1 Comparative Results

The proposed StRegA pipeline was compared against the three baseline models and an additional baseline—by combining the StRegA preprocessing steps with the ceVAE model. Figure 13.7 shows a qualitative comparison of the different methods. It can be observed that the proposed StRegA pipeline provides a more localised detection of the anomaly compared to the other methods. Tables 13.1 and 13.2 portray a quantitative comparison of the T1 and T2 models. The proposed method outperformed all the other models with statistical significance in both T1 and T2 cases while achieving 49% and 82% improvements in Dice scores over the baseline ceVAE for the task of segmenting tumours from T1w and T2w brain MRIs from the BraTS dataset. It can be observed that all the models perform better in the case of T2 than on T1ce. This can also be either because the tumours are easier to detect in T2 images than on T1ce images and could also be because of the segmentation performance of FSL in case of anomalous data, as FSL is typically used for anomaly-free images. Moreover, Fig. 13.4 compares the range of resulting Dice scores for the task of anomaly detection in BraTS T2 images for ceVAE, ceVAE with StRegA preprocessing, and StRegA, with the help of violin plots.

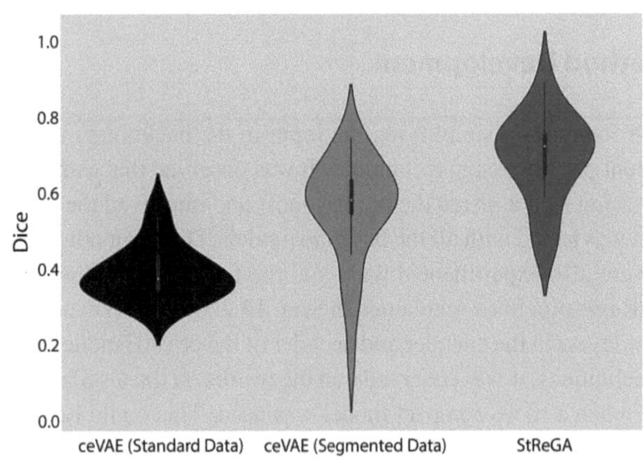

Fig 13.4 Violin plot comparing predictions on BraTS T2w data [pub:12]

It can be seen that the segmentation assists the ceVAE model in achieving better scores, while the complete StRegA pipeline manages to improve the scores even further. Finally, Figs. 13.9 and 13.10 show the output of the different stages of the pipeline for BraTS T1 and T2, respectively. Similarly, Fig. 13.8 shows the stages for synthetic anomalies.

13.3.2.2 StRegA as an Assistive Technology

The earlier comparisons have demonstrated the proposed pipeline's superiority over the baseline models. However, several cases of under-segmentation were observed. Detecting if there is any anomaly or providing the location of the anomaly might also be sufficient to improve the clinical workflow by assisting the radiologists in filtering anomalous cases and getting an approximate location of the same. Figures 13.5 and 13.6 show the potential of using StRegA as an assistive technology to localise the anomalies in the MRIs for BraTS T1ce and T2, respectively. This was achieved by creating bounding boxes around the prediction of StRegA, and they were compared against the bounding boxes generated from the ground-truth labels. It can be observed that the proposed method was able to properly locate the anomalies. As seen earlier, the T2 results were better than the T1ce ones. For T1ce, under-estimations were observed most of the time. However, for T2, the results are better, and most of the time, StRegA managed to mark them appropriately. Despite the under-estimations, the results show the potential of using this method as assistive technology—as a decision support system.

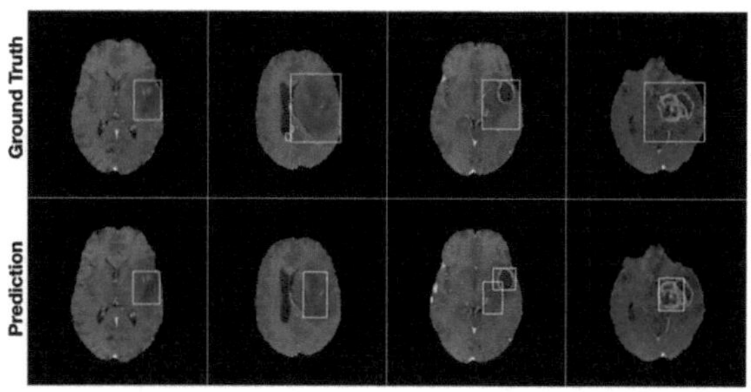

Fig 13.5 Anomaly localisation : BraTS T1ce [pub:12]

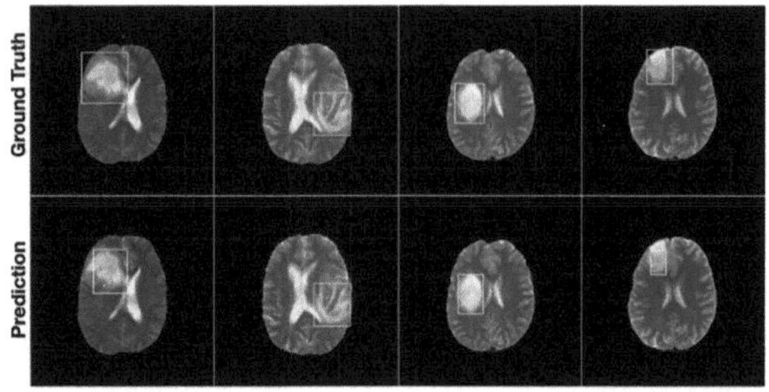

Fig 13.6 Anomaly localisation : BraTS T2 [pub:12]

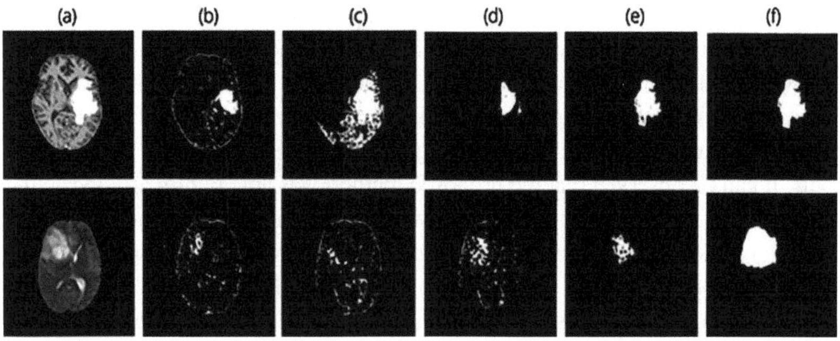

Fig 13.7 Qualitative comparison of the result of different methods. Top row: artificially generated anomalous data. Bottom row: BraTS T2w image. (a) A slice from the input volume, (b) ceVAE, (c) SkipVAE, (d) GMVAE, (e) StRegA and (f) ground truth [pub:12]

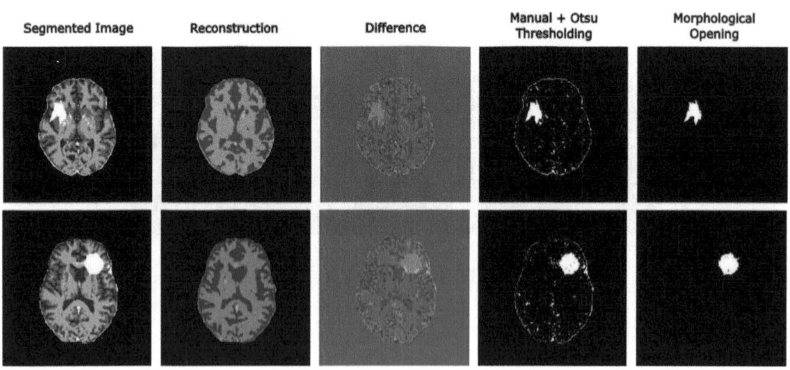

Fig 13.8 StRegA results on synthetically generated data [pub:12]

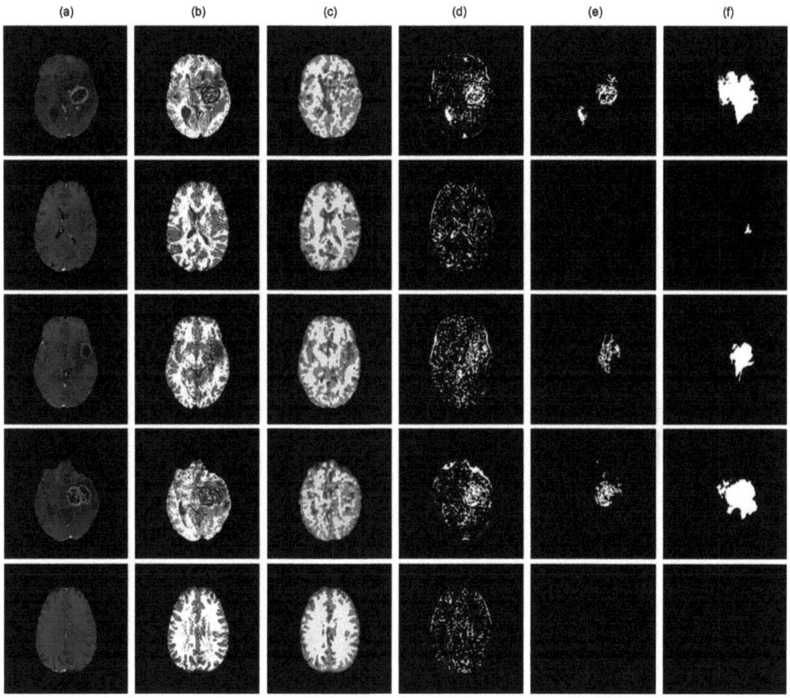

Fig 13.9 StRegA results on BraTS dataset T1ce images (a) Input Image (b) Segmented Image (c) Reconstruction (d) Manual and Otsu thresholding (e) Morphological Opening -> **Final prediction** (f) Ground Truth [pub:12]

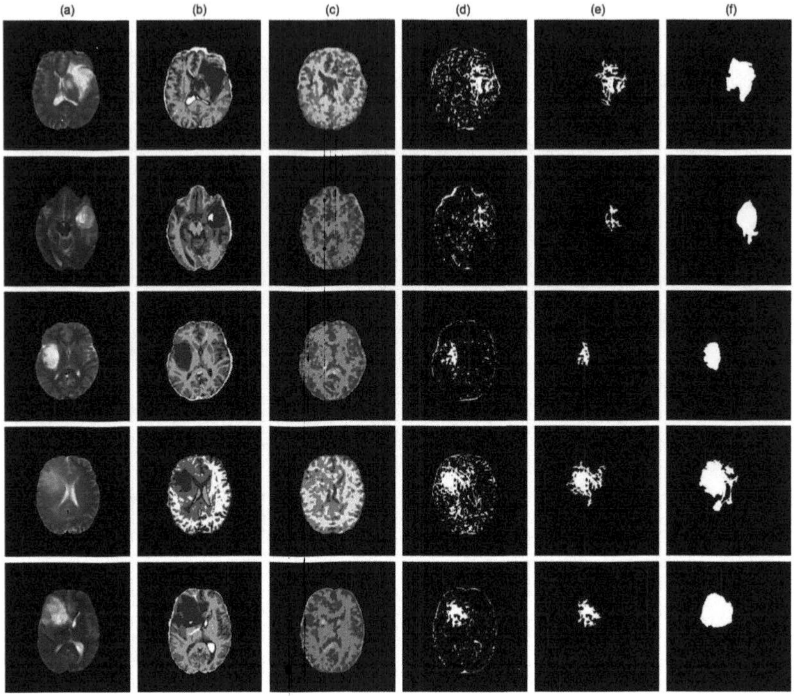

Fig 13.10 StRegA results on BraTS dataset T2w images (a) Input Image (b) Segmented Image (c) Reconstruction (d) Manual and Otsu thresholding (e) Morphological Opening **Final prediction** (f) Ground Truth [pub:12]

Table 13.1 Comparative quantitative results of StRegA with other models on BraTS T1ce data and other T1w data

Model	Dice Score ($\mu \pm \sigma$)		
	MOOD Toy Data	Synthetic Anomalous Data	BraTS
SkipAE	0.443±0.136	0.498±0.164	0.209±0.113
GMVAE	0.657±0.121	0.711±0.105	0.337±0.124
ceVAE	0.633±0.145	0.690±0.164	0.356±0.132
ceVAE (with StRegA pre-processing)	0.724±0.132	0.710±0.111	0.520±0.108
StRegA	0.737±0.098	0.723±0.134	0.531±0.112

Table 13.2 Comparative quantitative results of StRegA with other models on BraTS T2w data and other T1w data

Model	Dice Score ($\mu \pm \sigma$)		
	MOOD Toy Data	Synthetic Anomalous Data	BraTS
SkipAE	0.517±0.145	0.498±0.173	0.221±0.101
GMVAE	0.699±0.121	0.701±0.121	0.340±0.167
ceVAE	0.724±0.145	0.712±0.132	0.350±0.071
ceVAE (with StRegA pre-processing)	0.797±0.099	0.783±0.111	0.522±0.135
StRegA	0.856±0.098	0.859±0.112	0.642±0.101

13.3.3 Working with Undersampled MRIs

The test volumes were undersampled using MRUnder [url:8, code7] with five different sampling patterns: four Cartesian sampling patterns—1D Varden 30%, 1D Varden 10%, 2D Varden 10%, 2D Varden 5% and one radial sampling pattern—golden angle acquisition with 30 spokes, resulting in MR acceleration factors of 3, 10, 10, 20, and 17, respectively. The undersampled MRIs were then reconstructed slice-wise in 2D using the NCC1701 (Sect. 8.1) pipeline with the help of the already trained ReconResNet models for the respective sampling patterns—trained on brain-extracted T1w MRIs of the IXI dataset [url:12]. Then the reconstructed 2D slices were stacked together to obtain the final 3D volumes. The undersampled and the reconstructed volumes were then supplied to the StRegA pipeline.

The results of the main baseline model—ceVAE and the proposed StRegA pipeline on the fully-sampled, undersampled, and NCC1701 reconstructed datasets were compared using the Dice scores, shown in Table 13.3. It can be observed that both the methods were significantly affected by the undersampling artefacts, and the NCC1701 did not manage to increase the scores of the undersampled volumes similar to the fully-sampled ones. However, the reconstructed volumes manage to secure better scores than the undersampled volumes for both methods and for all the undersampling patterns. The amount drop in scores is more for StRegA than for the ceVAE model. This can be attributed to the fact that the StRegA heavily relies upon the preprocessing steps, as seen earlier, and the preprocessing steps include third-party tools—FSL, which might not be robust against artefacts. Nevertheless, the StRegA outperformed the baseline ceVAE model in all instances—for both undersampled as well as reconstructed MRIs for all undersampling patterns.

Table 13.3 Resulting Dice scores for the ceVAE baseline and the proposed StRegA pipeline on the BraTS T1w dataset, while working with the fully-sampled, undersampled, and NCC1701 reconstructed MRIs

Undersampling Pattern	Acc. Factor	Type	ceVAE	StRegA
–	1	Fully	0.356±0.132	0.531±0.112
1D Varden 30%	3	Under	0.262±0.120	0.282±0.120
		Recon	0.279±0.172	0.315±0.188
1D Varden 10%	10	Under	0.142±0.132	0.262±0.110
		Recon	0.199±0.156	0.303±0.099
2D Varden 10%	10	Under	0.251±0.103	0.262±0.110
		Recon	0.302±0.174	0.341±0.224
GA Radial 30sp	17	Under	0.179±0.119	0.179±0.119
		Recon	0.225±0.101	0.241±0.167
2D Varden 5%	20	Under	0.099±0.011	0.149±0.023
		Recon	0.136±0.206	0.215±0.131

For the acceleration factor of 10, StRegA managed to obtain an average Dice score of 0.341±0.224 after being reconstructed with NCC1701—which is comparable to the results of ceVAE on the original fully-sampled volumes (0.356±0.132).

13.4 Discussion

The results demonstrate that the proposed StRegA pipeline achieved significantly better scores than the baseline models in detecting brain tumours and artificial anomalies while being trained on anomaly-free brain MRIs. Even though StRegA has performed significantly better than the baseline models, it is to be noted that the predictions are not perfect—which is commonly seen with UAD techniques, and there is a strong dependency on the preprocessing steps, as can be seen from the results of ceVAE and ceVAE with StrRegA preprocessing. It is also worth mentioning as FSL segmentation is typically used on anomaly-free images; using them on anomalous data might have induced some bias in the training-testing of the models, which is not possible to quantify. Moreover, it was observed that the compact version of ceVAE proposed here performed better than the original ceVAE model, which might be attributed to the fact that the preprocessed images require a much simpler representation than images without preprocessing—consequently requiring less number of trainable parameters to learn the distribution of healthy-brain. Another interesting observation that can be made from the results is that the models performed 21% better on T2w images than T1ce from the BraTS dataset, even while being trained to focus on T1w images (T1ce was tested on models trained on purely T1w images, T2s were tested on models trained on a combination of T1 and T2). This can be due to the fact that the tumours are easier to distinguish in T2 than in T1ce due to the significant intensity difference between tumour and non-tumour tissues. It can also be due to some bias induced by the FSL's segmentation. Moreover, the FSL preprocessing might have also been the reason for the drop in performance while working with undersampled MRIs.

Undersegmentations can be observed in all the examples. This also resulted in a complete disappearance of a small anomaly as can be seen in Fig. 13.9 (row 2). Hence, one immediate point of concern is the size of the anomaly in the input image. As morphological opening would eliminate the prediction if it is too small, the StRegA post-processing steps might not be ideal for detecting small anomalies, e.g. multiple sclerosis. One way to deal with the small anomalies would be to reduce the area threshold. For synthetically generated anomalies, the pipeline performed almost perfectly, as can be seen in Fig. 13.8. Nevertheless, these models were trained in an unsupervised fashion—without any labelled training data. Hence, the requirement of large manually annotated training data can be mitigated. The anomalies detected by such a method can also be further used as weak labels to train weakly-supervised or semi-supervised models—which might improve the final segmentation quality without requiring much manual intervention.

StRegA as an assistive technology might help radiologists notice anomalies that might otherwise be overlooked. Diagnostics by radiologists remain undiscovered in up to 5–10% of cases [176]. The cause of these errors is due to unintentional bias, where radiologists focus more on the centre part of the MRI while neglecting peripheral findings. The ever-increasing number of radiological images for assessment increases the workload in radiology which in turn can result in more missed brain pathologies. The presented method might assist in localising anomalies (as seen in Fig. 13.6 and 13.6) and improve the overall diagnostic accuracy.

13.5 Summary and Outlook

This research addresses the challenge of anomaly detection in an unsupervised manner by presenting an unsupervised anomaly detection system named StRegA: **S**egmen**t**ation **Reg**ularised **A**nomaly, which combines a modified and compact version of the ceVAE model (cceVAE) with pre- and post-processing steps. The proposed method was trained using anomaly-free brain MRIs from two benchmark datasets and was then evaluated for the task of detection of brain tumours. The proposed method was compared against four baselines and outperformed all of them with statistical significance. StRegA achieved Dice scores of 0.531 ± 0.112 and 0.642 ± 0.101 while segmenting tumours, as well as 0.723 ± 0.134 and 0.859 ± 0.112 while segmenting artificial anomalies, from T1w and T2w MRIs, respectively—resulting in 49%, 83%, 5%, and 21% improvements, respectively in the four scenarios, over the best performing baseline model ceVAE, while achieving 2%, 23%, 2%, and 10% improvements over ceVAE with StRegA pre-processing. Undersegmentations were observed in most cases, however, the method could localise the anomalies properly. This work has shown the potential of using this method as part of a decision support system. When tested on undersampled and NCC1701 reconstructed volumes, a significant drop in the performance was observed. Nevertheless, the proposed StRegA pipeline outperformed the baseline ceVAE in all scenarios.

In the future, this method will be evaluated for the tasks of detecting other types of anomalies than brain tumours—to evaluate its generalisability to other pathologies. The current method faces difficulties while detecting small anomalies—this might be addressed by modifying the post-processing steps. Furthermore, techniques other than segmentation to simplify the input to the model will also be evaluated in the near future.

14 DS6: Vessel Segmentation

The original work of this vessel segmentation approach has been published in [pub:11], while some of the initial results were presented in [pub:26]. The code of this project is available publicly on GitHub [code:4].

14.1 Background and Motivation

Small vessels in the brain, such as the Lenticulostriate Arteries (LSA), which supply blood to the basal ganglia [76, 83], are the terminal branches of the arterial vascular tree. Pathology of these small vessels is associated with ageing, dementia and Alzheimer's disease [159, 248]. Segmentation and quantification of these small vessels is a critical step in the study of Cerebral Small Vessel Disease or CSVD [348]. 7-Tesla (7T) Time of Flight MRA is capable to depict such small vessels non-invasively [98, 109]. This research refers to vessels as small when they have an apparent diameter of only one to two voxels, as seen in 7T MRA with a resolution of 300 μm. Manual segmentation of small vessels in 7T MRA is reliable but also time-consuming and laborious. Manual and semi-automated techniques involve manual efforts in parameter tuning or annotating vessels, making it a time-consuming task requiring experience to achieve the required sensitivity to achieve robust detection of small vessels. Annotations created in a semi-automated fashion will be faster and much easier to create, but can be noisy. Fully-automated deep learning techniques might be capable of solving this problem, rendering time-consuming parameter tuning or manual intervention obsolete. However, they typically require large labelled training datasets for supervised learning. The aim of this research is to develop an approach that is capable of effectively segmenting small vessels by employing semi-supervised learning—that is capable of performing vessel segmentation by learning from a small noisy training dataset.

S. Chatterjee, *Deep Learning Innovations in MRI Reconstruction and Analysis*,
https://doi.org/10.1007/978-3-658-50741-1_14

14.1.1 Technical Background

In recent years, deep learning based architectures have been widely used for segmentation tasks [361, 371]. The UNet [186, 194] and Attention UNet [277] have proven to be promising techniques in biomedical image segmentation [231]. These methods can be trained using small-size datasets and have outperformed the traditional methods [186]. Therefore, these architectures are used as baselines for the proposed small vessel segmentation approach.

[247] proposed an improvement of the UNet architecture, by computing the overall loss as the sum of losses at each up-sampling scale of the UNet in its expansion path. This is known as deep supervision [219, 247] or Multi-Scale Supervision or MSS [338]. Backpropagating the losses calculated at different scales encourages the architecture to learn discriminative features at every level [219] and also facilitates more effective gradient flow to earlier stages [247], which enhances the learning of the network in comparison to UNet without deep supervision. The hypothesis behind choosing this is that learning features at different scales will help the model segment vessels of different sizes. Hence the UNet MSS was choosen as the backbone architecture.

[306] proposed an architecture for semi-supervised segmentation, where along with supervised learning, the network was made to segment consistently for a given class of transformation. The method was evaluated on a dataset consisting of both labelled and unlabelled chest X-ray images. The model was trained using a supervised loss from the labelled data, and an unsupervised consistency loss was computed between elastically transformed outputs of a Siamese network [34] made up of UNets. Elastic deformation is frequently used in neural networks as a data augmentation technique [67] as it can represent variations in the image realistically [257], including in the case of vessels. Since this paper dealt with a small dataset using UNet, this was considered as a reference and this strategy was leveraged to deal with the small dataset at hand. This research explores the possibility of creating a semi-supervised learning approach extending the earlier mentioned work, which is capable of learning from a small training set of 7T Time of Flight (ToF)-Magnetic Resonance Angiography (MRA) volumes—even when the training labels are noisy. Furthurmore, the method is to be evaluated on a manually annotated Region of Interest (ROI)—to evaluate whether or not the method is capable of overcoming the noise present in the training data. Finally, the method is to be evaluated on 1.5T and 3T Time of Flight (ToF)-Magnetic Resonance Angiography (MRA)—to evaluate if the 7T training is transferable to the lower field strengths.

14.2 Proposed Methodology

The backbone network is a modified version of UNet MSS [338]. The original UNet MSS [338] performs five times downsampling using strided convolution and transposed convolution for upsampling, whereas this modified version performs four times downsampling using Max-Pool and performs upsampling using interpolation. Moreover, the original UNet MSS uses Instance Normalisation and LeakyReLU in its convolution blocks; but this modified UNet MSS uses Batch Normalisation and ReLU. In addition to the final segmentation output, the output of two of the following layers (different scales) is also considered for loss calculation. These additional outputs were interpolated using nearest-neighbour to the size of the segmentation mask before loss calculation. Figure 14.1 portrays the modified version of the UNet MSS used in this research. Eq. 14.1 represents the loss function of the proposed UNet MSS. The m refers to the total up-sampling scales in the UNet, ℓ_{scale} is the loss at each up-sampling scale, α_i is the weight assigned to loss at a specific up-sampling level and network parameter θ.

$$\mathscr{L}_{MSS}(\theta) = \frac{1}{\sum_{i=1}^{m} \alpha_i} \sum_{i=1}^{m} \alpha_i \ell_{scale}^{i}(\theta) \tag{14.1}$$

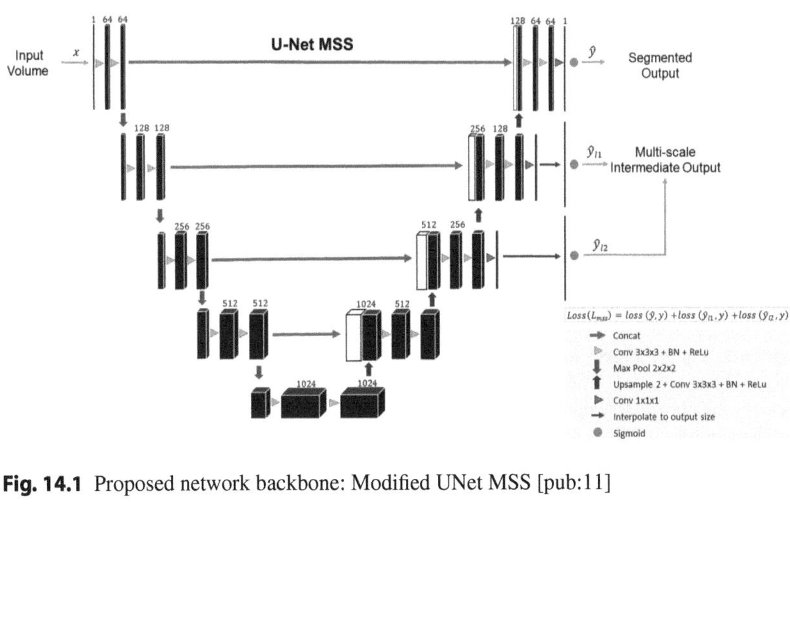

Fig. 14.1 Proposed network backbone: Modified UNet MSS [pub:11]

To enable the model to learn consistency under elastic deformations [67], the work by [306] was re-implemented and extended by using the modified UNet MSS in place of UNet, as shown in Fig. 14.2. Let $\mathscr{X}$ be the set of input volumes while $\mathscr{Y}$ is the set of corresponding labels and $\mathscr{T}$ be the set of elastic transformations. The proposed network uses a Siamese architecture to learn from the original data and deformed data using its two identical branches. The first branch is fed with the tuple (x,y), where $x \sim \mathscr{X}$, $y \sim \mathscr{Y}$, while the second branch is fed with the elastically transformed volume and label (t(x),t(y)), where $t \sim \mathscr{T}$. These tuples are passed through the UNet MSS to derive segmentation outputs $\hat{y}_1$ and $\hat{y}_2$ respectively. These outputs are compared with the corresponding labels to derive the supervised loss depicted in Eq. 14.2. Furthermore, the $\hat{y}_1$ is elastically transformed by t $\hat{y}_1$ to $\bar{y}$. Now the $\bar{y}$ is compared with $\hat{y}_2$ for computing the consistency loss in self-supervised manner as shown in Eq. 14.3. The network was trained to find the optimal value of θ that minimises the overall loss defined by the sum of the supervised and the consistency loss.

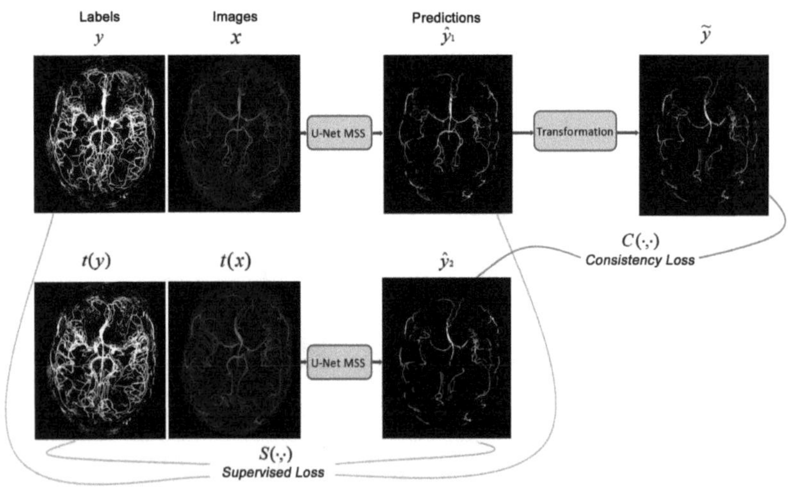

Fig. 14.2 The proposed network is based on Siamese architecture, using modified UNet MSS as the backbone. The original input is fed to the first branch while the second branch receives the elastically deformed version of the input, both branches use the same UNet MSS for segmentation. The segmentation output of the first branch is elastically deformed using a differentiable transformation layer and is then compared with the output of the second branch for consistency. (Images shown here are only representative and do not portray real results) [pub:11]

$$\mathscr{L}^{\tau}_{Sup}(\theta) = \frac{1}{\mid \mathscr{X} \mid} \sum_{x \in \mathscr{X}, t \in \tau} \mathscr{L}_{MSS}(x, y, \theta) + \mathscr{L}_{MSS}(t(x), t(y), \theta) \tag{14.2}$$

$$\mathscr{L}^{\tau}_{Cons}(\theta) = \frac{1}{\mid \mathscr{X} \mid} \sum_{x \in \mathscr{X}} \mathbb{E}_{t \sim \tau}[\mathscr{C}(t(f(x; \theta), f(t(x); \theta))] \tag{14.3}$$

It can be hypothesised that this proposed network can learn consistency under elastic transformations and will be able to learn from the available small dataset with noisy labels. The supervised loss $\mathscr{L}^{\tau}_{Sup}(\theta)$ trains the network to segment the images using the ground-truth labels, and the consistency loss $\mathscr{L}^{\tau}_{Cons}(\theta)$ ensures that the network is equivariant to the elastic transformation. This transformation can also be used as data augmentation for the models, but this was not applied in this research as it was already shown by [306] that the deformation consistency performs better than using it as data augmentation.

Elastic deformation of a given volume can be obtained by applying a displacement field to the volume, as shown in Eq. 14.4, where $\Delta x_{x,y,z}$, $\Delta y_{x,y,z}$ and $\Delta z_{x,y,z}$ are the displacement in x, y and z direction respectively; g denotes the grey values of each voxel of the original image, whereas g′ denotes the grey values of the deformed image [210]. The displacement fields were randomly generated by initialising a grid with random values within a specified limit, followed by interpolating the displacement at each voxel from the grid using cubic B-spline [56].

$$g'(x, y, z) = g\left(x + \Delta x_{x,y,z}, y + \Delta y_{x,y,z}, z + \Delta z_{x,y,z}\right) \tag{14.4}$$

14.3 Datasets and Labels

To be able to validate the proposed approach, a high-resolution Time of Flight (ToF)-Magnetic Resonance Angiography (MRA) dataset was needed. Images obtained with high spatial resolution on a 7T MR scanner contain more small vessels in comparison to a 3T MR scanner [205]. Hence, the dataset by [276] was chosen, which comprises of 3D Time of Flight (ToF)-Magnetic Resonance Angiography (MRA) of 11 subjects imaged at 7T. The data had an isotropic resolution of 300 µm and were acquired with prospective motion correction to prevent image blurring, hence preventing the loss of small vessels. For this dataset, no ground truth segmentation was available. The labels for these Magnetic Resonance Angiography (MRA) images were created semi-automatically using Ilastik [305] by a computer scientist (see Acknowledgement), who was initially trained and later, the segmentation quality was checked visually by a domain expert^{HM}. Ilastik is a widely used tool for pixel classification, which provides the capability to create custom pixel classification workflows. The annotator can visualise the volume and then can annotate the small vessels using points or scribbles by adjusting the pointer diameter. Ilastik uses these annotations and pixel features of the image to segment small vessels in the rest of the volume using the Random Forest algorithm. Further, the user can go through the rest of the slices and manually modify the segmentation. Alternatively, Ilastik can also be used for fully manual segmentation. For the label creation task on the dataset, Ilastik's semi-automatic option with manual interaction was used.

The quality of the labelled images can be visually assessed by comparing the Maximum Intensity Projection (MIP) of the original volume with the label, which is shown in Fig. 14.3. The label generated is not entirely perfect, as it is noisy, contains gaps in certain regions, and does not capture all small vessels, which can be attributed to the limitations of the tool. For a more accurate analysis of the method, the fully manual segmentation option of Ilastik was used to manually segment a small Region of Interest (ROI) of dimension 64^3, as can be seen in Fig. 14.7. For training and evaluation of the network, the dataset was randomly divided into training, validation and test sets in the ratio of 6:2:3. The performance of the methods was evaluated using 3-fold cross-validation. For further evaluation of the generalisation capabilities of the proposed network, publicly available Magnetic Resonance Angiography (MRA) images from the IXI dataset [url:12] were used. The data were collected on two different field strengths, 1.5T and 3T. The labels for these images were created using Ilastik in a semi-automated manner, as mentioned earlier.

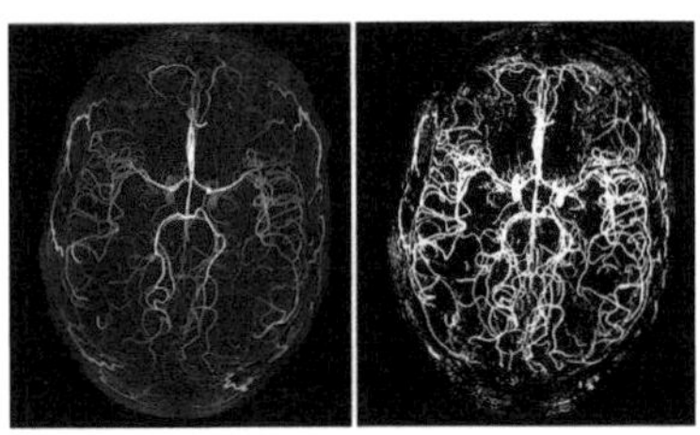

Fig. 14.3 Maximum Intensity Projection (MIP) of the actual volume and the segmentation labels created using Ilastik [pub:11]

14.4 Experimental Setup

The Siamese network used in this research consists of a UNet MSS [247] as the backbone architecture, as explained in the previous chapter. Every Magnetic Resonance Angiography (MRA) image volume was converted to 3D patches with dimensions 64^3, and with a stride of 32,32,16 across sagittal (width), coronal (height) and axial (depth), respectively, to get overlapped samples—to allow the network to learn inter-patch continuity. Having six image volumes for training and two volumes for validation, this configuration created 20,412 patches for training and 6,804 for validation. For additional trainings to judge the effect of the training set size on the models, three additional training sets with one, two and four volumes were also created. For the training of the deep learning models, 8000 patches were selected at random in each epoch and fixed 6000 patches for validation—to reduce the number of iterations per epoch, reducing the required training time. For evaluation, the three test volumes were converted to non-overlapping patches of size 64^3 using strides of 64 across all dimensions, this resulted in 1,012 patches. For the calculation of the supervised loss and the consistency loss, the Focal Tversky Loss or FTL [300] as in Eq. 4.26 was used and was optimised during training using the Adam optimiser [168] with a learning rate of 0.01. The range of γ is between 1 and 3. The Tversky Index (TI) was calculated using the Eq. 4.24, and α and β of the TI were chosen in this research after experimentation to be 0.7 and 0.75, respectively.

The implementation of the deep learning models was done using PyTorch [327]. A differentiable version of the elastic transformation was implemented by extending the random elastic transformation of TorchIO [394] by replacing its SimpleITK-based [151] implementation of the transformation with a modified version of the kernel transformation (using 3D B-spline kernel) of AirLab [283]. The number of control points along each dimension of the grid (see Fig. 14.2) was chosen randomly to be five, six or seven, to introduce different levels of deformation. The maximum displacement was set to 0.02, and two borders were locked. Elastic transformations, along with the selection of its number of control points, were performed randomly for each patch for each training iteration. In this manner, the network encountered differently deformed patches with different levels of deformations during the training process. The code of this implementation can be obtained from GitHub [code:4].

As all the models (proposed method and the baselines) converged between 20 and 30 epochs, the trainings were performed for 50 epochs with mixed precision [236, url:1] on Nvidia Tesla V100-SXM2-32GB GPU. Using Mixed precision, it was possible to train with a batch size of 20, in comparison to a batch size of eight with

Full precision. The lowest validation loss was noted, and the corresponding epoch's model weights were saved. These trained models were then made deformation-aware by further training for 50 more epochs with the proposed Siamese architecture.

14.5 Evaluation

In this section, the performance of the proposed method is compared against that of other methods. Two different non-deep learning (Non-DL) based methods were used for comparison—Frangi filter [43], which is one of the most widely used methods for vessel segmentation and the MultiScale Frangi Diffusion Filter (MSFDF) pipeline [254], which is one of the state-of-the-art methods for vessel segmentation for Susceptibility Weighted Imaging (SWI) and Time of Flight (ToF) image volumes. For both of these methods, the volumes were preprocessed with N4 bias field correction [119] to remove intensity inhomogeneities, which is a common effect in high field MR images resulting in spatial intensity variations [88]. Multiple experiments were performed with the manual tuning of the parameters of the Frangi filter to obtain a suitable parameter set. Finally, a three-scale (1,2,3) Frangi filter with a γ of 0.1 (Frangi correction constant) and a final threshold of 0.01 was chosen for this dataset and was applied on pre-normalised images (between 0–1). For the MSFDF pipeline, the official implementation [url:7] was used with all the default parameters as provided, except for the Otsu offset, which was set to 0.0 (default 0.2) to reduce noise contamination in the final segmentation. Besides the bias-field corrected image volumes, the method required a brain mask. To that end, the brain extraction tool or BET [63, 77] of the FSL [138] was used (fractional threshold set to 0.1).

The performance of the proposed model was compared against two established deep learning based baseline models—UNet [194], for being one of the most popular methods for segmentation and Attention UNet [277], which was proposed as an upgrade of UNet. The performance of the proposed method was compared with and without the deformation consistency; the Siamese network UNet MSS + Deformation (proposed method) and the modified UNet MSS (proposed backbone), respectively. The proposed method was also compared against a re-implemented version UNet + Deformation [306]. This re-implemented version had exactly the same setup as the proposed method, except for the network model. No pre-processing, such as bias field correction or brain extraction, was performed on the images before supplying them to the deep learning models, except for normalising the intensities (between 0–1).

Quantitative evaluation of the 3D segmentations was performed using the Sßrensen-Dice coefficient (Eq. 4.38) and the Intersection over Union (IoU, Eq. 4.39), also known as the Jaccard Index. For qualitative analysis, Maximum Intensity Projection (MIP) were computed to provide an overview of the entire 3D volume as a 2D representation.

The first set of evaluations was performed by comparing the results against the 7T Magnetic Resonance Angiography (MRA) dataset labels, which were created semi-automatically with Ilastik (see Sect. 14.3). Then the effect of the training set size on the results was evaluated. Moreover, given that the dataset labels are noisy, evaluations of the models were performed by comparing the results against a manually segmented ground-truth. Lastly, to evaluate the generalisation capabilities of the models, they were employed to segment 1.5T and 3T Magnetic Resonance Angiography (MRA) volumes.

14.5.1 On 7T Magnetic Resonance Angiography (MRA) Test Set

The first set of evaluations was performed for three out of eleven volumes of the 7T Magnetic Resonance Angiography (MRA) dataset with 300 μm isotropic resolution (test set separated from the dataset before training/validation).

As shown in Table 14.1, the Frangi filter and the [254] method did not perform competitively with Deep Learning methods quantitatively. Further, qualitatively it is evident from Fig. 14.4 that both Non-Deep Learning methods failed to detect small vessels in most of the cases. The Frangi filter failed to detect small vessels in columns three and four. In the [254] method, non-vessel regions got segmented as vessels, which is evident from the dominant blue in columns one and two. All deep

Table 14.1 Comparison of metrics for Non-DL and DL methods

Type	Method	Dice	IoU
Non-DL	Frangi Filter	51.81±3.09	35.00±2.85
Non-DL	MSFDF Pipeline	48.35±6.34	32.04±5.55
DL	Attention UNet	76.73±0.22	62.25±0.29
DL	UNet	76.19±0.17	61.54±0.22
DL	**UNet MSS**	79.35±0.35	65.81±0.47
DL	**UNet + Deformation**	79.44±0.89	65.97±1.23
DL	**UNet MSS + Deformation**	80.44±0.83	67.37±1.16

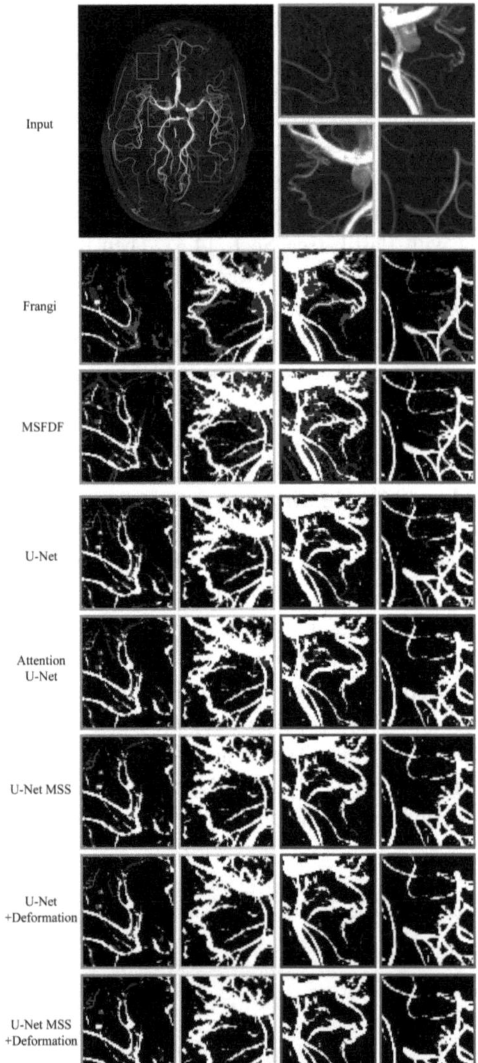

Fig. 14.4 Qualitative comparisons of the results of different non-DL and DL methods with the Maximum Intensity Projection (MIP) of the input volume. Red indicates false negatives and blue indicates false positives [pub:11]

neural networks outperformed traditional techniques by a large margin (51.81±3.09 for Frangi: the best non-DL baseline, 76.19±0.17 for baseline UNet: the worst DL baseline) even with imperfectly labelled data. Careful visual inspection of the MIPs showed that UNet MSS segmented the vessels better than the rest of the models. Further, on applying deformation-aware learning, a considerable improvement (4.27%) was observed with UNet and also with UNet MSS (1.37%). This can also be further noticed in evaluation with different training set sizes conducted on the same models, where deformation-aware learning improved the performance of the models.

14.5.1.1 Statistical Hypothesis Testing

The statistical significance of the improvements observed was analysed by means of an independent two-sample t-test. As the test set of this research contains only three volumes and the population size might be too small to perform a reliable hypothesis testing, an additional set of Dice scores was calculated by dividing the volumes on the x-axis into nine equal non-overlapping (independent) slabs—resulting in slabs with dimensions $80 \times 630 \times 195$. It was observed that deformation-aware learning resulted in significant improvement for both UNet and UNet MSS (p-values 0.004 and 0.013, respectively). Furthermore, the incorporation of multi-scale supervision with UNet resulted in a significant improvement (p-value 0.003). However, UNet MSS with deformation did not show any significant improvement over UNet with deformation (p-value 0.086).

14.5.2 On the Effect of the Training Set Size

Further experiments were performed with varying training set sizes to understand its influence on the performance of the models and to understand how a lower number of volumes can be used for training the models to archive adequate performance. Four different training set sizes were experimented with: Six, Four, Two and One. The sizes of the validation and test set were kept fixed (two and three, respectively). 3-fold cross-validation was used for each training set size.

Table 14.2 shows the results obtained by training the networks with different training set sizes, and Fig. 14.5 portrays the resultant Dice scores using violin plots to observe the data distribution. From the violin plots, it can be observed that the medians (the middle dotted lines inside the violins) are higher in every case for deformation-aware models. Moreover, it can be observed that the models trained with six volumes performed better than the models trained with smaller training sets and the models trained with one volume performed the worst. The result of the models trained with four and two volumes are very similar, but the results of trainings with two volumes are better than four. Given that the dataset labels are imperfect, it could have contributed to these counter-intuitive results, but requires further investigations.

Table 14.2 Quantitative evaluation of the effect of training set sizes on UNet and UNet MSS with and without deformation-aware learning

Model	Training set size	Without deformation		With deformation-aware learning	
		Dice	IoU	Dice	IoU
UNet	1	72.52±0.67	56.91±0.81	74.40±1.25	59.25±1.60
	2	76.99±0.21	62.73±0.36	78.13±0.47	64.28±0.61
	4	75.88±0.72	61.19±0.89	77.97±2.06	63.96±2.82
	6	76.19±0.17	61.54±0.22	79.44±0.89	65.97±1.23
UNet MSS	1	73.11±0.48	57.63±0.57	74.81±1.32	59.78±1.70
	2	75.52±0.78	60.76±1.07	77.84±2.35	63.95±3.19
	4	75.02±1.36	60.10±1.65	77.79±2.05	63.76±2.88
	6	79.35±0.35	65.81±0.47	80.44±0.83	67.37±1.16

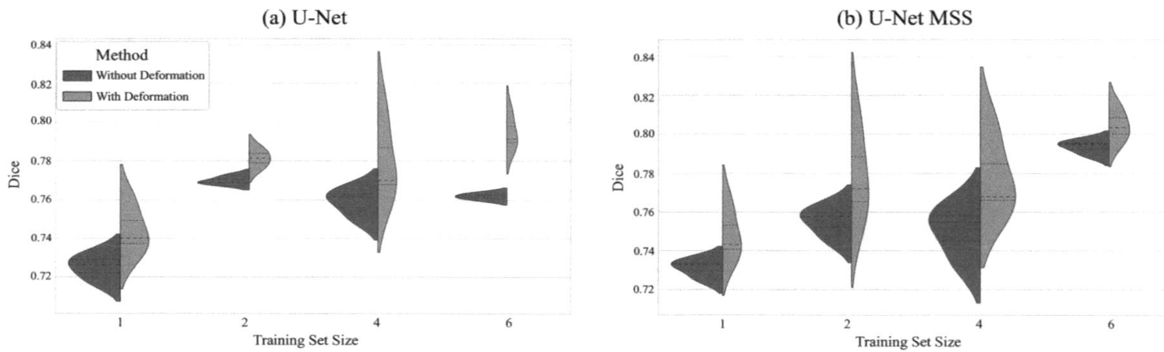

Fig. 14.5 Violin plots showing the quantitative results of the experiments with different training set sizes, for UNet (top) and UNet MSS (bottom) trained with and without deformation-aware learning [pub:11]

Models trained on a single volume still show good segmentation of vessels (74.81±1.32 Dice for UNet MSS + Deformation), which are comparable with the ones trained with six volumes even though typical deep learning models expect to have a large dataset. It can be further observed in Fig. 14.6 that UNet MSS trained with deformation-aware learning could maintain certain vessel continuity better than its counterpart, which was trained without deformation.

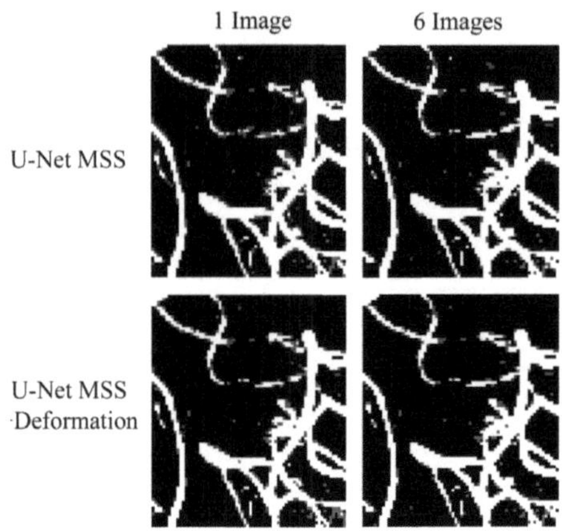

Fig. 14.6 Qualitative comparison of models trained with one volume and six volumes, for UNet MSS with and without deformation-aware learning. Red indicates false negatives and blue indicates false positives [pub:11]

14.5.3 On a Manually Segmented Region of Interest of 7T Magnetic Resonance Angiography (MRA)

The labels created semi-automatically using Ilastik were imperfect (as discussed in Sect. 14.3). Hence, a single 64^3 Region of Interest (ROI) was segmented manually from an input image of the 7T Magnetic Resonance Angiography (MRA) test set, and the performance of the methods was compared against this manually segmented ROI (considered perfect label, hence, ground truth).

Figure 14.7 shows a qualitative analysis comparing the manual segmentation against the prediction from all methods. The comparison shows that the semi-automated segmentation with Ilastik and all the baselines failed to detect the entire

Y-shaped small vessels at the centre of the ROI. On the contrary, the models trained with deformation were able to detect these Y-shaped small vessels and maintain vessel continuity. Further, for quantitative analysis, the Dice scores of the predictions against the manually segmented ROI were calculated (see Table 14.3). It can be observed that the training labels created with Ilastik achieved only a 50.21 Dice while being compared against the manually segmented ROI. Furthermore, UNet MSS resulted in 9.95% better Dice than UNet, and deformation-aware UNet MSS got 18.98% better Dice than the UNet MSS without deformation.

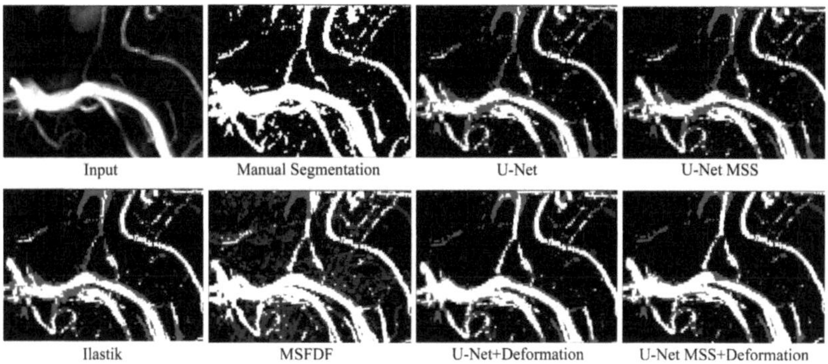

Fig. 14.7 Qualitative evaluation of various methods with the MIP of the input volume. Red colour indicates the false negative while blue represents false positive while being compared against manual segmentation [pub:11]

Table 14.3 Resultant Dice coefficients while comparing different methods against a manually segmented region

Type	Method	Dice
non-DL	MSFDF	52.39
non-DL	Frangi	57.59
Training Labels	Ilastik	50.21
DL	UNet	47.45
DL	UNet MSS	52.17
DL	UNet + deformation	59.81
DL	**UNet MSS + deformation**	**62.07**

14.5.4 On a Publicly Available Dataset, Having Lower Resolution than the Training set: IXI Magnetic Resonance Angiography (MRA)

To evaluate the generalisation capabilities of the model, it was applied to the publicly available IXI Magnetic Resonance Angiography (MRA) dataset. The data were acquired at 1.5T and 3T with a lower image resolution of 450 μm. Due to this reduced resolution, the IXI dataset is less sensitive to small vessels (based on the true vessel lumen and not the apparent vessel diameter in voxel). The resolution of this dataset (450 μm) is 1.5 times lower in each direction in comparison to the training dataset (300 μm) acquired using a 7T scanner. This difference in resolution motivated the experiments with two additional patch sizes (32^3 and 96^3) during inference apart from 64^3, which has been used during training. It can be observed that the patch size 96^3 gave the best performance, which is 1.5 times higher than the training patch size.

Table 14.4 summarises the quantitative performance of the models on 1.5T and 3T images. It shows that, in general, the UNet MSS model with deformation consistency outperforms the model without deformation consistency on both image sets (10.84% and 3.42% improvement with deformation for 1.5T and 3T, respectively, for patch size 96^3).

Table 14.4 Quantitative evaluation using Dice scores for 1.5T and 3T IXI dataset with different patch sizes.

Model	Patch	1.5T		3T	
		Dice	IoU	Dice	IoU
UNet MSS	32	36.97±0.48	22.75±3.69	37.74±0.42	23.31±3.17
	64	39.29±0.36	24.49±2.82	40.34±0.33	25.31±2.58
	96	52.51±0.82	35.88±7.44	48.25±0.16	31.81±1.47
UNet MSS + deformation	32	47.41±0.68	31.25±5.91	43.95±0.42	28.23±3.46
	64	43.94±0.39	28.22±3.31	45.96±0.23	29.86±1.97
	96	58.20±0.85	41.38±8.28	49.90±0.66	33.42±5.99

A qualitative analysis was performed on these image sets to gain further understanding of the segmentation performance(see Fig. 14.8). The Fig. 14.8 (a) and (b) shows the MIP and segmentation mask by UNet MSS and UNet MSS + deform with patch size 64 and 96 on 1.5T and 3T volumes respectively. The image shows that the predicted segmentation mask is in line with the MIP and shows various areas where the prediction is better than the provided imperfect dataset labels. With the increase in the patch size, there is an improvement in the predictions, but the level of noise increases as well.

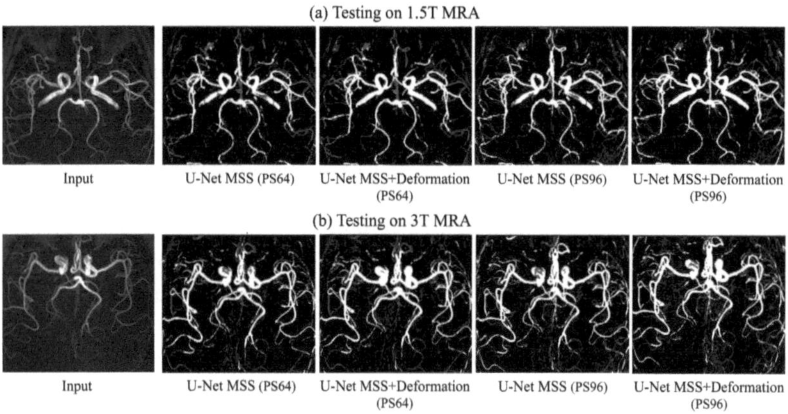

Fig. 14.8 Qualitative evaluation of UNet MSS with and without deformation-aware learning, while segmenting 1.5T and 3T Magnetic Resonance Angiography (MRA) volumes from the IXI dataset. The inference was performed using two different patch sizes: 64 and 96. Red indicates false negatives and blue indicates false positives, while being compared against the dataset labels created with Ilastik [pub:11]

14.5.5 Working with Undersampled MRAs

The initial fold of the cross-validation was used in this analysis. The test volumes were undersampled using MRUnder [url:8, code:7] with six different sampling patterns: five Cartesian sampling patterns—2D Varden 25%, 1D Varden 10%, Uniform 10-steps, 2D Varden 10%, 2D Varden 5% and one radial sampling pattern—golden angle acquisition with 30 spokes, resulting in MR acceleration factors of 4, 10, 10, 10, 20, and 17, respectively. The undersampled MRAs were then reconstructed slice-wise in 2D using the NCC1701 (Sect. 8.1) pipeline. Before that, the ReconResNet model of the NCC1701 was trained for the respective undersampling patterns using the same dataset—following the same splits for training, validation, and testing as DS6. Then the reconstructed 2D slices were stacked together to obtain the final 3D volumes. The undersampled and the reconstructed volumes were then supplied to the trained UNet MSS and the UNet MSS +

deformation models. As the dataset (all three splits—training, validation, testing) used in this research had noisy labels, additional comparisons were performed against a manually annotated test set (while still being trained with noisy training labels).

The results of the proposed UNet MSS and the UNet MSS + deformation models on the fully-sampled, undersampled, and NCC1701 reconstructed datasets were compared against the manually annotated dataset using the Dice scores, shown in Table 15.8. The NCC1701 reconstruction resulted in better Dice scores in all scenarios, while the biggest improvements for UNet MSS + deformation can be observed for Uniform 10-step from 17.06±1.29 to 39.26±1.53 and 16.08±0.77 to 36.90±1.56 for radial golden angle 30 spokes, for acceleration factors 10 and 17, respectively, while the biggest improvement for UNet MSS can be observed for 2D Varden 5% (acceleration factor 20) from 18.72±1.38 to 37.27±3.98. One noticeable fact from these results is that the result of the manually annotated dataset on the fully-sampled volumes is in favour of UNet MSS, but most of the results on the undersampled and reconstructed datasets were in favour of UNet MSS + deformation. This is because the models were trained with imperfect labels; the deformation-aware learning made the predictions close to the training labels, as shown in the earlier results. But on the other hand, the deformation-aware learning made the model more robust against noise, like undersampling artefacts. One surprising fact can be observed from the results of 2D Varden 25% of UNet MSS + deformation, which resulted in better Dice than the fully-sampled.

14.6 Discussion

This work presented a self-supervised approach using deformation-aware learning and has shown its application to small vessel segmentation task while being trained with a small, imperfectly labelled dataset.

In the above sections, it can be observed that the quantitative as well as the qualitative performance of UNet MSS is higher than UNet as seen in Table 14.1 and Fig. 14.4. This is in line with the literature (see also the summary of Sect. 7.3 and Sect. 14.2 Methodology) that the UNet MSS learns discriminative features as the loss is calculated at multiple levels, and this facilitates efficient gradient flow in comparison to UNet. Further, it can be observed that the deformation-based models outperform the models trained without deformation (see Table 14.1). This could be attributed to the deformation-aware training, where the model is made to learn the shape consistency of the vessels when they are deformed. The findings are in line with the findings of the previously published works [272, 306], where they have also shown that making the network transformation equivariant improved the performance. The proposed UNet MSS model with deformation outperforms all the other models in quantitative as well as qualitative aspects. The hypothesis behind the reason for this to be the efficient gradient flow of UNet MSS on top of the consistency learning that assists the model in learning discriminative features.

Given the imperfect annotation, the claim has been further strengthened performing the evaluation against manual segmentation. As seen in Table 14.3 and Fig. 14.7, the models trained with deformation-aware learning outperform the baselines. However, when the models were compared against a complete manually segmented dataset, the models without deformation-aware learning performed better, as can be seen in Table 14.5. However, it can also be observed that the deformation-aware learning made the models more robust against noise in the input volumes.

Table 14.5 Resulting Dice scores for the proposed UNet MSS and the UNet MSS + deformation models while compared with manually annotated test set, working with the fully-sampled, undersampled, and NCC1701 reconstructed MRAs

Undersampling Pattern	Acc. Factor	Type	UNet MSS	UNet MSS + deformation
–	1	Fully	66.55±0.67	64.77±0.80
2D Varden 25%	4	Under	49.90±2.23	52.22±2.09
		Recon	60.32±2.09	67.92±3.01
1D Varden 10%	10	Under	32.71±0.90	35.66±1.65
		Recon	40.17±2.61	42.63±1.22
2D Varden 10%	10	Under	45.37±2.28	43.35±1.59
		Recon	58.80±3.59	58.32±2.22
Uniform 10steps	10	Under	15.73±0.51	17.06±1.29
		Recon	39.74±0.53	39.26±1.53
GA Radial 30sp	17	Under	20.72±1.03	16.08±0.77
		Recon	27.65±1.65	36.90±1.56
2D Varden 5%	20	Under	18.72±1.38	33.03±4.04
		Recon	37.27±3.98	38.34±4.28

The performance of the model trained with and without deformation was analysed on lower resolution image volumes (450 μm isotropic resolution) of 1.5T and 3T where the model results improved with increasing patch size and was able to capture vessels that were not captured in the annotation, but the level of noise in the results increased as well. The network was trained using image volumes with 300 μm isotropic resolution with a patch size of 64, and it performed best for patch size 96 for the 450 μm image volumes. So it can be said that the performance of the model was dependent on the patch size, and a clear relationship can be observed between the image resolution and patch size as when the image resolution decreased by a factor of 1.5, the patch size had to be increased by a factor of 1.5 to get better results: which is counter-intuitive. This demands further in-depth investigations. It is to be noted that the models were trained on 7T Magnetic Resonance Angiography (MRA) without bias field correction.

The experiments performed with the different training set sizes have shown that the trainings with six volumes performed better than the trainings performed with smaller datasets. However, trainings performed with two and four volumes were inconclusive. They resulted in similar Dice scores for the proposed method (77.84±2.35 and 77.79±2.05 Dice respectively for UNet MSS + Deformation), but a training set size of two performed better than four for the rest of the models, which is completely counter-intuitive and further investigations are needed to determine the cause. Trainings performed with one volume yielded poorer, but still acceptable results (74.81±1.32 Dice for the proposed UNet MSS + Deformation). One possible future experiment can be performed by creating one perfect ground-truth for one volume and training the models with just that volume, and then comparing the performance against this research where up to six volumes (imperfectly labelled) were used for training.

The limitation of the proposed approach is the missing continuity of the small vessels in some of the regions, as seen in the last column of Fig. 14.4. This can be attributed to the fact that the model was trained using patches and lacks the context of continuity within the whole volume, which leads the model to predict stray points with a higher intensity as small vessels, which could be avoided if continuity is taken into consideration. Prior knowledge of vessels being cylindrical in nature is disrupted with deformation to a certain extent. Also, since the Dice coefficient is calculated concerning an imperfect labelled image, some of the inherently detected vessels in the proposed model are considered as over-segmentation. Moreover, the shortcoming of the quantitative evaluation of the results is that the method has been evaluated by applying metrics (Dice and IoU) on the whole volume, and has not been evaluated for different vessel diameters specifically—which was not possible to perform due to the presence of noise in the labels.

14.7 Summary and Outlook

The proposed network uses a modified version of UNet Multi-Scale Supervision (MSS) architecture with deformations, which makes it equivariant to elastic deformations. The evaluation of this proposed approach was performed for the task of small vessel segmentation from MR angiograms. The training and testing of the model were performed using a 7T Magnetic Resonance Angiography (MRA) dataset with a resolution of 300 μm. Segmentation of 7T Magnetic Resonance Angiography (MRA) is highly relevant as it can help in the detection of cerebral small vessel disease. The proposed method was compared against non-machine learning methods, such as the Frangi filter and the MSFDF pipeline; as well as against Deep learning methods such as UNet and Attention UNet. The proposed method outperformed all these baseline methods while being trained using a small, imperfectly labelled (noisy) dataset. It could be observed that the models with deformation consistency significantly outperformed the ones without. When the models were trained with only one volume, deformation-aware models could maintain vessel continuity better than their counterparts without deformation-aware learning. As the dataset labels were imperfect, the improvements achieved by the proposed model (UNet MSS + Deformation) can be observed clearer when the results were compared against manually segmented ROI, where UNet MSS achieved 52.17 and 62.07 Dice scores, with and without deformation respectively. This research has shown that the proposed modified UNet MSS with deformation-aware learning outperformed the baseline methods, and deformation-aware learning outperformed models without deformation-awareness. However, while comparing the models against a completely manually annotated dataset, it was observed that the model without deformation-aware learning performed better, while the model with deformation-aware learning was more robust against noise in the input volume. Furthermore, it can be said that the current trained network of the proposed method can be employed to perform vessel segmentation on 7T Magnetic Resonance Angiography (MRA). The network can be fine-tuned and also used for 1.5T and 3T Magnetic Resonance Angiography (MRA) images.

Future work is to explore further generalisability of the model by performing a mixed training with different resolutions (150 μm, 300 μm and 450 μm) and with different field strengths (1.5T, 3T and 7T). Moreover, trainings will be performed using MIP as an additional loss term to try to improve the continuity of vessels. Experiments can also be performed by training the network from scratch by using the network's predicted output as labels as they are better than the original imperfect annotations. Furthermore, introspection on how deformation-aware learning is helping the models learn different shapes of vessels is also an interesting future direction of work.

MICDIR: Image Registration 15

The original work of this image registration approach has been published in [pub:35], while some of the initial results were presented in [pub:35] and [pub:29]. The code of this project is available publicly on GitHub [code:6].

15.1 Background and Motivation

Image registration is the process where two given images are aligned to an identical geometrical coordinate system. Medical image registration tasks can be divided into two categories: intramodal—registering images of the same modality (e.g. MRI-T1 or MRI-T2 or CT) of different subjects into a standard coordinate system, and intermodal—registering images of different modalities (e.g. MRI-T1 to MRI-T2, or MRI to CT) of the same subject. Traditional approaches focus on directly optimising pairs of images based on some criteria or function. More modern approaches include machine learning, deep learning based solutions that try to understand the problem as a whole rather than focusing on individual pairs of images. Such approaches have far-reaching advantages, e.g. they can be very fast, and they can have the capability of accommodating a wide array of deformations. Deep learning based solutions try to map the moving images to fixed images by learning a generalisable representation of the displacement vector field that represents the change in the two images. Several DL-based techniques have been proposed in the past—some of them are discussed in Sect. 7.4. However, they do not explicitly encode global dependencies (i.e. the overall anatomical view of the supplied image) and therefore cannot track large deformations. The aim of this research is to develop a solution using unsupervised deep learning that is capable of registering images for both intermodal and intramodal cases, which can better integrate structural information into the solution and is able to handle small as well as large deformations. For achieving this

S. Chatterjee, *Deep Learning Innovations in MRI Reconstruction and Analysis*,
https://doi.org/10.1007/978-3-658-50741-1_15

goal, this research aims at performing multi-scale image registration—to improve the performance in case of small as well as large deformations, incorporating the Self-Constructing Graph (SCG) with the model—to support the network to learn and encode the minute structural co-relations of the given image-pairs, and cycle consistency—to make the method inverse-consistent (i.e. capable of registering both images to each other—improving the overall image registration quality). The proposed method has been employed for the task of intramodal and intermodal image registration of brain MRIs and has been compared against two non-deep learning and three deep learning baselines. However, as the focus of NCC1701 (the framework which is used as a precursor for the reconstruction of undersampled MRI) is on T1w images, only the intramodal registration results using T1w image are shown here in this chapter. Intermodal results are presented in the original publication [pub:5]. Finally, the research shows the result of an ablation study to understand the contributions of the three aforementioned components.

15.2 Methodology

This section explains the proposed methodology—MICDIR and its nuances. It starts by explaining the different background methods utilised for building the model, then furnishing the details about the proposed method—including the hypotheses behind choosing the different components, the proposed network architecture, training procedure; explains the baseline methods against which the proposed method has been compared, mentions the dataset that has been utilised, and finally elucidates the different evaluation methods.

15.2.1 Proposed Method: MICDIR

Deep learning based deformable image registration techniques such as Voxelmorph [303] have been successful in capturing fine changes and providing smooth deformations. However, Voxelmorph, as well as ICNet [295] and FIRE [312] do not explicitly encode global dependencies (e.g. overall anatomical view of the supplied image in case of medical images) and track large deformations (i.e. large differences between the images). This research tries to address these issues by employing a self-constructing graph network (SCGNet) [356] to encode semantics—which might improve the learning process of the model and help the model to generalise better, multi-scale supervision to be able to work well in case of small as well as large deformations, and cycle consistency (also known as inverse consistency or IC) for making the deformations consistent in both forward (moving-fixed) and inverse (fixed-moving) directions.

15.2.1.1 Hypotheses

The MSCGUNet tries to capture global dependencies using SCGNet [356], wherein encoded CNN features are used to identify semantics. It is hypothesised that these dependencies could help the model to learn relationships even between two distantly located structures in the fixed and moving image and help generate better deformations. SCGNet can find clusters of similar information in the image using latent space features [356]. If the model knows about the landmarks, then it would help improve the deformation field.

[pub:11] uses multi-scale supervised UNet to perform segmentation of vessels, where the multi-scale supervision has been employed to handle a different amount of deformations, where deformations are predicted, and losses are calculated for both original image and a downsampled image. It is hypothesised that deformations on a downsampled image would cover for larger deformations while the deformations on the original image would produce finer deformations. The multi-scale supervision also allows for faster convergence and brings stability to the training as there would be multiple channels through which gradients can flow.

An adapted version of the cycle consistency loss, inspired by [214], was employed to maintain consistent deformations. Cycle consistency, also known as Inverse Consistency (IC), helps to maintain flow correspondence between fixed and moving images, thereby drastically reducing the number of possible deformation fields to achieve the same deformation in both forward (deformation from fixed to moving) and inverse (deformation from moving to fixed) directions.

15.2.1.2 Method Overview

Let f, m be fixed and moving nD images where they belong to domain $\Omega \subset \mathbb{R}^n$. The network model parameterises the function G_θ that calculates the deformation field ϕ given a pair of fixed f and moving m images. The deformation field will have n-channels as it has to represent displacement in each of the n-axes. If it is a 3D image (also known as volume), then the deformation field will have three channels where each channel represents displacement in x, y and z axes. The proposed network model MSCGUNet takes in the fixed f and moving m image along with its downsampled counterparts f_d and m_d and computes deformation fields ϕ and ϕ_d. Both pairs are passed through the network, and corresponding deformation fields are then applied to $m \circ \phi$ and $m_d \circ \phi_d$. Spatial transformers are used to warp the moving image, and a similarity metric is used to compare f with m and f_d with

m_d. Gradient descent is used to minimise the loss and find optimal parameters $\hat{\theta}$. Fig 15.1 provides an overview of the MICDIR method.

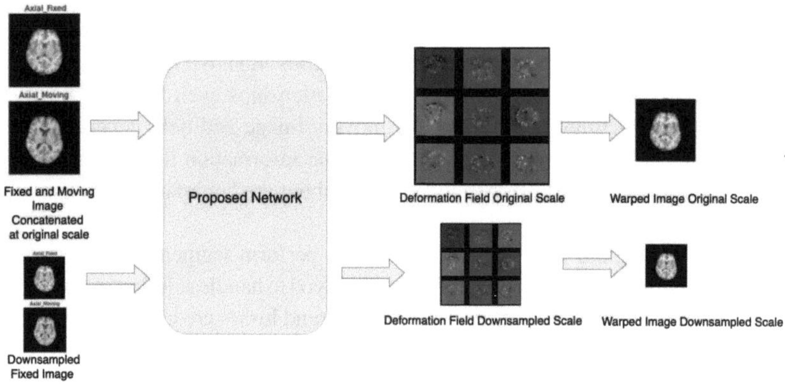

Fig. 15.1 Proposed Method Overview: MICDIR [pub:5]

15.2.1.3 Architecture of MSCGUNet—the Proposed Network

The fixed f and moving m images are concatenated on the channel dimension and supplied together to the network. The proposed model uses a UNet structure to encode and decode the image information. For encoding, it uses convolution with a stride of 2 and LeakyReLU activation to downsample the image and learn relevant features. Encoders near the latent space learn more high-level features, whilst encoders at the beginning learn low-level features that act as an image pyramid. The encoded information is fed into the latent space of MSCGUNet, which is a self-constructing graph neural network (SCGNet). The SCGNet first parameterises the encodings to latent embeddings with mean and standard deviation and derives an undirected graph through an inner product of latent embeddings. The output is then fed into a fully-connected Graph Convolution Network (GCN) which attempts to learn the semantics of the image (e.g. brain anatomy in case of brain MRIs). The GCN output is upsampled through trilinear interpolation and supplied to a CNN decoder, which also takes input from the respective encoder as skip connections. Each decoder block scales up the spatial dimensions of the image and reduces the channels by a factor of 2. Skip connections help improve gradient backpropagation and training convergence. It also supplies the corresponding encoder information that provides location information that would be missing from the decoder. Fixed image f and f_d are concatenated to the UNet output to maintain the contrast of the fixed image [308]. Finally, deformation fields are smoothed by convolution layers which help learn deformation at finer scales. Figure 15.2 illustrates the model architecture.

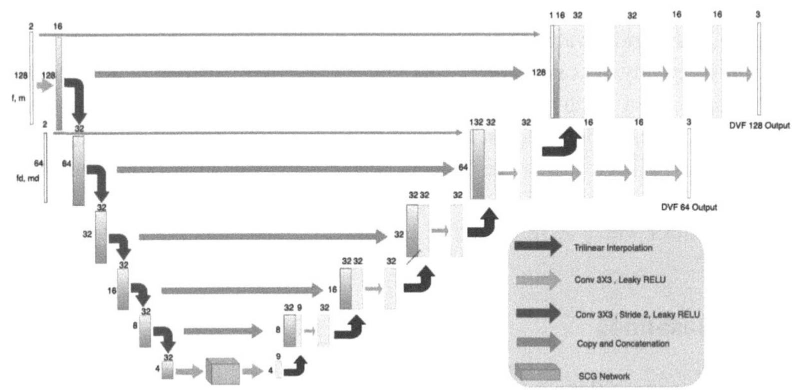

Fig. 15.2 Architecture of MSCGUNet: **M**ulti-scale **UNet** with **S**elf-**C**onstructing **G**raph Latent [pub:5]

15.2.1.4 Training Procedure

Fixed f and moving m images, along with their downsampled counterparts f_d and moving m_d, are used to train the network. f and m are given as input to the MSCGUNet, and it provides deformation fields at both original and downsampled scale, ϕ and ϕ_d, respectively. They are then applied on m and m_d to obtain the warped images w and w_d, and are finally compared against f and f_d to compute the loss. This process is repeated for the same image pair by swapping fixed and moving images—to obtain the inverse consistency loss. Finally, both losses are added up and backpropagated through the network, as shown in Eq. 15.1.

$$\begin{aligned} l_{total} = \alpha(l_{sim_{f->m}} + l_{sim_{m->f}}) + \alpha_d(l_{sim_{f_d->m_d}} + l_{sim_{m_d->f_d}}) \\ +\beta(l_{sm_{f->m}} + l_{sm_{m->f}}) + \beta_d(l_{sm_{f_d->m_d}} + l_{sm_{m_d->f_d}}) \\ +\lambda(l_{scg_{f->m}} + l_{scg_{m->f}}) \end{aligned} \tag{15.1}$$

Normalised Cross-correlation (NCC) and Normalised Mutual Information (NMI) were used as the similarity measure l_{sim}, for intramodal and intermodal registration, respectively. Smoothness loss l_{sm} [17] acts as a regulariser that penalises sharp deformations, and SCG loss l_{scg} helps to learn similar features in the input image pairs. α, α_d, β, β_d, and λ in the equation are hyper-parameters—coefficients to adjust the impact of each component in the final optimisation process. After experimenting with various values, the final values were set to be -1.2, -0.6, 0.5, 0.25, and 5, respectively. It is noteworthy that this set of hyper-parameters was suitable for the experiment setup (dataset—imaging modalities, preprocessing steps used, etc.) and for other setups—different sets of hyper-parameters might result in better performance. In this research, the network was trained using 200 volumes for 1000 epochs using the aforementioned loss (Eq. 15.1), which was optimised using the Adam optimiser with a learning rate of $3e-4$.

15.2.2 Baselines

The performance of the proposed model was compared against a non-deep learning based baseline, a direct optimisation technique, and three deep learning based baselines. Those baselines have been elucidated in 7.4, and their implementation configurations are explained here.

15.2.2.1 Non-Deep Learning Baseline: ANTs

Advanced Normalization Tools (ANTs) [100] is regarded as one of the state-of-the-art medical image registration and segmentation toolkit and is considered to be a gold standard in comparative studies. ANTs, a non-deep learning framework, offers several types of linear and non-linear transforms and different choices of optimisation metrics. The typical optimisation metrics include mutual information, cross-correlation, mattes and demons. For deformable intramodal registration, symmetric normalisation (SyN) [97] transform with cross-correlation optimisation metric was used, while for intermodal, the same transform was used with mutual information as optimisation metric.

15.2.2.2 Naive Baseline: Direct Optimisation

The previous sections have explained the technique of parameterising a function using large deep learning models to calculate deformation fields. For the direct optimisation idea, the usage of deep learning models is ditched, and the deformation field is directly optimised using gradient descent algorithms without any model parameters. The rationale behind direct optimisation is to experiment on registration performance without any deep learning model or complex mathematical optimisation procedure. This idea is mathematically supported by [55], which describes various techniques based on gradients to estimate optical flow, and [366], which discusses using multi-scale, iterative, numerical approaches and using additional speckle information to solve registration problems.

In these experiments, an optimiser is directly applied to a deformation field of size $3 \times 128 \times 128 \times 128$ for a pair of images, and then the deformation field is optimised using image similarity metrics such as NCC or NMI and smoothness loss. Equation 15.2 explains the loss function used to optimise the deformation field. f and m represents fixed and moving image, $l_{sim_{m->f}}$ represents similarity loss and $l_{sm_{m->f}}$ represents smoothness loss to regularise deformation field. This research used the PyTorch implementations [url:3] of four different optimisers: Adam [168], AdamW [233], RMSProp [136, 148], and SGD [10, 156].

$$l_{total} = \alpha(l_{sim_{m->f}}) + \beta(l_{sm_{m->f}}) \tag{15.2}$$

15.2.2.3 Deep Learning Baseline: ICNet

For preprocessing pipeline 1 intramodal registration, the number of iterations had been set as 50, batch size as 2, and the total number of epochs as 2000. For pipeline 2 intramodal registration, the number of iterations was set as 50, total epochs as 1000, and batch size as 2. For intermodal registration with the dataset being preprocessed with pipeline 1, batch size was set to 2, iterations to 100, and epochs equal to 1000. Hyperparameter (taken from the original paper) for the inverse-consistent constraint was set to 0.05, the anti-folding constraint to 100000 and the smoothness constraint to 0.5. The loss was optimised for training using the Adam optimiser with a learning rate of 0.0005.

15.2.2.4 Deep Learning Baseline: ADMIR

As the focus of the current research is on deformable registration, the affine registration module of ADMIR was not utilised. Only the deformable part of ADMIR was used to perform registration, and consequently, the loss function was adjusted accordingly. The network was trained using the Adam optimiser with a learning rate of 0.0001 for 1000 epochs. Every image in the training set was successively selected as the moving image and concatenated with the fixed image as a whole to feed into the deformable ConvNet in each epoch.

15.2.2.5 Deep Learning Baseline: Voxelmorph

The network was trained with $\lambda = 1$ with a batch size of 5 and was trained for a total of 1000 epochs using the Adam optimiser with a learning rate of 0.0001. For intramodal registration, NCC and for intermodal, NMI was employed as the loss function.

15.2.3 Dataset

The IXI Dataset [url:12] consists of nearly 600 MRIs from healthy subjects. For each subject, the dataset consists of five MR image acquisition protocols, among which T1w and T2w MRIs were used in this research. Intramodal registrations were performed among the different subjects of the T1w images, while the intermodal registrations were performed by taking pairs of T1–T2 MRIs of each subject. The T1 weighted images are with voxel ordering AIL (A–P within I–S within L–R) consisting of 150 slices in each volume, with an in-plane matrix size of 256×256 pixels, having a resolution of $0.9375 \times 0.9375 \times 1.2000 m^3$/voxel. The T2 weighted images are with RPI (R-L within P–A within I–S) voxel ordering, comprising 116 to 130 slices in each volume, with the same matrix size and resolution as the T1 weighted images.

15.2.3.1 Data Preprocessing

Data preprocessing has always been an integral part of machine learning—going hand in hand with the GIGO (Garbage in Garbage out) concept. These techniques may involve the removal or reduction of noise and artefacts and address the intensity differences of images from different scanners using techniques like image filtering, resampling, and intensity normalisation.

Preprocessing techniques employed in this research were aimed at 3D brain MRIs. Two different preprocessing pipelines were created—the first pipeline was used as the main pipeline—the models were trained and tested by preprocessing the dataset using this pipeline. The second pipeline was also applied to the test set—to evaluate the generalisation capabilities of the networks while encountering data preprocessed differently than the training set. These pipelines were roughly based on existing research papers with some modifications/adaptations to improve the registration performance. As this research focuses on deformable registration, affine registration was performed as part of the preprocessing steps.

Pipeline 1: FreeSurfer Preprocessed Dataset The full preprocessing stream of FreeSurfer [48] cortical reconstruction process named *recon-all*, consists of distinct

31 preprocessing stages, divided into sub-directives. Following Voxelmorph [303], this research uses the first sub directive, *autorecon1* which consists of stages as shown in Fig. 15.3 and the stages are described below along with the approximate time required for their executions on an AMD Opteron 64bit 2.5GHz processor.

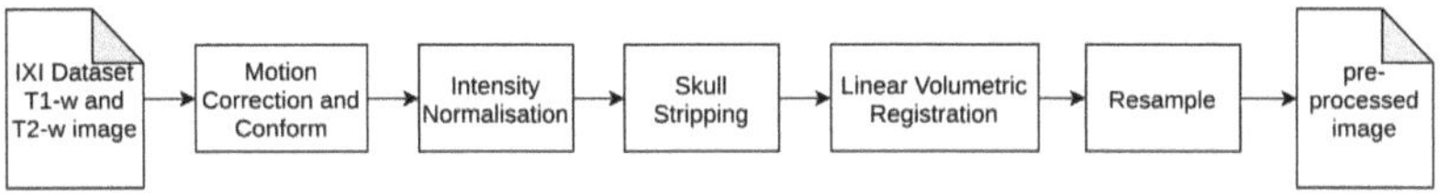

Fig. 15.3 Pre-processing Pipeline 1 [pub:5]

Motion Correction and Conform: The motion correction step takes multiple source volumes as input and corrects for small motions between these volumes, and averages them together. Processing time for the Motion Correction step is estimated to be less than five minutes.

NU Intensity Correction: Variations in scanners or parameters during MR image acquisition often results in intensity non-uniformity in MR data. Hence the non-parametric non-Uniform normalisation (N3) step is incorporated to account for these intensity differences. This step makes relatively few assumptions about the data (non Parametric) and performs four iterations of normalisation. NU intensity correction takes approximately three minutes to process.

Talairach: Talairach is a FreeSurfer script that computes affine transform from the input volume to the MNI305 atlas. The transform computation internally uses the MINC program *mritotal* and these coordinates are used as seed points in subsequent stages of the cortical reconstruction process. Talairach computation is performed approximately within a minute.

Normalisation: The normalisation step scales the intensities of all voxels in such a way that the mean intensity of white matter becomes 110, and this step takes approximately three minutes for a volume.

Skull Strip: Eliminating extra-cranial and non-brain tissues might be an important step to ensure better segmentation of brain regions for many clinical applications and analyses. The skull strip step removes the skull from normalised input volume and generates a brain mask. *mri_watershed* program is then executed. The whole procedure takes under a minute to complete. The *autorecon1* directive of the FreeSurfer cortical reconstruction process end with this stage.

Resample: The affine registered images were further resampled to a $128 \times 128 \times 128$ grid with 1mm isotropic voxels. The preprocessed dataset was split into 200 image pairs for the train and 50 image pairs for the test.

Pipeline 2 The pipeline 2 follows steps similar to the preprocessing pipeline as performed in ICNet [295]. Individual stages of the pipeline are depicted in Fig. 15.4

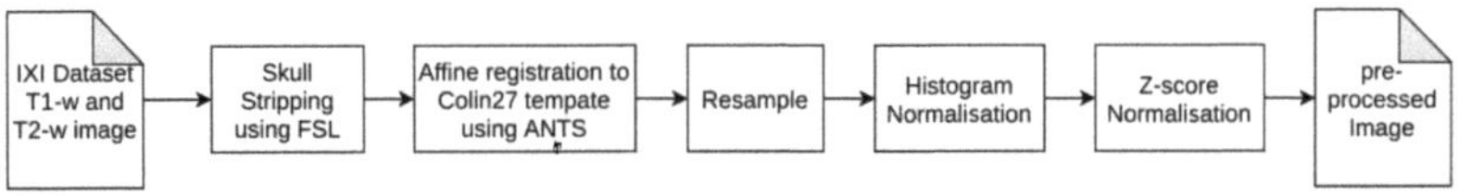

Fig. 15.4 Pre-processing Pipeline 2 [pub:5]

Skull Strip using FSL: Brain Extraction Tool (BET) of FSL [63, 77] is used to extract the brain by removing non-brain tissues from the whole head.

Affine registration to Colin27 template: The skull-stripped input volumes are then affine registered to the Colin27 [44] atlas. Colin27 atlas with high SNR and structure definition is an average of 27 T1 weighted MRI scans of the same individual. This was performed using the Advanced Normalization Tools (ANTs) [100] and took less than a minute to complete.

Resample: After affine registration with Colin27 atlas, the image volumes are of dimensions $181 \times 218 \times 181$. To ensure consistency with the pipeline 1 dataset, they were resampled to a $128 \times 128 \times 128$ grid having isotropic voxels of 1mm. Resampling is performed with the help of nibabel [346].

Histogram and Z-score Normalisation: Intensity normalisation is performed by matching the intensity histogram of each brain MRI to the Colin27 template using the Histogram matching algorithm. Furthermore, Z-score normalisation was performed to ensure the mean intensity of each image is zero and ensure the standard deviation is one.

15.2.4 Evaluation

The evaluation of the different methods was performed by comparing the registered images against the fixed images quantitatively and qualitatively. For quantitative evaluation of both intramodal and intermodal registration, Pearson correlation coefficient, SßrensenDice coefficient (Eq. 4.38), and Kullback-Leibler distance were employed. For intramodal registration, Structural Similarity Index and mean-squared error were additionally used. These two metrics were not used for the intermodal registration as they are intensity-based techniques and would not work for evaluation with different modalities. The statistical significance of the changes in all the metric values observed for the different methods was measured using

2-tailed T-tests with a standard significance level of 0.05. Apart from these quantitative evaluation measures, images were also compared visually.

Pearson Correlation Coefficient: Pearson Correlation Coefficient (PCC) is a statistical measure used to understand the departure of two random variables from independence. The Pearson correlation is calculated as

$$PCC = 1/N - 1\sum_{i=1}^{N}((x_i - \mu_x)/\sigma_x)((y_i - \mu_y)/\sigma_y) \tag{15.3}$$

where x_i and $\_i$ are realisations of random variables X and Y and μ_x and μ_y are the means of X and Y respectively, σ_x and σ_y are standard deviations of X and Y respectively and N is the number of sample pairs.

Dice Coefficient: Dice coefficient (Eq. 4.38) or simply Dice is an overlap-based metric extensively used while evaluating image segmentation methods. To compare the registration methods using the Dice score, first, the fixed image and the registered image were segmented into four segments; namely, cerebrospinal fluid (CSF), white matter (WM), grey matter (GM) and the background, using the ANTs atropos segmentation [122]. Then the Dice scores for individual segments were computed between the fixed image X and registered image Y and were averaged to get the Dice score for the whole volume. It is to be noted that the segmentation accuracy is a major factor affecting the Dice scores as the evaluation relies entirely on the segmented results.

Kullback-Leibler Distance: Kullback-Leibler Distance (KLD) is an evaluation metric used belonging to information theory evaluation measures that gives a statistical distance measure between two joint intensity distributions. Given observed and expected intensity distributions as $p_o(x, y)$ and $p_s(x, y)$ the KLD is computed as

$$KLD = \sum_{x\in\chi}\sum_{y\in\chi} p_o(x, y)log(p_o(x, y)/p_e(x, y)) \tag{15.4}$$

Structural Similarity Index Measure: The SSIM values were calculated between the registered (warped) image and the fixed image for intramodal registration—to evaluate perceptual image similarity. This was calculated using Eq. 4.35.

15.3 Results and Discussion

The results of the proposed method were compared against the baseline methods quantitatively using the metrics discussed in Sect. 15.2.4 and also qualitatively. Initially, different optimisers were compared for direction optimisation (see Sect. 15.2.2.2), and the best performing optimiser was then used for further comparisons. Comparisons have been performed for the task of intramodal registration (T1w MRIs of different subjects) and intermodal registration (pairwise T1–T2 registration of each subject). However, this section only presents the intramodal registration results as the focus of the undersampling pipeline NCC1701 (Sect. 8.1) has been on T1w images.

15.3.1 Direct Optimisation

Four different optimisers were compared using their PyTorch implementations: Adam, AdamW, RMSProp, and SGD. Intramodal results have been reported quantitatively in Table 15.1. In three out of five metrics for Intramodal (while the other two favour ADAM), RMSPROP performed better than the other optimisers. Hence, RMSPROP was chosen as the optimiser for direction optimisation for further comparison against the other methods. The optimisation was performed for 1500 epochs and took 3.47 minutes each for deformable registration on a 12 GB GPU. 1500 was observed to be the best for the optimisation techniques—both quantitatively and qualitatively. Fewer epochs were not enough, while a higher number of epochs resulted in deformation fields that were too smooth or unrealistic. It could mean that constraints should be added to better update the deformation field during the optimisation process (Fig. 15.5).

Table 15.1 Intramodal quantitative evaluation for direct optimisation

Optimiser	IntraModal				
	SSIM	Pearson Correlation	Dice Score	MSE	KL Distance
ADAM	**0.9762± 0.0081**	0.9886± 0.0041	0.7824± 0.0346	0.0010± 0.0004	**0.0578± 0.0224**
ADAMW	0.9753± 0.0082	0.9868± 0.0051	0.7836± 0.0355	0.0010± 0.0004	0.0589± 0.0225
RMSPROP	0.9729± 0.0076	**0.9912± 0.0017**	**0.7926± 0.0195**	**0.0007± 0.0001**	0.0649± 0.0183
SGD	0.9315± 0.0088	0.9584± 0.0073	0.6044± 0.0229	0.0036± 0.0005	0.0665± 0.0223

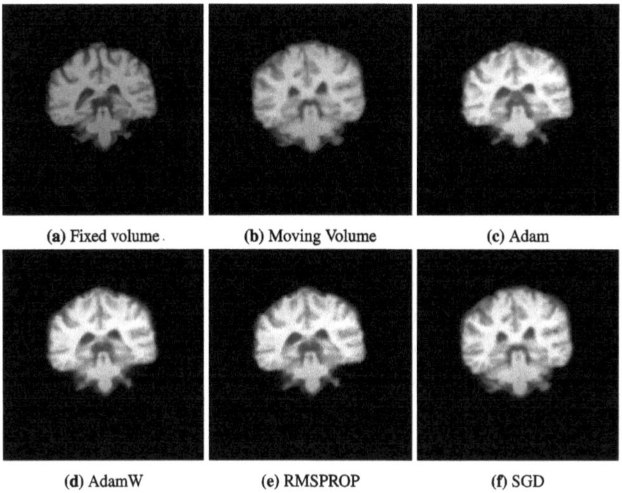

Fig. 15.5 Intramodal T1 weighted registration using direct optimisation, Coronal view [pub:5]

15.3.2 Comparative Results

The performance of the proposed method was compared quantitatively and qualitatively against three deep learning baseline models: ICNet [295], Voxelmorph [303], and ADMIR [370], direct optimisation of the deformation field using the RMSPROP optimiser, and the gold standard—ANTs SyN registration.

Table 15.2 shows the resulting metric values for the different methods for the task of intramodal registration while being preprocessed with pipeline 1 (Sect. 15.2.3.1) —the same preprocessing pipeline as the training dataset. The deep learning methods, with the exception of ICNet, consistently seem to outperform the ANTs SyN registration algorithm when evaluated on all the metrics. Among the experimented deep learning baselines, ADMIR was found to be the best performing one for the intramodal task. Another interesting observation was that the direct optimisation using RMSPROP performed better than the ANTs registration on all metrics while being even better than the ADMIR model on three out of five metrics. The proposed MICDIR outperformed all the baseline models on four out of five metrics with statistical significance, while securing the same score as direct optimisation on the last metric—while obtaining statistically significant improvement over the other baselines. The figures 15.6 and 15.7 provide visual comparisons of the registration results of the different methods with interesting observable areas of the marked for coronal and sagittal orientation, respectively. The qualitative comparisons agree with the quantitative evaluation that the proposed MICDIR resulted in better similarities with the fixed image than the baselines. Finally, the Deformation Vector

Table 15.2 Intramodal quantitative evaluation on dataset preprocessed with pipeline 1

Algorithm	IntraModal				
	SSIM	PCC	Dice Score	MSE	KL Distance
Ants(SyN)	0.9611± 0.0065	0.9792± 0.0042	0.7166± 0.0245	0.0017± 0.0003	0.0734± 0.0825
RMSPROP	0.9729± 0.0076	0.9912± 0.0017	0.7926± 0.0195	**0.0007± 0.0001**	0.0649± 0.0183
ADMIR	0.9775± 0.0064	0.9894± 0.0031	0.7860± 0.0310	0.0009± 0.0002	0.0551± 0.0250
ICNet	0.9271± 0.0129	0.9542± 0.0110	0.5996± 0.0350	0.0040± 0.0010	0.1294± 0.2857
VoxelMorph	0.9711± 0.0052	0.9878± 0.0029	0.7747± 0.0260	0.0010± 0.0002	0.0535± 0.0246
MICDIR	**0.9822± 0.0047**	**0.9916± 0.0020**	**0.8070± 0.0234**	**0.0007± 0.0001**	**0.0506± 0.0222**

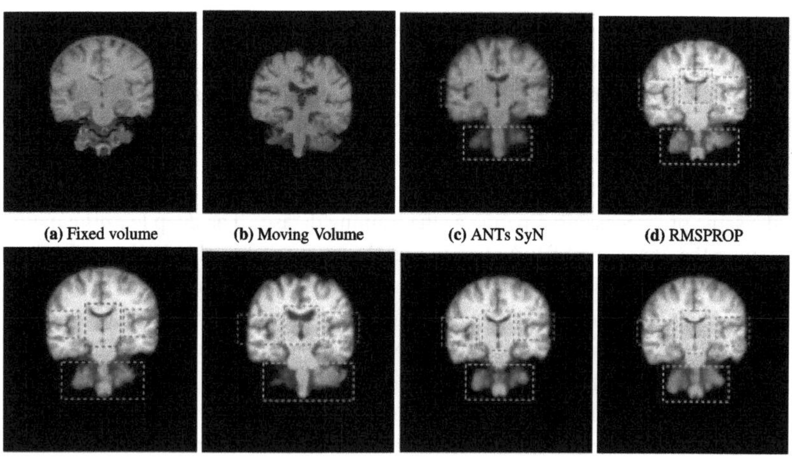

(a) Fixed volume (b) Moving Volume (c) ANTs SyN (d) RMSPROP

(e) ADMIR (f) ICNet (g) Voxelmoprh (h) MICDIR

Fig. 15.6 Intramodal T1 weighted registration, in coronal view [pub:5]

Fields (DVF) generated by the ADMIR and Voxelmoprh are compared qualitatively against the one generated by MICDIR, shown in Fig. 15.8.

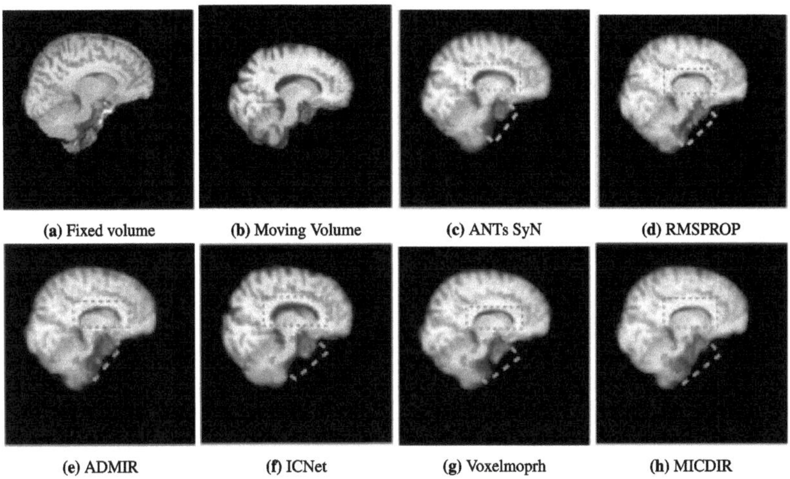

Fig. 15.7 Intramodal T1 weighted registration, in sagittal view [pub:5]

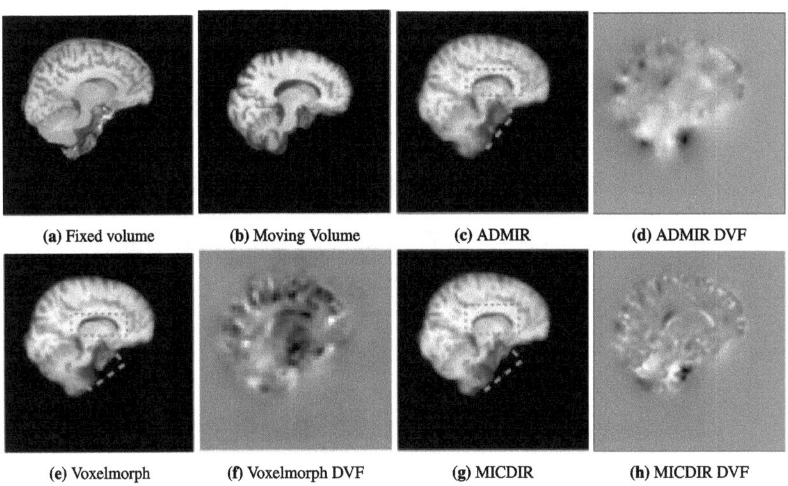

Fig. 15.8 Intramodal T1 weighted registration—comparison of the deformation fields, in sagittal view [pub:5]

The performance of these networks was further evaluated on data preprocessed with pipeline 2 while being trained on data preprocessed with pipeline 1—to compare the generalisation performance of these networks on data preprocessed differently than training. As observed from table 15.3, the MICDIR showed significant improvement in all four metrics when compared to gold standards ANTs SyN and deep learning networks ICNet and Voxelmorph. These results strongly suggest that the MICDIR can also generalise much better than the other deep learning networks and provides better performance even with respect to the ANTs SyN registration algorithm, even when trained with different preprocessing techniques.

Table 15.3 Intramodal quantitative evaluation on dataset preprocessed with pipeline 2

Algorithm	IntraModal			
	SSIM	PCC	Dice Score	MSE
Ants(SyN)	0.8734±0.0326	0.9702±0.0189	0.7103±0.0616	0.0051±0.0037
ICNet	0.7879±0.0576	0.9371±0.0243	0.5598±0.1181	0.0098±0.0062
VoxelMorph	0.8851±0.0260	0.9724±0.0130	0.6999±0.0580	0.0048±0.0030
MICDIR	**0.9025±0.0236**	**0.9761±0.0116**	**0.7445±0.0447**	**0.0041±0.0037**

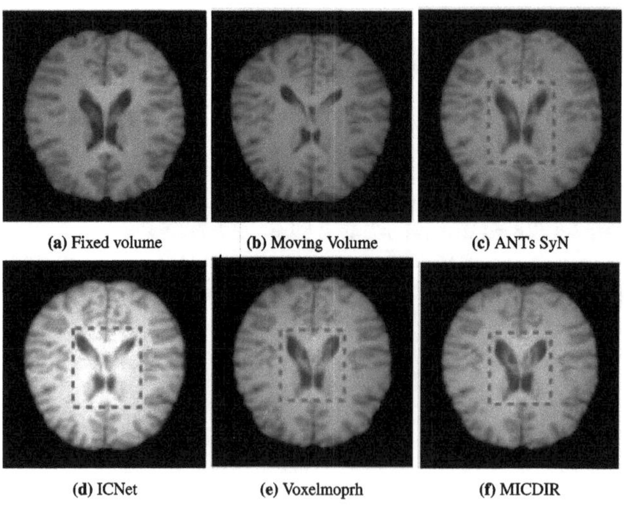

Fig. 15.9 Intramodal T1 weighted registration with data preprocessed using pipeline 2, in axial view [pub:5]

Figures 15.9 and 15.10 show the superiority of the proposed MICDIR over the baselines qualitatively in terms of the registration quality and the DVFs, respectively.

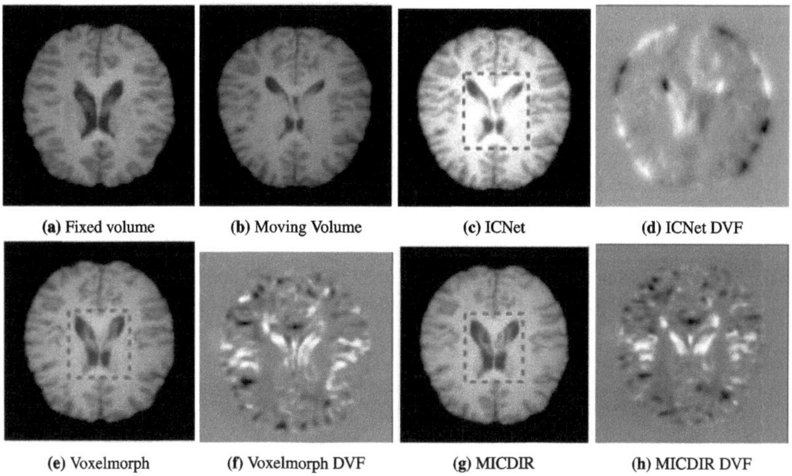

Fig. 15.10 Intramodal T1 weighted registration with data preprocessed using pipeline 2—comparison of the deformation fields, in axial view [pub:5]

15.3.2.1 Template Registration:

One use-case scenario for intramodal registration is to register a template image as the moving image to a given target image. After obtaining the DVF, it can then be applied to the segmentation mask of the template image to transform it into the segmentation mask of the target image—to perform template-based segmentation. This use-case was evaluated by using the Colin27 template [44] as the moving image and registering it against the images from the test set as fixed images. The images were processed using preprocessing pipeline 1. Table 15.4 presents the resultant metric values, and the proposed method outperformed all the baseline methods in four out of five metrics. Figure 15.11 shows an example registration for visual comparison, where the qualitative results corroborate the results of the quantitative evaluation.

Table 15.4 Intramodal quantitative valuation on dataset preprocessed with pipeline 1 registered to Colin27 template

Algorithm	IntraModal				
	SSIM	PCC	Dice Score	MSE	KL Distance
Ants(SyN)	0.9641± 0.0058	0.9818± 0.0028	0.7420± 0.0178	0.0015± 0.0002	0.0567± 0.0184
RMSPROP	0.9774± 0.0048	0.9895± 0.0020	0.7957± 0.0138	0.0009± 0.0002	**0.0445± 0.0121**
ADMIR	0.9803± 0.0034	0.9909± 0.0017	0.8071± 0.0147	0.0007± 0.0001	0.0496± 0.0167
VoxelMorph	0.9779± 0.0042	0.9885± 0.0023	0.7901± 0.0159	0.0009± 0.0002	0.0463± 0.0192
MICDIR	**0.9840± 0.0030**	**0.9923± 0.0014**	**0.8221± 0.0134**	**0.0006± 0.0001**	0.0448± 0.0173

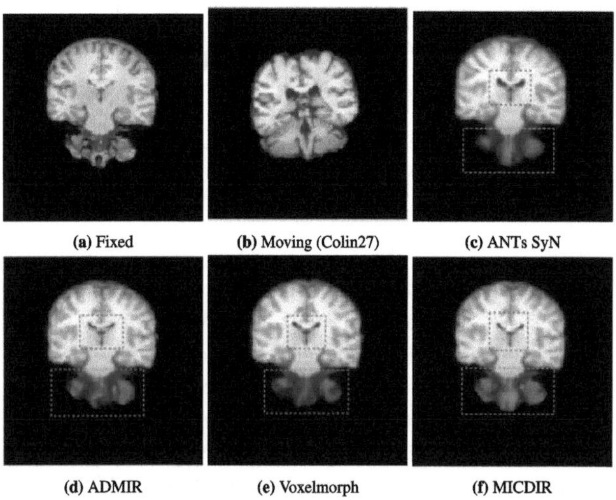

Fig. 15.11 Intramodal T1 weighted registration with Colin27 template, in coronal view

15.3.2.2 Inverse-Consistency:

MICDIR employs the cycle consistency loss, which makes the MSCGUNet inverse-consistent—making it possible for the network to register in both directions: moving to fixed—the intended direction, also used during training all the models (the proposed, as well as the baseline models), as well as fixed to moving. Evaluations were performed to assess the inverse-consistency of the model. For these evaluations, MICDIR and the baselines received the fixed as moving and moving as fixed. It is to be noted that the deep learning models (MICDIR, as well as the baselines) were trained to register moving images to the fixed images, while the baseline deep learning models were not inverse-consistent and might not be a fair comparison as, without inverse-consistency, this swap of fixed-moving images might be difficult to handle for those models. However, the non-deep learning baselines are unaffected by this change as they do not involve any training. In this scenario as well, MICDIR demonstrated statistically significant improvements over all the baselines, quantitative results are presented in Table 15.5, while the qualitative comparison has been shown in Fig. 15.12. Statistical tests were also performed for each method comparing the registration quality of moving to fixed (original registration order, results are in Table 15.2) and fixed to moving (flipped testing, presented in Table 15.5). The proposed MICDIR, as well as the non-deep learning baselines (ANTs and RMSPROP) showed no statistically significant difference, while the differences in the metric values of the deep learning baselines (ADMIR and VoxelMorph) were statistically significant. This evaluation conclusively demonstrates the inverse-consistency of MICDIR.

Table 15.5 Quantitative results for dataset preprocessed with pipeline 1 with fixed and moving images flipped

Algorithm	IntraModal				
	SSIM	PCC	Dice Score	MSE	KL Distance
Ants(SyN)	0.9613± 0.0071	0.9805± 0.0045	0.7361± 0.0247	0.0018± 0.0006	0.0486± 0.0920
RMSPROP	0.9740± 0.0087	0.9876± 0.0050	0.7935± 0.0315	0.0012± 0.0007	0.0417± 0.0128
ADMIR	0.9664± 0.0074	0.9796± 0.0041	0.7822± 0.0289	0.0010± 0.0005	0.0360± 0.0744
VoxelMorph	0.9637± 0.0076	0.9771± 0.0047	0.7670± 0.0313	0.0012± 0.0005	0.0420± 0.0806
MICDIR	**0.9825± 0.0054**	**0.9922± 0.0024**	**0.8101± 0.0212**	**0.0008± 0.0005**	**0.0341± 0.0093**

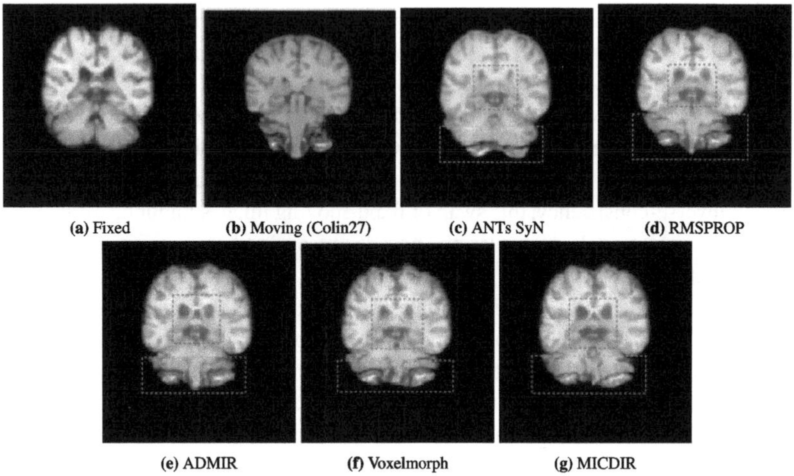

Fig. 15.12 Intramodal T1 weighted registration with fixed and moving images flipped (original moving as the new fixed and the original fixed as the new moving), in coronal view [pub:5]

15.3.3 Ablation Study of MICDIR

As described in detail in Sect. 15.2.1, MICDIR extends the UNET architecture by adding the concepts of encoding global dependencies by employing Self-Constructing Graph (SCG), Multi-Scale Supervision (MSS) for faster convergence and better attention to detail in different scales, and finally Inverse Consistency (IC)—which ensures pair of images are symmetrically deformed towards one another. Hence, it is interesting to evaluate the contribution of the different modifications. An ablation study was conducted by adding each of the modifications separately and observing the changes in the performance in terms of SSIM and Dice. The models were trained and tested on data preprocessed with pipeline 1. It can be observed in Tables 15.6 and 15.7 for intermodal and intramodal, respectively, that each of the modifications (SCG, MSS, IC) improved the performance of the base model—while the configurations involving MSS and IC significantly improved the registration. A combination of MSS and IC further improved the performance over both models with MSS and IC, while the final MICDIR comprising of all thee three modifications (SCG, MSS, IC) resulted in the best performing model.

Table 15.6 Ablation Study: Intramodal registration performance evaluation quantitatively for test set preprocessed with pipeline 1 T1 weighted volumes for different network configurations

Configuration			Evaluation Metric	
SCG	MSS	IC	SSIM	Dice
x	x	x	0.9713±0.0048	0.7738±0.0243
x	x	✓	0.9767±0.0047	0.7866±0.0228
x	✓	x	0.9782±0.0053	0.7902±0.0211
✓	x	x	0.9757±0.0051	0.7760±0.0232
x	✓	✓	0.9802±0.0047	0.7980±0.0233
✓	✓	✓	**0.9822±0.0047**	**0.8070±0.0234**

Table 15.7 Ablation Study: Intramodal registration performance evaluation for Test set preprocessed with pipeline 2 T1 weighted volumes with and without SCG in latent

Configuration			Evaluation Metric	
SCG	MSS	IC	SSIM	Dice
x	x	x	0.8794±0.0495	0.6768±0.1305
✓	x	x	**0.8875±0.0504**	**0.6905±0.1335**

From this ablation study, it was observed that the SCG component marginally improves the scores when compared to the base network. So an additional experiment was performed by comparing the base network and the base network with SCG by testing on data preprocessed with pipeline 2. In this experiment, a significant improvement can be observed with SCG in terms of both SSIM and Dice. Hence, it can be concluded that the SCG component helps improve the network's generalisation capabilities.

15.3.4 Working with Undersampled MRIs

The initial fold of the cross-validation was used in this analysis. The test volumes were undersampled using MRUnder [url:8, code:7] with five different sampling patterns: four Cartesian sampling patterns—1D Varden 30%, 1D Varden 10%, 2D Varden 10%, 2D Varden 5% and one radial sampling pattern—golden angle acquisition with 30 spokes, resulting in MR acceleration factors of 3, 10, 10, 20, and 17, respectively. The undersampled MRIs were then reconstructed slice-wise in

2D using the NCC1701 (Sect. 8.1) pipeline with the help of the already trained ReconResNet models for the respective sampling patterns—trained on T1w MRIs of the IXI dataset [url:12] as a part of the original research of NCC1701. Then the reconstructed 2D slices were stacked together to obtain the final 3D volumes. The undersampled and the reconstructed volumes were then supplied to the trained MSCGUNet model of the MICDIR pipeline.

The results of the proposed MICDIR pipeline on the fully-sampled, undersampled, and NCC1701 reconstructed datasets were compared using the Dice scores, shown in Table 15.8. The performance of the model was considerably affected by the undersamplings. The resulting Dice score of 0.79±0.11 dropped to the minimum of 0.51±0.13 for the acceleration factor of 20, while the NCC1701 improves that to 0.65±0.20. For the second-highest acceleration factor experimented with in this research, i.e. 17, the model resulted in an average Dice of 0.60±0.17, which the NCC1701 improved to 0.70±0.17 with its reconstruction.

Table 15.8 Resulting Dice scores for the task of intramodal registration preprocessed with pipeline 1, while working with the fully-sampled, undersampled, and NCC1701 reconstructed MRIs

Undersampling Pattern	Acc. Factor	Type	MICDIR
–	1	Fully	0.79±0.11
1D Varden 30%	3	Under	0.65±0.14
		Recon	0.74±0.09
1D Varden 10%	10	Under	0.63±0.22
		Recon	0.72±0.08
2D Varden 10%	10	Under	0.66±0.07
		Recon	0.68±0.24
GA Radial 30sp	17	Under	0.60±0.17
		Recon	0.70±0.17
2D Varden 5%	20	Under	0.54±0.13
		Recon	0.65±0.20

15.4 Summary and Outlook

This research proposed MICDIR—multi-scale inverse-consistent deformable image registration using the novel MSCGUNet—a multi-scale UNet with a self-constructing graph at its latent space and showed its application for the task of intramodal image registration. The proposed network allows to explicitly encode global dependencies and semantics present in the given image pairs, i.e. structure and overall view of the anatomy in the supplied image, by incorporating a self-constructing graph network in the latent space of a UNet model. The multi-scale architecture helps track different amounts of deformations, and the inverse consistency constraint ensures that the deformations are consistent. The proposed method was compared against three deep learning based deformable registration techniques, direct optimisation baseline—employing optimisers used in deep learning to optimise the deformation field directly, and against a gold standard registration pipeline. For the task of intramodal registration where T1w MR volumes of different subjects were co-registered to the same coordinate space, the proposed method outperformed all the baseline models, and the improvements observed were statistically significant. Furthermore, an ablation study of the proposed method was performed, which shows how much the different modifications are helping in improving the performance. Experiments have revealed that the MICDIR not only outperforms the baselines, but also generalises better—making it more robust against variations in the input. Finally, while working with undersampled MRIs reconstructed with NCC1701, the method resulted in average Dice scores of 0.70 ± 0.17 and 0.65 ± 0.20 for acceleration factors 17 (radial) and 20 (Cartesian), respectively, whereas securing 0.79 ± 0.11 while working with the fully-sampled volumes. The performance of the proposed method, in general, might further be improved by designing a loss function that targets transformer loss—just like the SCG loss, and also by integrating an affine network preceding the deformable network and training them end-to-end. The deformation field could be optimised better by adding other losses such as anti-folding loss.

Part V
Tackling the Motion

Summary
This pre-final part of this thesis presents the application of the methods proposed in Part III in the field of retrospective motion correction. The first section proposes a prior knowledge based motion correction in the image space using the ReconResNet model and the second section demonstrates the applicability of the hybrid spaces models (Sect. 10.2) while performing deep learning based retrospective motion correction.

Retrospective Motion Correction 16

Motion is yet another source of image degradation in MRI. Initially, several models were evaluated for estimating the level of motion present in a given MRI and also for performing motion correction, which was presented in [pub:47], and [pub:46], [pub:44, pub:2] presented a reference-free technique to predict the amount of motion present in a given MRI by predicting what would have been its SSIM value if that motion would have been artificially induced on a motion-free image. Based on the predicted SSIM value, they were further grouped into different corruption classes.

This chapter deals with this problem using two approaches—correction in the image space with the help of prior knowledge and correction in the hybrid space of image and k-space. The models used in this chapter were proposed initially for undersampled MRI reconstruction in earlier chapters. Apart from attempting to solve the problem of motion correction, this chapter also demonstrates the application generalisability of the proposed models. This work has been performed in collaboration with Alessandro Sciarra.

16.1 Correction in the Image Space

Retrospective Motion Correction (RMC) modifies the MR image data during the reconstruction process [197]. An increasing number of attempts using deep learning techniques have shown that it is possible to remove or reduce motion artefacts retrospectively [260, 278, 282, 317, 320, 328, 372]. It has also been observed that supplying additional prior knowledge, such as image priors, can help improve the performance of deep learning based reconstructions [204, 367]. This work aims to remove motion artefacts from corrupted brain images by improving existing deep learning models, by exploiting available non-corrupted image priors. This work has been published in [pub:27].

S. Chatterjee, *Deep Learning Innovations in MRI Reconstruction and Analysis*, https://doi.org/10.1007/978-3-658-50741-1_16

16.1.1 Methodology

16.1.1.1 Data Preparation

T1, T2, and PD images of 100 subjects (for each—training, testing, and validation) from the publicly available IXI Dataset [url:12] were used in this study. T2-weighted images were artificially corrupted with motion using a modified version of the RandomMotion transformation of TorchIO [394] (v0.17.45). The experiments were performed with 10 simulated movements with a rotation in the range of −1.75 to +1.75 degrees without any translation. In this modified version of the RandomMotion function, in-plane motion corruption was randomly performed in X or Y direction. An example of different levels of motion is given in Fig. 16.1.

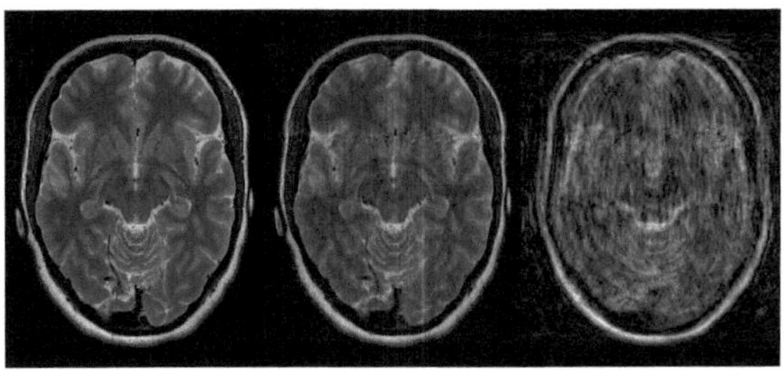

Fig. 16.1 Different levels of motion corruption. From left to right: (a) No corruption, (b) Mildly corrupted, (c) Heavily corrupted

16.1.1.2 Image Priors

Supplying additional images as prior knowledge along with the corrupted image can help improve the performance of deep learning models [204, 367]. In this research, experiments were performed using two different types of image priors—similar slices from different subjects of the same MRI contrast and different MRI contrasts of the same subject.

Similar Slices: During the motion correction, ten similar (same slice position) slices of the same MRI contrast were randomly chosen from different subjects and supplied as prior along with the motion corrupted image. The motivation behind this type of prior is that while performing motion correction on a certain image, images of the same contrast but of different subjects, which are not corrupted by motion, can readily be available. In these experiments, only T2-weighted images from the IXI dataset were used.

Different Contrasts of the Same Subject: Multiple contrasts of the same subject are usually acquired during a clinical routine. If one of those various contrasts is corrupted by motion, then that image can be corrected by using the other contrasts of the same subject as prior. All three available contrasts of the IXI Dataset were co-registered against the T2-weighted images. T2-weighted images were artificially corrupted; T1 and PD images were used as priors during the correction process.

16.1.1.3 Network Architectures

UNet [186] and ReconResNet [pub:8] were used as the baselines for this work. ReconResNet has been introduced in previous a section (Sect. 8.1 for undersampled MRI reconstruction in the image space, while UNet has also been employed for that same task in the past [269]. By employing them for the task of motion correction, the task-generalisability of those networks has also been evaluated here. These baseline networks were then modified such that they could receive priors. Two different methods of supplying priors were tested.

Multi-Channel Network: Each motion corrupted image, and the priors were concatenated on the channel dimension and were supplied to the network as multi-channel input. The baselines were receiving only one channel image as input; whereas for the multi-channel approach, the models received $1 + n\_prior$ channel images as input (Fig. 16.2 and 16.3).

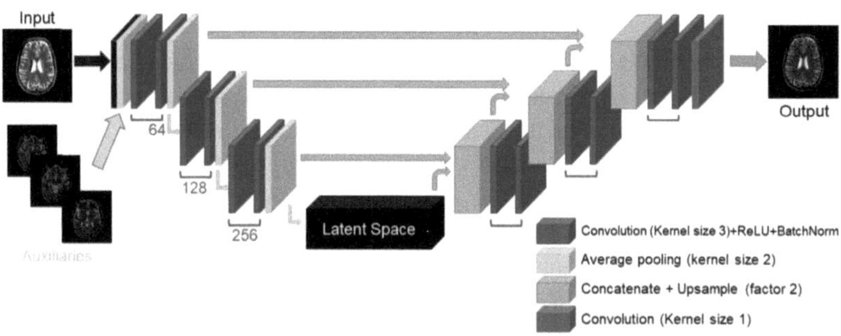

Fig. 16.2 Network architecture of the proposed Multi-channel UNet

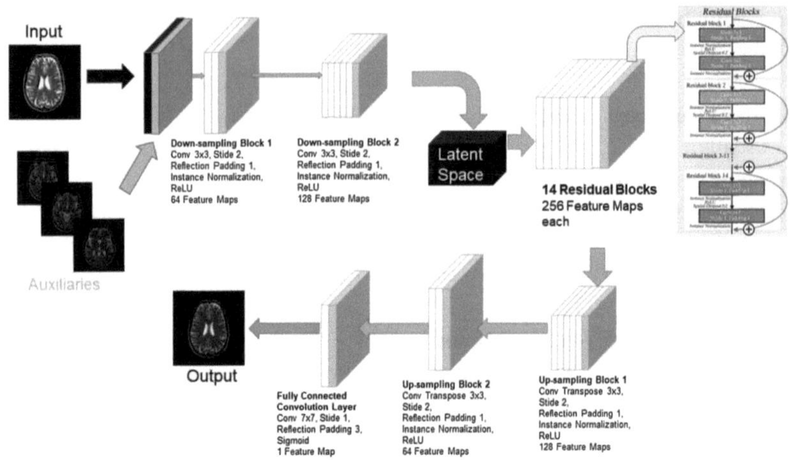

Fig. 16.3 Network architecture of the proposed Multi-channel ReconResNet

Dual-Branch Network: For this approach, modified versions of the baselines were created by adding an additional branch to the baselines for the priors. The main branch was supplied with the motion corrupted image, and the priors were supplied to the auxiliary branch. For the UNet, the auxiliary branch was identical to the contraction path and the latent space of the UNet except for the number of input channels. The skip-connections were supplied only from the main branch of the network, and no skip-connections were supplied from the auxiliary branch. For the ResNet, the auxiliary branch was identical to the downsampling blocks of the ResNet up to the residual blocks, with the only difference being the number of input channels. For both network models, the main branch and the auxiliary branch generated two different latent space representations. These latent representations were combined and forwarded for generating the final output. Two different methods were considered for combining the latent spaces—by simple addition or by concatenation and convolution with a kernel size of one—to generate the final combined latent space. This combined latent space was forwarded to the expansion path of the UNet. In the case of ResNet, this latent representation was forwarded to the residual blocks for further operation (Fig. 16.4 and 16.5).

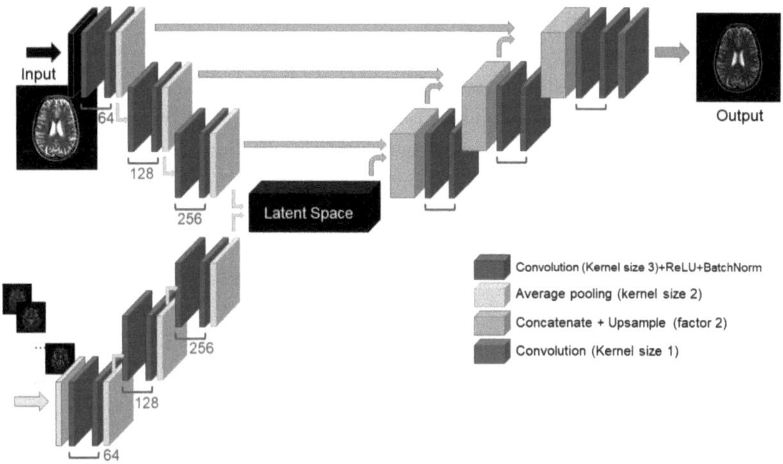

Fig. 16.4 Network architecture of the proposed Dual-Branch UNet

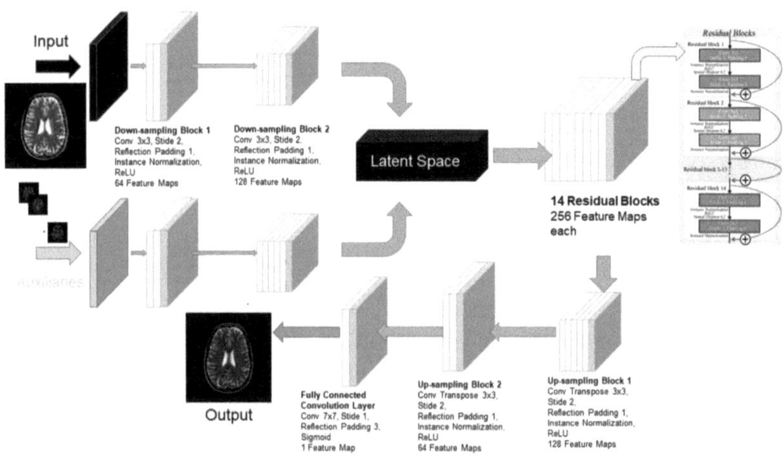

Fig. 16.5 Network architecture of the proposed Dual-Branch ReconResNet

16.1.2 Results and Discussion

Figure 16.6 shows the performance of the different methods based on the SSIM [74] values and Fig. 16.7 shows a representative example result. Between the two different types of priors, it was observed that supplying ten similar slices of the same contrast but of different subjects did not improve the motion correction. Nonetheless, supplying different contrasts of the same subject improved the motion correction significantly for most of the tests carried out. ResNet's performance improved for both types of prior supply methods—multi-channel and dual-branch. For UNet however, only the multi-channel approach has shown significant improvement.

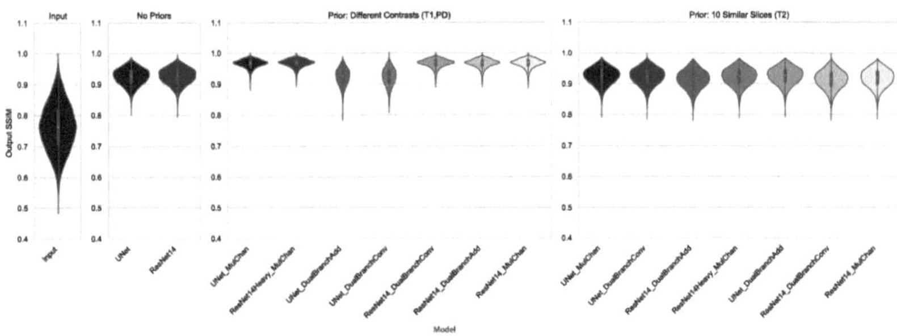

Fig. 16.6 Plots showing the performance of the various methods, based on SSIM [pub:27]

16.1.3 Summary and Outlook

The experiments presented here have shown promising results for supplying other contrasts as prior for deep learning based motion correction (*[RQ:1.1]*). Both initial models were originally used for the task of undersampled MRI reconstruction, and even without any prior, they managed to perform motion correction to a certain extent—with further improved with the help of prior knowledge (*[RQ:1.1]*, (*[RQ:7]*)). Between the two different types of priors experimented here, it was observed that supplying similar slices of the same contrast but of different subjects did not improve the performance, but supplying different contrasts of the same subject did (*[RQ:1]*). Moreover, in terms of the method of supplying prior for ResNet, the multi-channel and the dual-network approach both have shown promising results. However, for UNet only the multi-channel approach has shown improvements over the baseline. The reason for the failure could be attributed to the lack of skip connections from the auxiliary branch, but the skip connections

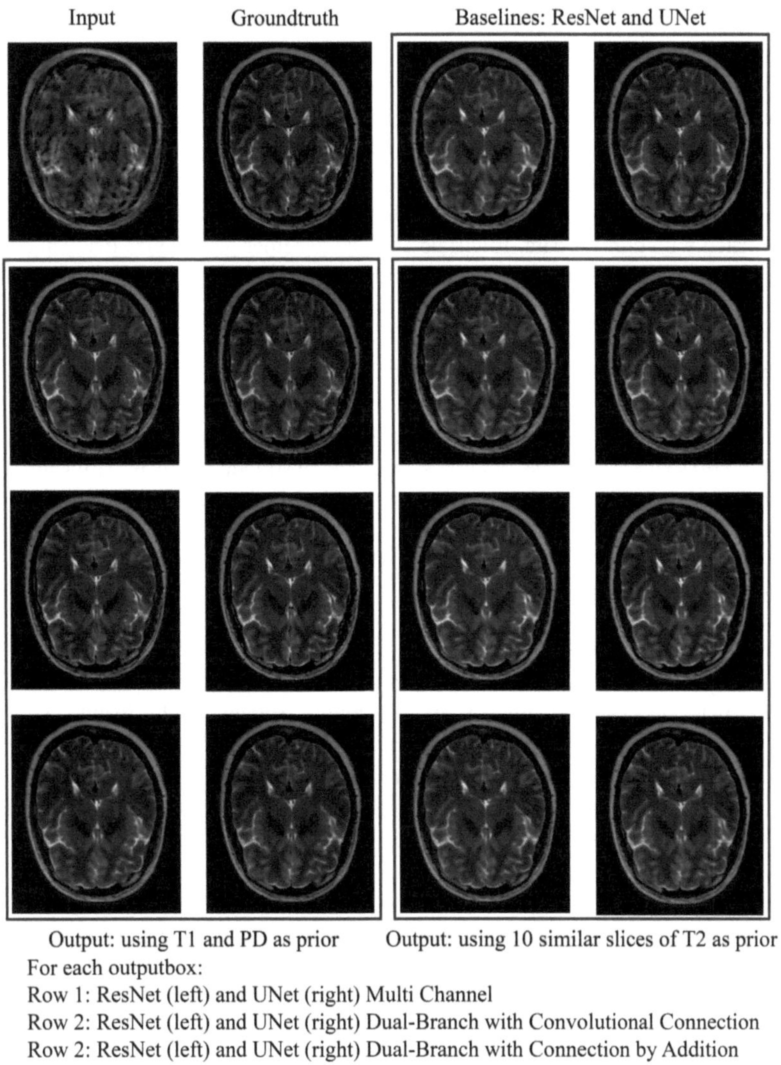

Fig. 16.7 One example slice to show the motion correction performance of the various methods [pub:27]

will make it similar to the multi-channel approach. Further investigations will be performed to determine the cause of the failure and also other strategies for supplying the priors to the UNet. Further experiments will also be performed to check the performance of these approaches when the corrupted volumes and the prior volumes are not co-registered. Supplying ten similar slices as prior did not show any improvement for any of the networks. Further investigations will be performed to determine the reasons and possible solutions. Furthermore, the strategies will be validated with clinical data. It will be assessed whether pathologies that can only be seen in a corrupted contrast are preserved after performing the correction.

16.2 Correction in the Hybrid Space

Section 10.2 has demonstrated the advantage of deep learning models working in the hybrid space for the task of undersampled reconstruction. This research aims at implementing the models proposed in that section for the task of motion correction.

16.2.1 Methodology

16.2.1.1 Data Preparation

The IXI dataset [url:12] was used in this study that contains T1, T2, PD, MRA, and DTI MRIs collected from three different hospitals—Hammersmith Hospital, Guy's Hospital, and Institute of Psychiatry, acquired at 3T and 1.5T field strengths. For this research, T1w volumes from Hammersmith Hospital and Guy's Hospital were utilised in this research—which were divided into train, validation, and test sets, with 200, 135, and 150 volumes. The volumes were artificially corrupted by randomly modifying the k-space lines with rotation. With this, the different lines of the k-space are rotated at different angles—to simulate the real motion corruption as discussed in Sect. 2.5.2. For each slice, a random sigma σ between 1.0 and 2.0 was selected with uniform distribution. Then for each phase-encoding line in the k-space of that image, a random rotation degree ϕ was chosen by multiplying σ randomly with either -1, 0, or $+1$ with each option having equal probabilities of occurrence. Then that phase-encoding line was rotated by $\phi°$. In this manner, every phase-encoding line is rotated differently as it might happen in a real motion corruption. During training, the probability of a slice getting corrupted by this method was set to 80%, and the same slice was provided with different σ in each epoch—ntroducing more randomness in the dataset.

Contrast Augmentation To make the models more robust against against differences in the image contrast during inference than the training set, a series of contrast augmentation techniques was introduced in this research, jointly with [pub:2]. Four different methods constitute this set of contrast augmentation:-

- Gamma adjustment [45]
- Logarithmic adjustment [38]
- Sigmoid adjustment [47]
- Adaptive histogram adjustment [24]

Gamma adjustment was performed by choosing a random floating point gamma value between 0.75 and 1.75, logarithmic adjustment by choosing a random float between -0.5 and $+0.5$ for the gain, sigmoid adjustment by choosing a random floating point cutoff between 0.01 and 0.75 and an integer gain value between 1 and 4, and finally, the adaptive histogram adjustment was performed with 512 bins by randomly choosing kernel size between 25 and 100 with a clip limit of 0.01. Within the set of contrast augmentation techniques, the probability of adaptive histogram adjustment was set to 10%. While for the other three, it was set to 30%. This was performed as the time required to apply adaptive histogram adjustment is considerably more than the other three. One of these four adjustment methods was randomly chosen for each slice during each epoch of the training. The probability of this set being applied on a particular slice was set to 75%.

16.2.1.2 Network Architectures

Fourier-PD and Fourier-PDUNet models, originally proposed in Sect. 10.2 for undersampled reconstruction, were employed here for the task of motion correction in the hybrid space. The primal blocks of these networks are responsible for correcting the artefacts in the image space, while the dual blocks are responsible for correcting the k-space. Both these models use four complex-valued convolution layers separated by $\mathbb{C}$ReLU activation functions inside each of the dual blocks. Each of the primal blocks of Fourier-PD employs four convolutional layers separated by PReLU activation functions, while Fourier-PDUNet employs a UNet [186] model in each of its primal blocks. Inverse Fast Fourier Transform (iFFT) was applied on the corrupted images to obtain their corresponding k-spaces, and they were supplied to the models to obtain the corrected image as output—shown in Fig. 16.8. There was one difference in the Fourier-PDUNet model here for motion correction from the one proposed earlier in Sect. 10.2 for undersampled reconstruction - the number of iterates. Here Fourier-PDUNet uses two iterates, while for undersampled reconstruction, it used ten. This was decided after evaluating the different number of iterates and their influence as for any iterative reconstruction algorithm, an optimal (usually minimal) number of iterations has to be found for the network to converge. The performance of these models were compared against a fully image space model—ReconResNet (Sect. 8.1.1.2 [pub:8]).

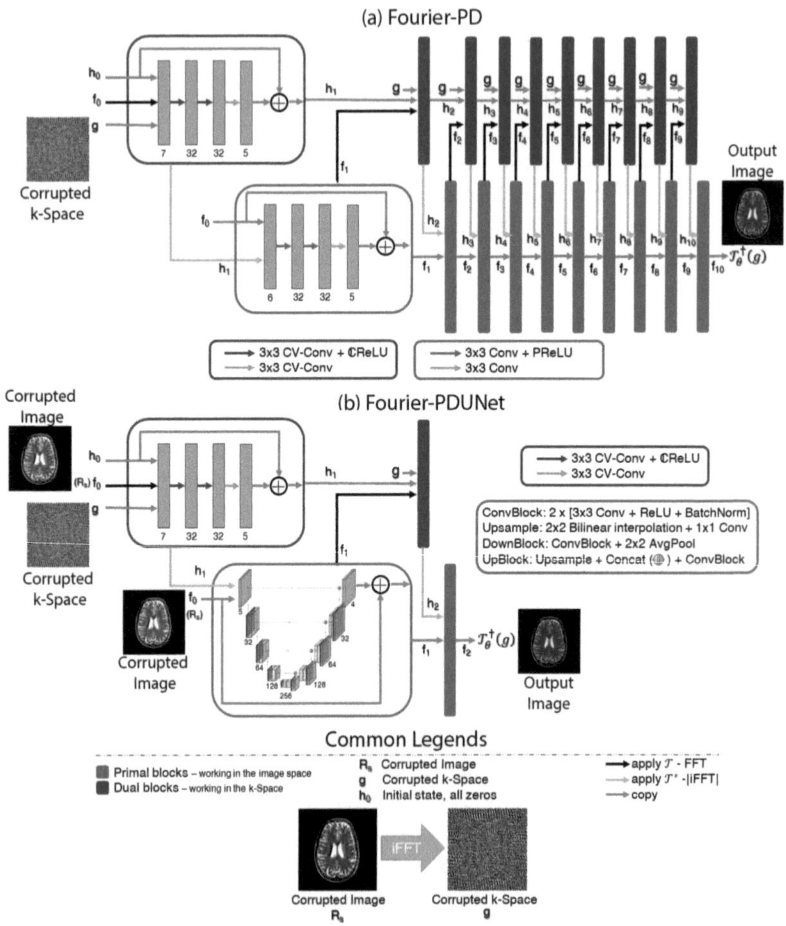

Fig. 16.8 Fourier-PD and Fourier-PDUNet models for Motion Correction

16.2.1.3 Implementation and Training

The models were trained by calculating the loss in the image space by comparing the final output of the model (output of the final primal block in case of the Fourier-PD and Fourier-PDUNet models) against the ground-truth fully-sampled images with the help of L1 loss (Eq. 4.28), and it was minimised using the Adam optimiser (Initial learning rate 0.001, decayed by 10 after every 25 epochs; $\beta_1 = 0.9$, $\beta_2 = 0.999$, $\epsilon = 1e - 09$) with batch size one for 50 epochs. This network was implemented using PyTorch [327] with the help of pytorch-complex [code:10] for the complex-valued operations and was trained using Nvidia RTX 2080Ti GPUs. The codes of these models are available as part of the NCC1701 code on GitHub [code:8].

16.2.2 Results

The results of the models were compared against the ground-truth uncorrupted MRIs using SSIM and NRMSE, for which Fig. 16.9 shows the results. It can be observed that all three models performed very similarly. However, Fourier-PD outperformed both ReconResNet and Fourier-PDUNet in terms of SSIM with statistical significance—which was computed using Mann-Whitney U test [7], while ReconResNet achieved the best NRMSE values. To compare the models further, UQI was employed—results are portrayed in Fig. 16.10. UQI, however, tells a different story. According to UQI, Fourier-PDUNet outperformed both the other models with statistical significance, while Fourier-PD outperformed ReconResNet. Furthermore, the models were compared qualitatively in Fig. 16.11. Qualitative comparison agreed

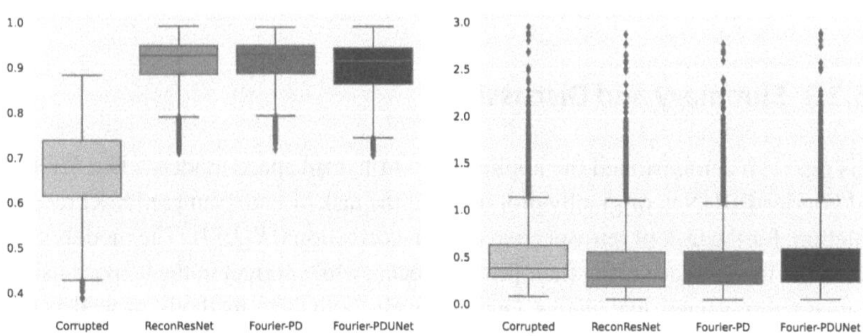

(a) SSIM **(b)** NRMSE

Fig. 16.9 Quantitative comparison of the models using SSIM and NRMSE

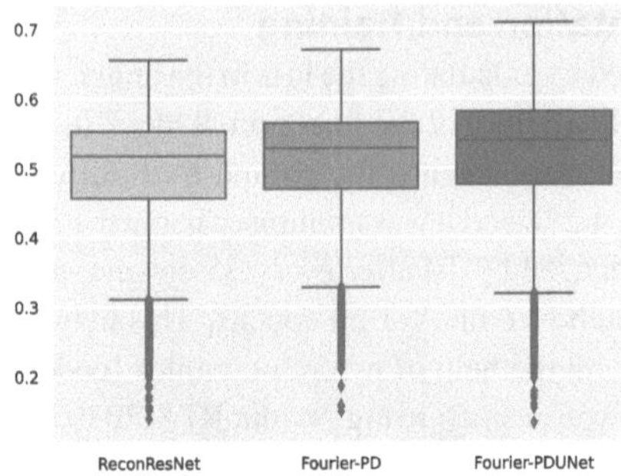

Fig. 16.10 Quantitative comparison of the models using UQI

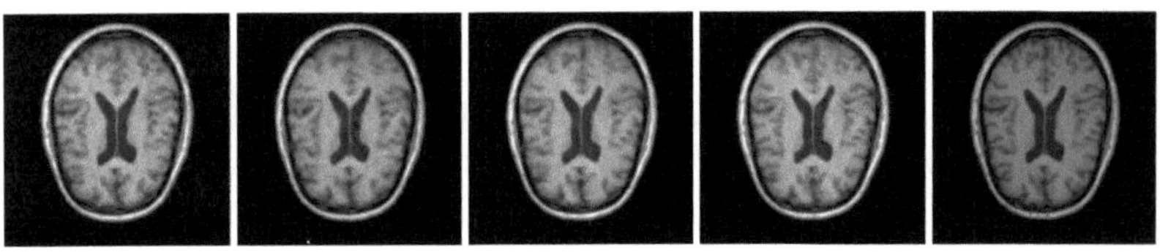

(a) Corrupted Input (b) ReconResNet (c) Fourier-PD (d) Fourier-PDUNet (e) Ground-truth

Fig. 16.11 Qualitative comparison of the models against the ground-truth

more with the UQI than the other metrics. It can be seen clearly that the Fourier-PD and Fourier-PDUNet performed considerably better than the ReconResNet model. However, it is not easy to distinguish the difference between Fourier-PD and Fourier-PDUNet results, except for the fact that the Fourier-PDUNet result is a bit sharper.

16.2.3 Summary and Discussion

This research demonstrated the applicability of hybrid space models - Fourier-PD and Fourier-PDUNet, originally proposed for the task of undersampled MRI reconstruction, for the task of retrospective motion correction (*[RQ:7]*). The models correct in the image space using their primal blocks, while correct in the k-space using their complex-valued dual blocks. Quantitatively, both these methods, as well as the image space model ReconResNet, resulted in similar scores. The highest average SSIM score was 0.912ś0.05, obtained by the Fourier-PD model while improving the corrupted images, with an average SSIM of 0.672ś0.083. Qualitatively, the hybrid space models performed better than the ReconResNet model. This research demonstrated the advantage of using hybrid space for motion correction than correcting solely in the image space—which is in line with the findings presented before from undersampled MRI reconstruction in Sect. 10.2.3.

Part VI
Conclusion and Outlook

Summary
The final concluding part summarises the research presented in this thesis, provides conclusions drawn from the experiments, and finally presents the directions for future research. The initial chapter of this part provides the answers to the research questions presented in the introductory section, obtained by the research presented in this thesis. The second chapter provides a summary of the main technical and application-orientated contributions of this thesis. The final section presents different ideas for future research.

Concluding Remarks

17

This thesis studies various deep learning techniques for the tasks of undersampled MRI reconstruction and retrospective motion correction. The different proposed methods aim to answer the different research questions furnished in Sect. 1.2. This thesis proposed two different methods for artefact reduction in the image space of undersampled MRIs, while proposing another method aimed at the reconstruction of undersampled radial MRIs, which unifies the problem of sparse CT and undersampled radial MRIs. Furthermore, the problem of undersampled MRI reconstruction has been dealt with as a Super-Resolution problem, and three different methods have been proposed—one of which attempts to learn both spatial and temporal relationships to perform undersampled dynamic MRIs. Eventually, the problem of undersampled reconstruction was addressed directly at k-space and hybrid space (mix of image and k-space). Afterwards, this thesis proposed five different automatic deep learning based MRI processing pipelines and evaluated them while working with undersampled and reconstructed data. Finally, the models proposed for undersampled reconstruction (ReconResNet, Fourier-PD, Fourier-PDUNet) were employed for the task of retrospective motion correction. In this chapter, Sect. 17.1 provides concise answers to the research questions, and Sect. 17.2 provides a comprehensive list pointing out the main contributions of this thesis.

17.1 Answers to the Research Questions

This thesis addresses nine research questions and five sub-research questions. The outcome of those is summarised here.

[RQ:1] The term "prior knowledge" can be defined in various ways. Knowledge about the data and using the training data to train the deep learning models can

S. Chatterjee, *Deep Learning Innovations in MRI Reconstruction and Analysis*,
https://doi.org/10.1007/978-3-658-50741-1_17

be considered as prior knowledge. All the methods presented in this thesis are deep learning based and use training data to learn, and they have performed better than other non-deep learning based techniques. Hence, prior knowledge has been exploited successfully in this manner and has improved the results.

This thesis also addresses the point of prior knowledge more explicitly—by incorporating it into the network models. For the task of super-resolution of dynamic MRIs, the static subject-specific prior scans, which are typically acquired prior to any interventional procedures—so-called a planning scan or a preinterventional MRimaging—were considered in this thesis as prior knowledge, to be precise, as image priors. Sect. 9.3.1 fine-tuned trained models using this image prior, and that improved the performance of super-resolution reconstruction. Moreover, Sect. 9.3.2 proposed DDoS-UNet model, which utilises this same prior image (i.e., subject-specific prior scan) as the starting point of the reconstruction process, termed as the "Antipasto" phase, and continues by exploiting the temporal relationship present between the different time points of a dynamic MRI. This temporal information can also be considered as prior knowledge, as the super-resolved previous time point acts as a prior image while reconstructing the current time point. This thesis has shown that both of these image priors have significantly improved the super-resolution reconstruction performance.

Two other types of image priors have been explored for the task of retrospective motion correction—similar slices from different subjects of the same MRI contrast and different MRI contrasts of the same subject. The motivation behind the first one is that while performing motion correction on a certain image, images of the same contrast but of different subjects, which are not corrupted by motion, might readily be available. On the other hand, multiple contrasts of the same subject are usually acquired during clinical routines. If one of those various contrasts is corrupted by motion, then that image might be corrected by using the other contrasts of the same subject as prior—motivation behind the second type of image prior. The experiments shown in Sect. 16.1 have demonstrated that using different contrasts of the same subject as image priors improves the performance of motion correction. However, using the same contrast from different subjects as prior did not show any noticeable improvement ***[RQ:1.1]***.

[RQ:2] Section 8.1 proposed the ReconResNet model, which is a ResNet-based model that uses dropout-regularised residual blocks to perform undersampled MRI reconstruction in the image space. It has outperformed the established model for undersampled MRI reconstruction in the image space—UNet, with statistical significance. Experiments have revealed that the proposed model performed well up to an acceleration factor of 20 for 2D random Variable density (Varden) Cartesian sampling.

[RQ:3] Section 9.2 proposed the ShuffleUNet model, which is inspired by the tight frame UNet that addresses the problems of blurred or smoothed output and chequerboard artefacts by replacing the strided convolutions with lossless pooling layers (pixel unshuffle), by replacing deconvolution (transposed convolution or interpolation) with a sub-pixel convolution operation (pixel shuffle) and by removing the batch normalisation layers. The proposed model outperformed the UNet model with statistical significance for the task of Single-Image Super-Resolution of static diffusion-weighted brain MRIs. Section 9.3.2 proposed the DDoS-UNet model for super-resolution of dynamic MRIs, which also outperformed UNet with statistical significance.

Furthermore, the DDoS-UNet proposed in Sect. 9.3.2, works with dynamic MRIs while using the static planning scan as a prior image. The network is provided with pairs of high and low-resolution MRIs to super-resolve the low-resolution MRI while using the high-resolution image as a prior image. The network learns two different representations: the temporal relationship between $\hat{H}R_{TP(n)}$ and $\hat{H}R_{TP(n-1)}$, the Super-Resolution relationship between $LR_{TP(n)}$ and $\hat{H}R_{TP(n)}$ ***[RQ:3.1]***. The proposed network was trained using an artificially created dynamic dataset using a static benchmark dataset and then tested using dynamic MRIs comprising 25 time points. It was observed that even though the network was trained using a dataset with MRI acquisition parameters very different from the test set, the network was able to super-resolve the given input images better than the model working without incorporating the temporal information as the prior image with statistical significance.

[RQ:4] A comparative study of real-valued CNN and Complex-valued Convolutional Neural Network (CV-CNN) is presented in Sect. 10.1.1 for the task of brain tumour classification and segmentation. It has been demonstrated that the CV-CNN models outperform the CNN models having the same number of features, as well as models having the same number of trainable parameters ***[RQ:4.1]***. Hence, CV-CNNs were employed while working in the k-space. Two models working in the hybrid space—Fourier-PD and Fourier-PDUNet have been proposed in Sect. 10.2. These models extend PD network and PDUNet by employing CV-CNNs as the dual blocks to work in the k-space, while the primal blocks worked in the image space. The proposed model optimises in both image and k-space—attempts to remove artefacts in the image space using the primal blocks and predict the missing frequencies in the k-space using the dual blocks. Both these hybrid space models outperformed the pure image space model ReconResNet. The improvement achieved by Fourier-PDUNet was statistically significant over both Fourier-PD and ReconResNet models ***[RQ:4.2]***.

[RQ:5] The ReconResNet model was not only applied for the task of undersampled Cartesian MRI reconstruction but also for radial MRI. Section 8.1 has also proposed a data consistency step of radial MRIs as a part of the NCC1701 framework and has been shown to successfully reconstruct till an acceleration factor of 17 while reconstructing radial golden-angle samplings.

Section 8.2 presented a unified deep learning framework for the reconstruction of sparse CT and undersampled radial MRIs ***[RQ:5.1]***. This research shows the applicability of using different sinogram upsampling based sparse CT reconstruction techniques for reconstructing undersampled radially-acquired MRI by performing spoke-wise 1-dimensional Inverse Fast Fourier Transform as a preprocessing step, followed by applying Filtered Backprojection. PD-UNet model has been proposed in this research which outperformed the established image space correction model UNet, as well as other CT-inspired radial MRI reconstruction baselines with statistical significance. It is also noteworthy that all the sinogram upsampling based baseline models outperformed the conventional NUFFT reconstruction of the undersampled radial MRI with statistical significance.

[RQ:6] All but one of the existing deep learning models, which are discussed in Sect. 5.2 have shown experiments on undersampled MRIs with acceleration factors 4 and 8, while one of the methods (a GAN-based method) has been shown till an acceleration factor of 10. The limitations of the proposed NCC1701 framework and the ReconResNet model were evaluated within the so-called "Limbo" experiment in Sect. 9. It has been demonstrated that the framework was able to reconstruct undersampling factors as high as 20 for 2D Varden sampling and 10 for 1D Varden sampling. These findings were re-confirmed on another dataset in Sect. 8.1.2.2. When it comes to radial undersampling, the framework worked successfully while working with only 30 spokes—resulting in an acceleration factor of 17, shown in Sect. 8.1.2.1 and 8.1.2.2.

The PD-UNet (Sect. 8.2) has been seen to perform successfully while reconstructing undersampled radial MRIs up to an acceleration factor of 16. In Sect. 9.2, ShuffleUNet has been employed for a theoretical acceleration factor of 8 (calculated from the amount of data utilised) while super-resolving static MRIs. When it comes to the super-resolution of undersampled dynamic MRI, a prior knowledge-based fine-tuning approach has been demonstrated to work up to a Cartesian acceleration factor of 16, while the DDoS-UNet model performed successfully till a factor of 25.

In summary, this thesis presented different methods for dealing with highly undersampled MRIs (most methods worked for acceleration factors 16 or higher). The highest level of undersampling experimented with successfully in this thesis for Cartesian MRI dynamic MRI is an acceleration factor of 25, for Cartesian static MRI is 20, and for radial static MRI is 17.

[RQ:7] Part V has shown the applicability of ReconResNet, Fourier-PD, and Fourier-PDUNet models for the task of retrospective motion correction, which were initially proposed for undersampled MRI reconstruction. Sect. 16.1 demonstrated that the ReconResNet model performed better than the UNet model for retrospective motion correction—which is in line with the findings of the experiments of the undersampled MRI reconstruction in the image space. The incorporation of prior knowledge further improved the performance of the models, which has been discussed earlier. Moreover, Sect. 16.2 has demonstrated that the Fourier-PD and Fourier-PDUNet models also performed better than the ReconResNet model for the task of motion correction in hybrid space, which is another outcome in line with the findings of the undersampled reconstruction experiments.

[RQ:8] Part IV has proposed five different automatic MRI processing pipelines—supervised spatiospatial models for the classification of brain tumours, GP-models for supervised classification, and weakly-supervised segmentation of brain tumours using inherently explainable models, deformation-aware semi-supervised learning (DS6) for small vessel segmentation in 7T MRAs using noisy training data, multi-scale inverse-consistent deformable image registration of intramodal brain MRIs using multi-scale UNet with self-constructing graph latent (MICDIR), and unsupervised anomaly detection in brain MRIs using the StRegA pipeline.

While evaluating the pipelines on undersampled data, it was observed that all the models' performance was impacted by undersampling. Incorporating NCC1701 as a precursor to those pipelines improved the performance in every single scenario. However, in most scenarios, the improved performance on the NCC1701 reconstructed images was still poorer than those pipelines' original performance on fully sampled data.

17.2 Contributions

The main technical and application-oriented contributions of the different sections of this thesis are summarised as follows:-

- Section 8.1 proposed NCC1701—an undersampled MRI reconstruction framework for Cartesian and radial MRI, which utilises the proposed ReconResNet model—a modified ResNet-based model with regularised residual blocks as the network backbone of the framework, combined with data consistency steps.
  - This research introduced a data consistency step for radial MRIs.
  - Demonstrated successful reconstruction up to an acceleration factor of 20 for Cartesian and 17 for radial MRIs.
  - The proposed approach was evaluated on data with different pathologies—brain tumour and Alzheimer's disease.
  - Additional experiments have shown the network's performance by training different sampling patterns together and by reconstructing different orientations than those used for training orientation.

- Section 8.2 proposed the Primal-Dual UNet model, which optimises in the sinogram and image space, achieving better results in terms of accuracy and speed while reconstructing radial MRIs up to an acceleration factor of 16.
  - This research shows the applicability of using sparse CT reconstruction techniques for reconstructing undersampled radially-acquired MRI by performing spoke-wise 1-dimensional Inverse Fast Fourier Transform as a preprocessing step—making it possible to have a unified deep learning based CT and radial MRI reconstruction method.
  - Apart from improving the overall results, the method improved the reconstruction quality of regions of interest—liver, kidneys, and spleen.

- ShuffleUNet has been proposed in Sect. 9.2, which is inspired by the tight frame UNet that addresses the problems of blurred or smoothed output and chequerboard artefacts by replacing the strided convolutions with lossless pooling layers (pixel unshuffle), by replacing deconvolution (transposed convolution or interpolation) with a sub-pixel convolution operation (pixel shuffle) and by removing the batch normalisation layers.
  - The method has been applied for the task of Single-Image Super-Resolution (SISR) of static volumetric diffusion-weighted MRIs for a theoretical MR acceleration factor of 8.

- Section 9.3.1 presents a method to incorporate prior knowledge in deep-learning based super-resolution and its application in dynamic MRI.
  - This research addresses the trade-off between the spatial and temporal resolution of dynamic MRI by incorporating a static high-resolution scan as prior knowledge while performing spatial super-resolution on low-resolution

  images, effectively reducing the required scan time per volume—which in turn will improve the temporal resolution.
  - A 3D UNet model was first trained on a benchmark dataset and was fine-tuned using a subject-specific prior planning scan.
  - Evaluations have been shown till an acceleration factor of 16.

- DDoS-UNet has been proposed in Sect. 9.3.2 which incorporates the temporal information into the network model for performing super-resolution of dynamic MRIs.
  - The proposed method super-resolves the low-resolution dynamic MRI with the help of a static prior scan and by exploiting the temporal relationship between the different time points.
  - The proposed network was trained using an artificially created dynamic dataset from a static abdominal benchmark dataset and then tested using dynamic MRIs comprising 25 time points.
  - The method has been evaluated using Cartesian undersampling by taking the centre k-space data up to an acceleration factor of 25.

- A comparative study of real-valued CNNs and complex-valued CNNs (CV-CNNs) is presented in Sect. 10.1.1 for the task of brain tumour classification and segmentation.
  - It has been demonstrated that the CV-CNN models outperform the CNN models having the same number of features, as well as models having the same number of trainable parameters.
  - Serves as the motivation behind using CV-CNNs while working with complex k-space data.

- Section 10.1.2 proposed KSPResNet—a CV-CNN version of the ReconResNet model to work directly in the k-space and attempts to predict the missing k-space frequencies, and preliminary experiments have been shown for the reconstruction of undersampled MRIs with an acceleration factor of 4.
- Fourier-PD and Fourier-PDUNet models have been proposed in Sect. 10.2, which extends the PD network and PDUNet models using complex-valued operations.
  - To be able to work with the complex-valued k-space data, the dual blocks of both these networks were modified by replacing the convolution layers with complex-valued convolution layers and PReLU activation functions with Complex Rectified Linear Unit ($\mathbb{C}$ReLU) activations.

  - The proposed model optimises in both image and k-space—attempts to remove artefacts in the image space using the primal blocks and predict the missing frequencies in the k-space using the dual blocks.
  - Experiments have shown that the models perform better than pure image-space model ReconResNet up to an acceleration factor of 16.

- Part IV demonstrated that incorporating of NCC1701 as a precursor to five different automatic MRI processing pipelines improved their performance on undersampled MRIs.
- Part V has shown the applicability of ReconResNet, Fourier-PD, and Fourier-PDUNet models for the task of retrospective motion correction, which were originally proposed for undersampled MRI reconstruction.
  - Section 16.1 has shown that using other contrasts of the same subject, supplied along with the corrupted contrast MRI as prior knowledge, improves the performance of motion correction.
  - Fourier-PD and Fourier-PDUNet attempt to correct the corrupted k-space using the dual blocks, while attempting to remove artefacts using the primal blocks—shown in Sect. 16.2.
  - Section 16.2 has demonstrated, similar to the undersampled MRI reconstruction, that performing correction in both k-space as well as image space results in better performance than only correcting in the image-space.

Moreover, this thesis has further contributions in terms of five different automatic deep learning based MRI processing pipelines, which are summarised as follows:-

- Chapter 11 demonstrated the possibility of applying spatiotemporal models which are typically used for video classification tasks—ResNet(2+1)D and ResNet Mixed Convolution as "spatiospatial" models by treating one dimension (slice dimension) differently than the other two spatial dimensions of the 3D volumetric images and applied for the task of brain tumour classification, achieving better results than a pure 3D model.
- Three inherently explainable classifiers have been presented in Chap. 12 and have been employed for the task of brain tumour classification-segmentation. The localisation heatmaps provided by the networks—representing the models' focus areas and that are also used in classification decision-making—can be directly interpreted without requiring any post-hoc methods to derive information for model explanations. The models are trained by using the input image and only the classification labels as ground-truth in a supervised fashion—without using

any information about the location of the region of interest (i.e. the segmentation labels), making the segmentation training of the models weakly-supervised through classification labels.

- Chapter 13 presented an unsupervised anomaly detection pipeline, StRegA—Segmentation Regularised Anomaly, which combines the proposed Compact Context-Encoding Variational Autoencoder (cceVAE) with preprocessing steps to simplify the input to the model and post-processing steps to improve the prediction of the model. The proposed method has been trained on anomaly-free brain MRI datasets and then was evaluated on the task of brain tumour detection.
- Chapter 14 proposes a semi-supervised learning approach, deformation-aware learning DS6, which can learn to perform volumetric segmentation from a small training set. Furthermore, a modified version of UNet Multi-scale Supervision (UNet MSS) has been proposed here as the network backbone. The proposed DS6 approach combines elastic deformation with the network backbone using a Siamese architecture. The proposed method has been applied for the task of vessel segmentation from 7T ToF-MRA volumes.
- MICDIR—**M**ulti-scale **I**nverse-**C**onsistent **D**eformable **I**mage **R**egistration, which utilises the proposed MSCGUNet—**M**ulti-scale **UNet** with **S**elf-**C**onstructing **G**raph Latent, for deformable image registration for 3D volumetric images—has been proposed in Chap. 15. The method performs image registration at two different scales (i.e., original resolution as well as down-scaled resolution) using the multi-scale UNet architecture, tries to encode the semantics to improve generalisation using the self-constructing graph network as the latent space of the said UNet, and attempts to make the method inverse-consistent using the cycle consistency loss. The proposed method has been employed for the task of intramodal image registration of brain MRIs.

Directions for Future Research 18

Even after taking significant steps aimed at addressing the problems presented here, several questions are still unanswered. This chapter discusses the limitations of the proposed approaches and points toward different directions of future research.

18.1 Enhancing the Training Process

There are several ways the training process might be improved for the models. The models for undersampled reconstruction were trained for each acceleration factor separately. Mixing the different acceleration factors might improve the training performance of the models by introducing an augmentation effect and might also eliminate the need for training for each acceleration factor separately. This might also help the network learn how undersampling the data influences the creation of artefacts and ultimately learn how to suppress these artefacts independent of the acceleration factor—which might make the model applicable to unseen undersampling patterns as well. Similarly, increasing the range of motion corruption shown to the models might make them more robust against those artefacts in general and also increase their applicability in real workflows. Furthermore, trainings can be performed by combining motion and undersamplings—which might make the models more generally applicable in dealing with different types of artefacts. Another research direction to improve the training process is with the help of loss engineering. For training the undersampled reconstruction or motion correction models, this thesis employed MAE, SSIM as loss functions, while also used PLN in certain cases. Further improvements to the loss function might improve the training of the model, effectively improving the performance [386].

S. Chatterjee, *Deep Learning Innovations in MRI Reconstruction and Analysis*, https://doi.org/10.1007/978-3-658-50741-1_18

18.2 Improvement of the k-space Model

This thesis has shown some preliminary experiments of working directly and only in the k-space using the proposed CV-CNN model: KSPResNet. Even though the model has shown promising results, some remaining artefacts were observed in the results, which demands further investigations. One hypothesis for the results can be that the small Fourier kernels can only produce low frequencies and must be replaced by larger kernels. The variance among the different k-spaces might also have some negative impact on the results. Moreover, due to the inherent circular relationship within the k-space, these models might benefit from developing certain circular complex-valued convolution kernels. These ideas demand further investigations. Another research direction that can be taken for the k-space model is the direction of Recurrent Neural Network (RNN). Such a model has been proposed to learn to transform the given k-space into the image space [392]. It will be interesting to investigate such a model for k-space only operations and will also be interesting to integrate technologies such as bidirectional LSTM and CV-CNN with that model. By improving the k-space model, the hybrid space model presented in this thesis might also be improved—by incorporating those same ideas to improve the dual blocks working in the k-space.

18.3 Integration with Parallel Imaging

The methods proposed in this thesis for undersampled reconstruction, as well as motion correction, work with single-channel input. However, in most MR scanners, multi-channel coils are available—making it possible to use Parallel Imaging (PI). Established PI methods, such as SENSitivity Encoding (SENSE) [52] or Generalized Auto-calibrating Partially Parallel Acquisitions (GRAPPA) [62], which are not necessarily slow, can also help reduce the scan time. The methods presented in this thesis try to complement PI techniques. The proposed models (undersampled reconstruction, as well as motion correction) can be extended by making them multi-channel to work alongside PI techniques. This can not only further speed up the acquisition but also might assist the network in acquiring more knowledge from the different coil images.

18.4 Detailed Evaluation of the Results

Quantitative metrics and qualitative visualisation—the methods employed in this thesis for evaluating the results are also the same path followed by most other researchers. However, the ultimate goal of these deep learning based undersampled reconstruction or motion correction approaches is to be able to use for diagnostics or to be used as a part of processing pipelines. The NCC1701 was employed as a precursor to five different processing pipelines in this thesis, and it improved the performance of those pipelines while encountering undersampled data. However, a more thorough and comparative study should be conducted—comparing the different undersampled reconstruction techniques proposed in this thesis (e.g. PDUNet, Fourier-PDUNet, ShuffleUNet, DDoS-UNet, etc.) and also other state-of-the-art methods. Furthermore, those pipelines should also be evaluated on motion corrupted data, and the applicability of the proposed, as well as other state-of-the-art motion correction methods as a precursor to those pipelines to deal with motion artefacts, should also be examined.

Furthermore, the reconstruction results or motion correction results themselves should also be compared in a clinical context. Quantitative evaluations do not tell much about the things of clinical interest. Radiologists would be able to visually compare the results qualitatively and can provide feedback on the applicability of these methods in a clinical workflow [343]. A blind paired comparative study, similar to [pub:15], can be performed with the help of clinicians to compare the different proposed methods and also other state-of-the-art techniques.

The same goes for the automatic MRI processing pipelines proposed here. Those pipelines were also compared quantitatively and qualitatively in this thesis. Nevertheless, they should be evaluated by clinicians—to evaluate their added value as decision support systems.

As a final note, the models trained for undersampled MRI reconstruction and motion correction were trained on lesion-free MRIs only. Evaluation of NCC1701 has been shown here with pathological data, but the other methods should also go through a similar evaluation.

18.5 Black-box Problem

Clinicians are often very sceptical about applying automatic image processing approaches, especially deep learning based methods, in practice. One main reason for this is the black box nature of these approaches and the inherent problem of missing insights into the automatically derived decisions. The methods proposed in this thesis for undersampled MRI reconstruction and motion correction, as well as MRI processing pipelines, have shown impressive results. Nonetheless, they (all except one of the proposed models, which is inherently explainable) are mostly "black box" in nature, meaning the inner working mechanism, the complex reasoning done by these models, is difficult for human beings to comprehend. Getting a better understanding of these models using interpretability and/or explainability can also expose the possible failure scenarios of these models. Moreover, by doing uncertainty quantification, the possible errors in the result can be estimated without any available ground-truth. Another aspect to consider is that some of the models proposed here can be more or less interpretable or explainable than the other proposed models or even other state-of-the-art models, and some models might do better reasoning than some other models even if the result might be worse—making them less prone to failures during their application. It has been seen that a model might perform better than another model, while the reasoning of the model might not be that accurate [pub:3,pub:1]. Hence, an in-depth study of different aspects of interpretability, explainability, and uncertainty should be performed involving clinicians. This can not only improve trust in these methods but also might unveil ways to improve them.

18.6 Integration into the Clinical Workflow

This thesis's contributions can broadly be grouped into two groups: improvement of MR acquisition in terms of image quality and automatic MRI processing pipelines. All the experiments shown here regarding the first group are performed by simulating the artefacts (i.e., artificially performing undersampling or inducing motion). Implementing them as part of the acquisition workflow (i.e., integrating with the MR scanner) might face difficulties, such as computational limitations and unforeseen differences between real and artificially simulated artefacts. Hence, these methods should be evaluated by integrating them into the clinical workflow. For the second group of experiments, they should also be integrated into the clinical workflow, for example, tumour classification-segmentation or anomaly detection—to evaluate their performance given the computational limitations or other unforeseen challenges, but also to evaluate how much value they add to the workflow to improve the decision-making process of the radiologists.

Interventional MRI: One of the important workflows for dynamic MRIs is the MR-guided interventions. The methods proposed for dynamic MRIs, as well as the other methods, can be candidates for such a workflow for the inherent fast nature of execution of these methods—a hypothesis that should be evaluated by integrating them into interventional workflows. Moreover, clinical interventions are performed with devices, such as needles and catheters, which are not present in the training data. Evaluations should be performed on images with such devices to check the models' applicability in interventional scenarios.

18.7 End-to-End Training of Undersampled Reconstruction and MRI Processing Pipelines

This thesis has shown that using deep learning based undersampled MRI reconstruction as a precursor to other MRI processing pipelines improves the performance of those pipelines on undersampled data in all experimented scenarios. However, the pipelines did not manage to reach the level of performance of the fully sampled MRIs. This result opens up two different directions of research. If the accuracy of the undersampled reconstruction can be improved—the result of the pipelines might also improve. However, another interesting research direction is to perform end-to-end training of both methods. If the primary goal of a certain MR acquisition is to be used for an MR processing pipeline, for example, MRA for vessel segmentation, the undersampled reconstruction model can be added in front of the vessel segmentation model, and then both the models can be trained end-to-end. In this manner, the performance of the vessel segmentation models might improve on undersampled MRIs, but it might also improve the reconstruction quality of the undersampled reconstruction model, as this might act as a type of perceptual loss for the undersampled reconstruction model.

Web References

[url:1] Nvidia. *Nvidia Apex*. 2018. URL: https://github.com/NVIDIA/apex (visited on 06/13/2022).

[url:2] PyTorch. *PyTorch Interpolation*. 2018. URL: https://pytorch.org/docs/stable/generated/torch.nn.functional.interpolate.html (visited on 06/13/2022).

[url:3] PyTorch. *PyTorch Optimisers*. 2018. URL: https://pytorch.org/docs/stable/optim.html (visited on 06/13/2022).

[url:4] PyTorch. *PyTorch: BCEWithLogitsLoss*. 2018. URL: https://pytorch.org/docs/stable/generated/torch.nn.BCEWithLogitsLoss.html (visited on 06/13/2022).

[url:5] PyTorch. *PyTorch: Conv1D*. 2018. URL: https://pytorch.org/docs/stable/generated/torch.nn.Conv1d.html (visited on 06/13/2022).

[url:6] PyTorch. *Torchvision models*. 2018. URL: https://pytorch.org/vision/stable/models.html#video-classification (visited on 06/13/2022).

[url:7] Micha l Bernier. *Braincharter Vasculature Notebook*. 2020. URL: https://github.com/braincharter/vasculature_notebook (visited on 06/13/2022).

[url:8] Soumick Chatterjee. *soumickmj/MRUnder: Initial Release*. Version v0.1. June 2020. 10.5281/zenodo.3901455. URL: https://github.com/soumickmj/MRUnder.

[url:9] Google DeepMind. *Kinetics-400 Dataset*. 2020. Project URL: https://deepmind.com/research/open-source/kinetics and can be downloaded from: https://github.com/cvdfoundation/kinetics-dataset (visited on 06/13/2022).

[url:10] Christopher Thomas. *Deep learning image enhancement insights on loss function engineering*. 2020. URL: https://medium.com/data-science/deep-learning-image-enhancement-insights-on-loss-function-engineering-f57ccbb585d7 (visited on 06/13/2022).

S. Chatterjee, *Deep Learning Innovations in MRI Reconstruction and Analysis*,
https://doi.org/10.1007/978-3-658-50741-1

[url:11] ADNI. *Alzheimers Disease Neuroimaging Initiative (ADNI) database*. 2021. URL: https://adni.loni.usc.edu (visited on 06/13/2022).

[url:12] IXI. *IXI Dataset*. 2021. URL: https://brain-development.org/ixi-dataset/ (visited on 06/13/2022).

[url:13] MRI Questions. *Relative Field Strength*. 2021. URL: https://mriquestions.com/how-strong-is-30t.html (visited on 06/13/2022).

List of Codes

Available publicly on GitHub:

[code:1] *Brain Tumour Classification using GP-Models*. URL: https://github.com/soumickmj/GPModels.

[code:2] *Brain Tumour Classification using Spatiospatial Models*. URL: https://github.com/farazahmeds/Classification-of-brain-tumor-using-Spatiotemporal-models.

[code:3] DDoS-UNet. URL: https://github.com/soumickmj/DDoS.

[code:4] DS6. URL: https://github.com/soumickmj/DS6.

[code:5] Fine-tuning dynamic DL-SR-MRI. URL: https://github.com/soumickmj/FTSuperResDynMRI.

[code:6] MICDIR. URL: https://github.com/soumickmj/MICDIR.

[code:7] MRUnder. URL: https://github.com/soumickmj/MRUnder.

[code:8] NCC1701. URL: https://github.com/soumickmj/NCC1701.

[code:9] PD-UNet. URL: https://github.com/phernst/pd-unet.

[code:10] PyTorch Complex. URL: https://github.com/soumickmj/pytorch-complex.

[code:11] ShuffleUNet, as a part of NCC1701. URL: https://github.com/soumickmj/NCC1701/tree/main/Bridge/WarpDrives/ShuffleUNet.

[code:12] StRegA. URL: https://github.com/soumickmj/StRegA.

[code:13] Tricorder. URL: https://github.com/soumickmj/Tricorder.

List of Publications

Journal Articles (as first author or co-first author)

[pub:1] Soumick Chatterjee, Hadya Yassin, Andreas Nürnberger, and Oliver Speck. "Weakly-supervised segmentation using inherently-explainable classification models and their application to brain tumour classification". In: *Neurocomputing* 682 (2026), p. 133460. https://doi.org/10.1016/j.neucom.2026.133460.

[pub:2] Alessandro Sciarra, Soumick Chatterjee, Max Dünnwald, Andreas Nürnberger, Oliver Speck, and Steffen Oeltze-Jafra. "Automated SSIM Regression for detection and quantification of Motion Artifacts in Brain MR Images". In: *arXiv preprint arXiv:2206.06725* (in review). arXiv: 2206.06725.

[pub:3] Soumick Chatterjee, Fatima Saad, Chompunuch Sarasaen, Suhita Ghosh, Valerie Krug, Rupali Khatun, Rahul Mishra, Nirja Desai, Petia Radeva, Georg Rose, et al. "Exploration of Interpretability Techniques for Deep COVID-19 Classification Using Chest X-ray Images". In: *Journal of Imaging* 10.2 (2024), p. 45. https://doi.org/10.3390/jimaging10020045.

[pub:4] Soumick Chatterjee, Chompunuch Sarasaen, Georg Rose, Andreas Nürnberger, and Oliver Speck. "Ddos-unet: Incorporating temporal information using dynamic dual-channel unet for enhancing super-resolution of dynamic mri". In: *IEEE Access* 12 (2024), pp. 99122–99136. https://doi.org/10.1109/ACCESS.2024.3427674

[pub:5] Soumick Chatterjee, Himanshi Bajaj, Istiyak H Siddiquee, Nandish Bandi Subbarayappa, Steve Simon, Suraj Bangalore Shashidhar, Oliver Speck, and Andreas Nürnberger. "MICDIR: Multi-scale inverse-consistent deformable image registration using UNetMSS with self-constructing graph latent". In: *Computerized Medical Imaging and Graphics* 108 (2023), p. 102267. https://doi.org/10.1016/j.compmedimag.2023.102267.

[pub:6] Philipp Ernst, Soumick Chatterjee, Georg Rose, Oliver Speck, and Andreas Nürnberger. "Sinogram upsampling using Primal-Dual UNet for undersampled CT and radial MRI reconstruction". In: *Neural Networks* 166 (2023): 704–721. https://doi.org/10.1016/j.neunet.2023.08.004.

[pub:7] Hana Haselji, Soumick Chatterjee, Robert Frysch, Vojtch Kulvait, Vladimir Semshchikov, Bennet Hensen, FrankWacker, Inga Brüsch, ThomasWerncke, Oliver Speck, et al. "Liver segmentation using Turbolift learning for CT and cone-beam C-arm perfusion imaging". In: *Comput-*

ers in Biology and Medicine 154 (2023), p. 106539. https://doi.org/10.1016/j.compbiomed.2023.106539.

[pub:8] Soumick Chatterjee, Mario Breitkopf, Chompunuch Sarasaen, Hadya Yassin, Georg Rose, Andreas Nürnberger, and Oliver Speck. "ReconResNet: Regularised residual learning for MR image reconstruction of Undersampled Cartesian and Radial data". In: *Computers in Biology and Medicine* 143 (2022), p. 105321. ISSN: 0010-4825. https://doi.org/10.1016/j.compbiomed.2022.105321.

[pub:9] Soumick Chatterjee, Arnab Das, Chirag Mandal, Budhaditya Mukhopadhyay, Manish Vipinraj, Aniruddh Shukla, Rajatha Nagaraja Rao, Chompunuch Sarasaen, Oliver Speck, and Andreas Nürnberger. "TorchEsegeta: Framework for Interpretability and Explainability of Image-based Deep Learning Models". In: *Applied Sciences* 12.4 (2022), p. 1834. https://doi.org/10.3390/app12041834.

[pub:10] Soumick Chatterjee, Faraz Ahmed Nizamani, Andreas Nürnberger, and Oliver Speck. "Classification of brain tumours in MR images using deep spatiospatial models". In: *Scientific Reports* 12.1 (2022), pp. 1–11. https://doi.org/10.1038/s41598-022-05572-6.

[pub:11] Soumick Chatterjee, Kartik Prabhu, Mahantesh Pattadkal, Gerda Bortsova, Chompunuch Sarasaen, Florian Dubost, Hendrik Mattern, Marleen de Bruijne, Oliver Speck, and Andreas Nürnberger. "DS6, Deformation-Aware Semi-Supervised Learning: Application to Small Vessel Segmentation with Noisy Training Data". In: *Journal of Imaging* 8.10 (2022), p. 259. https://doi.org/10.3390/jimaging8100259.

[pub:12] Soumick Chatterjee, Alessandro Sciarra, Max Dünnwald, Pavan Tummala, Shubham Kumar Agrawal, Aishwarya Jauhari, Aman Kalra, Steffen Oeltze-Jafra, Oliver Speck, and Andreas Nürnberger. "StRegA: Unsupervised Anomaly Detection in Brain MRIs using a Compact Context-encoding Variational Autoencoder". In: *Computers in Biology and Medicine* 149 (2022), p. 106093. ISSN: 0010-4825. https://doi.org/10.1016/j.compbiomed.2022.106093.

[pub:13] Chompunuch Sarasaen, Soumick Chatterjee, Mario Breitkopf, Georg Rose, Andreas Nürnberger, and Oliver Speck. "Fine-tuning deep learning model parameters for improved super-resolution of dynamic mri with prior-knowledge;". In: *Artificial Intelligence in Medicine* 121 (2021), p. 102196. https://doi.org/10.1016/j.artmed.2021.102196.

[pub:14] Soumick Chatterjee, Pramod George Jose, and Debabrata Datta. "Text Classification Using SVM Enhanced by Multithreading and CUDA". In:

International Journal of Modern Education & Computer Science 11.1 (2019). https://doi.org/10.5815/ijmecs.2019.01.02.

Journal Articles
(as a co-author)

[pub:15] Alessandro Sciarra, Hendrik Mattern, Renat Yakupov, Soumick Chatterjee, Daniel Stucht, Steffen Oeltze-Jafra, Frank Godenschweger, and Oliver Speck. "Quantitative evaluation of prospective motion correction in healthy subjects at 7T MRI". In: *Magnetic resonance in medicine* 87.2 (2022), pp. 646–657. https://doi.org/10.1002/mrm.28998.

[pub:16] A Emre Kavur, Soumick Chatterjee, Philipp Ernst, Oliver Speck, Andreas Nürnberger, M. Alper Selver, et al. "CHAOS challenge-combined (CT-MR) healthy abdominal organ segmentation". In: *Medical Image Analysis* 69 (2021), p. 101950. https://doi.org/10.1016/j.media.2020.101950.

Conference Full Articles
(as first author or co-first author)

[pub:17] Soumick Chatterjee, Sowmya Prakash, and Andreas Nürnberger. "[Workshop Paper] Flavours of Convolution for Unsupervised Aspect Extraction and Aspect-based Sentiment Analysis". In: *Proceedings of the Seventh Workshop on Natural Language for Artificial Intelligence (NL4AI 2023) co-located with 22th International Conference of the Italian Association for Artificial Intelligence (AI* IA 2023)*. Vol. 3551. CEUR-WS, 2023, p. 7. URL: https://ceur-ws.org/Vol-3551/paper7.pdf.

[pub:18] Soumick Chatterjee, Alessandro Sciarra, Max Dünnwald, Raghava Vinaykanth Mushunuri, Ranadheer Podishetti, Rajatha Nagaraja Rao, Geetha Doddapaneni Gopinath, Steffen Oeltze-Jafra, Oliver Speck, and Andreas Nürnberger. "ShuffleUNet: Super resolution of diffusion-weighted MRIs using deep learning". In: *2021 29th European Signal Processing Conference (EUSIPCO)*. IEEE. 2021, pp. 940–944. https://doi.org/10.23919/EUSIPCO54536.2021.9615963.

Conference Full Articles
(as a co-author)

[pub:19] Domenico Iuso, Soumick Chatterjee, Rob Heylen, Sven Cornelissen, Jan De Beenhouwer, and Jan Sijbers. "Evaluation of deeply supervised neural networks for 3D pore segmentation in additive manufacturing". In: *Developments in X-Ray Tomography XIV*. Vol. 12242. International Society for Optics and Photonics. SPIE, 2022. https://doi.org/10.1117/12.2633318.

[pub:20] Dhanunjaya Mitta, Soumick Chatterjee, Oliver Speck, and Andreas Nürnberger. "Upgraded W-Net with Attention Gates and Its Application in Unsupervised 3D Liver Segmentation". In: *Proceedings of the 10th International Conference on Pattern Recognition Applications and Methods—ICPRAM*. INSTICC. SciTePress, Feb. 2021, pp. 488–494. ISBN: 978-989-758-486-2. https://doi.org/10.5220/0010221504880494.

[pub:21] Chompunuch Sarasaen, Soumick Chatterjee, Mario Breitkopf, Domenico Iuso, Georg Rose, and Oliver Speck. "Breathing deformation model-application to multi-resolution abdominal MRI". In: *2019 41st Annual International Conference of the IEEE Engineering in Medicine and Biology Society (EMBC)*. IEEE. 2019, pp. 2769–2772. https://doi.org/10.1109/EMBC.2019.8857706.

[pub:22] Rupali Khatun and Soumick Chatterjee. "Machine learning approach for segmenting glands in colon histology images using local intensity and texture features". In: *2018 IEEE 8th International Advance Computing Conference (IACC)*. IEEE. 2018, pp. 314–320. https://doi.org/10.1109/IADCC.2018.8692135.

Conference Short Articles (as first author or co-first author)

[pub:23] Soumick Chatterjee, Chompunuch Sarasaen, Georg Rose, Andreas Nürnberger, and Oliver Speck. "DDoS-UNet: Incorporating temporal information using Dynamic Dual-channel UNet for enhancing super-resolution of dynamic MRI". In: *Medical Imaging with Deep Learning*. 2022. URL: https://openreview.net/forum?id=S7S6gPtbKU4.

[pub:24] Alessandro Sciarra, Soumick Chatterjee, Max Dünnwald, Giuseppe Placidi, Andreas Nürnberger, Oliver Speck, and Steffen Oeltze-Jafra. "Reference-less SSIM Regression for Detection and Quantification of Motion Artefacts in Brain MRIs". In: *Medical Imaging with Deep Learning*. 2022. URL: https://openreview.net/forum?id=24cqMfboXhH.

[pub:25] Soumick Chatterjee, Mario Breitkopf, Chompunuch Sarasaen, Hadya Yassin, Georg Rose, Andreas Nürnberger, and Oliver Speck. "ReconResNet: Regularised Residual Learning for MR Image Reconstruction of Undersampled Cartesian and Radial Data". In: *Medical Imaging with Deep Learning*. 2021. URL: https://openreview.net/forum?id=KNEKu-W4Avz.

[pub:26] Soumick Chatterjee, Kartik Prabhu, Mahantesh Pattadkal, Gerda Bortsova, Chompunuch Sarasaen, Florian Dubost, Hendrik Mattern, Marleen de Bruijne, Oliver Speck, and Andreas Nürnberger. "DS6, Deformation-aware Semi-supervised Learning: Application to Small Vessel Segmentation with Noisy Training Data”. In: *Medical Imaging with Deep Learning*. 2021. URL: https://openreview.net/forum?id=2t0_AxD1otB.

[pub:27] Soumick Chatterjee, Alessandro Sciarra, Max Dünnwald, Steffen Oeltze-Jafra, Andreas Nürnberger, and Oliver Speck. "Retrospective Motion Correction of MR Images using Prior-Assisted Deep Learning”. In: *Medical Imaging meets NeurIPS 2020*. Dec. 2020.

Conference Short Articles (as a co-author)

[pub:28] Philipp Ernst, Soumick Chatterjee, Georg Rose, and Andreas Nürnberger. “Primal-Dual UNet for Sparse View Cone Beam Computed Tomography Volume Reconstruction”. In: *Medical Imaging with Deep Learning*. 2022. URL: https://openreview.net/forum?id=RVKcDeJ2fCi.

Conference Abstracts

[pub:29] Soumick Chatterjee, Himanshi Bajaj, Steve Simon, Suraj Shashidhar, Oliver Speck, and Andreas Nürnberger. “Multi-scale UNet with Self-Constructing Graph Latent for Deformable Image Registration”. In: *Joint Annual Meeting ISMRM-ESMRMB 2022*. May 2022, p. 5563.

[pub:30] Soumick Chatterjee, Alessandro Sciarra, Max Dünnwald, Anitha Bhat Talagini Ashoka, Steffen Oeltze-Jafra, Oliver Speck, and Andreas Nürnberger. “Uncertainty quantification for ground-truth free evaluation of deep learning reconstructions”. In: *Joint Annual Meeting ISMRM-ESMRMB* 2022. May 2022, p. 5631.

[pub:31] Soumick Chatterjee, Alessandro Sciarra, Max Dünnwald, Pavan Tummala, Shubham Agrawal, Aishwarya Jauhari, Aman Kalra, Steffen Oeltze-Jafra, Oliver Speck, and Andreas Nürnberger. “StRegA: Unsupervised Anomaly Detection in Brain MRIs using Compact Context-encoding Variational Autoencoder”. In: *Joint Annual Meeting ISMRM-ESMRMB* 2022. May 2022, p. 5531.

[pub:32] Soumick Chatterjee, Hadya Yassin, Florian Dubost, Andreas Nürnberger, and Oliver Speck. “Learning to segment brain tumours using an explainable classifier”. In: *Joint Annual Meeting ISMRM-ESMRMB* 2022. May 2022, p. 5592.

[pub:33] Philipp Ernst, Soumick Chatterjee, Georg Rose, Oliver Speck, and Andreas Nürnberger. "Sinogram upsampling using Primal-Dual UNet for undersampled CT and radial MRI reconstruction". In: *IEEE International Symposium on Biomedical Imaging (ISBI)* 2022. Mar. 2022.

[pub:34] Chompunuch Sarasaen, Soumick Chatterjee, Georg Rose, Andreas Nürnberger, and Oliver Speck. "Motion-robust dynamic abdominal MRI using k-t GRASP and dynamic dualchannel training of super-resolution U-Net (DDoS-UNet)". In: *Joint Annual Meeting ISMRM-ESMRMB* 2022. May 2022, p. 2965.

[pub:35] Soumick Chatterjee, Himanshi Bajaj, Suraj Shashidhar, Sanjeeth Busnur Indushekar, Steve Simon, Istiyak Siddiquee, Nandish B S, Oliver Speck, and Andreas Nürnberger. "A Comparative Study of Deep Learning Based Deformable Image Registration Techniques". In: *2021 ISMRM & SMRT Annual Meeting & Exhibition.* May 2021, p. 2436.

[pub:36] Soumick Chatterjee, Arnab Das, Chirag Mandal, Budhaditya Mukhopadhyay, Manish Bhandari Vipinraj Bhandari, Aniruddh Shukla, Oliver Speck, and Andreas Nürnberger. "Interpretability Techniques for Deep Learning based Segmentation Models". In: *2021 ISMRM & SMRT Annual Meeting & Exhibition.* May 2021, p. 2400.

[pub:37] Soumick Chatterjee, Chompunuch Sarasaen, Alessandro Sciarra, Mario Breitkopf, Steffen Oeltze-Jafra, Andreas Nürnberger, and Oliver Speck. "Going beyond the image space: undersampled MRI reconstruction directly in the k-space using a complex valued residual neural network". In: *2021 ISMRM & SMRT Annual Meeting & Exhibition.* May 2021, p. 1757.

[pub:38] Soumick Chatterjee, Alessandro Sciarra, Max Dünnwald, Shubham Agrawal, Pavan Tummala, Disha Setlur, Aman Kalra, Aishwarya Jauhari, Steffen Oeltze-Jafra, Oliver Speck, and Andreas Nürnberger. "Unsupervised reconstruction based anomaly detection using a Variational Auto Encoder". In: *2021 ISMRM & SMRT Annual Meeting & Exhibition.* May 2021, p. 2399.

[pub:39] Vishwesh Nath, ..., Soumick Chatterjee, Alessandro Sciarra, Max Dünnwald, Steffen Oeltze-Jafra, Andreas Nürnberger, Oliver Speck, ..., and Jana Hutter. "Resolving to super resolution multi-dimensional diffusion imaging (Super-MUDI)". In: *2021 ISMRM & SMRT Annual Meeting & Exhibition.* May 2021, p. 0103.

[pub:40] Chompunuch Sarasaen, Soumick Chatterjee, Andreas Nürnberger, and Oliver Speck. "DDoS: Dynamic Dual-Channel U-Net for Improving Deep Learning Based Super-Resolution of Abdominal Dynamic MRI".

In: *ESMRMB 2021 Book of Abstracts, published in Magnetic Resonance Materials in Physics, Biology and Medicine*. Vol. 34. Sept. 2021, S6.O3. https://doi.org/10.1007/s10334-021-00947-8.

[pub:41] Chompunuch Sarasaen, Soumick Chatterjee, Fatima Saad, Mario Breitkopf, Andreas Nürnberger, and Oliver Speck. “Fine-tuning deep learning model parameters for improved super-resolution of dynamic MRI with prior-knowledge”. In: *2021 ISMRM & SMRT Annual Meeting & Exhibition*. May 2021, p. 1778.

[pub:42] Soumick Chatterjee, Praveen Putti, Andreas Nürnberger, and Oliver Speck. “Wavelet filtering of undersampled MRI using trainable wavelets and CNN”. In: *ESMRMB 2020 Book of Abstracts, published in Magnetic Resonance Materials in Physics, Biology and Medicine*. Oct. 2020, p. L01.106. https://doi.org/10.1007/s10334-020-00876-y.

[pub:43] Hendrik Mattern, Alessandro Sciarra, Max Dünnwald, Soumick Chatterjee, Ursula Mueller, Steffen Oeltze-Jafra, and Oliver Speck. "Contrast prediction-based regularization for iterative reconstructions (PROSIT)”. In: *2020 ISMRM & SMRT Annual Meeting & Exhibition*. Aug. 2020, p. 3462.

[pub:44] Dhanunjaya Mitta, Soumick Chatterjee, Oliver Speck, and Andreas Nürnberger. “Unsupervised learning for Abdominal MRI Segmentation using 3D Attention W-Net”. In: *2020 ISMRM & SMRT Annual Meeting & Exhibition*. Aug. 2020, p. 3556.

[pub:45] Chompunuch Sarasaen, Soumick Chatterjee, Andreas Nürnberger, and Oliver Speck. “Super Resolution of Dynamic MRI using Deep Learning, Enhanced by Prior-Knowledge”. In: *ESMRMB 2020 Book of Abstracts, published in Magnetic Resonance Materials in Physics, Biology and Medicine*. Vol. 33. Oct. 2020, S03.04. https://doi.org/10.1007/s10334-020-00874-0.

[pub:46] Alessandro Sciarra, Soumick Chatterjee, Max Dünnwald, Oliver Speck, and Steffen Oeltze- Jafra. “Evaluation of Deep Learning Techniques for Motion Artifacts Removal”. In: *2020 ISMRM & SMRT Annual Meeting & Exhibition*. Aug. 2020, p. 3370.

[pub:47] Alessandro Sciarra, Max Dünnwald, Soumick Chatterjee, Oliver Speck, and Steffen Oeltze- Jafra. “Classification of Motion Corrupted Brain MR Images using Deep Learning Techniques”. In: *ESMRMB 2020 Book of Abstracts, published in Magnetic Resonance Materials in Physics, Biology and Medicine*. Vol. 33. Oct. 2020, S03.09. https://doi.org/10.1007/s10334-020-00874-0.

[pub:48] Soumick Chatterjee, Mario Breitkopf, Chompunuch Sarasaen, Georg Rose, Andreas Nürnberger, and Oliver Speck. "A deep learning approach for reconstruction of undersampled Cartesian and Radial data". In: *ESMRMB Congress* 2019. Oct. 2019.

[pub:49] Soumick Chatterjee, Mario Breitkopf, Chompunuch Sarasaen, Georg Rose, Andreas Nürnberger, and Oliver Speck. "Comparison between the usage of same and different variable density undersampling patterns for Deep Learning based MRI Reconstruction". In: *4th Image-Guided Interventions Conference* (IGIC). Nov. 2019.

[pub:50] Chompunuch Sarasaen, Soumick Chatterjee, Mario Breitkopf, Georg Rose, and Oliver Speck. "Generating breathing deformation model from low resolution 4D MRI". In: *4th Image-Guided Interventions Conference (IGIC)*. Nov. 2019.

Furthermore, the following masters theses of students that I had the pleasure of supervising, some of which have substantially contributed to this work

[stud:1] Adarsh Kuzhipathalil (co-supervised with Jiahua Xu). "Road Scene Understanding with Multi-Modal Sensor Data". Master's Thesis. Otto von Guericke University Magdeburg, 2022.

[stud:2] Hadya Sa'ad Abdalwadoud Yassin. "Brain Tumor Segmentation using Inherently Explainable Classification Network". Master's Thesis. Otto von Guericke University Magdeburg, 2022.

[stud:3] Jayanth Varma Dantuluri. "Sensor Fusion and AI for the Detection of Humans in Indoor Environments". Master's Thesis. Otto von Guericke University Magdeburg, 2021.

[stud:4] Faraz Ahmed Nizamani. "Classification of Brain Tumour In MR Images using Deep Learning Methods with a focus on Model Interpretability". Master's Thesis. Otto von Guericke University Magdeburg, 2021.

[stud:5] Sowmya Prakash. "Unsupervised aspect extraction and Aspect based sentiment analysis". Master's Thesis. Otto von Guericke University Magdeburg, 2021.

[stud:6] Adithya Viswanathan. "Comparison of Iterative vs Deep-learning based reconstruction approach in Undersampled Magnetic Resonance Imaging". Master's Thesis. Otto von Guericke University Magdeburg, 2021.

[stud:7] Dhanunjaya Mitta. "Unsupervised MR Abdominal Image Segmentation by Using Deep Neural Networks". Master's Thesis. Otto von Guericke University Magdeburg, 2019.

Credits are given in the respective chapters

References

1. Thomas Bayes. "LII. An essay towards solving a problem in the doctrine of chances. By the late Rev. Mr. Bayes, FRS communicated by Mr. Price, in a letter to John Canton, AMFR S". In: *Philosophical transactions of the Royal Society of London* 53 (1763), pp. 370–418.
2. Pierre-Simon de Laplace. *Théorie analytique des probabilités; Par M. le comte Laplace,....* Mme Ve Courcier, imprimeur-libraire pour les mathématiques et la marine ..., 1812.
3. Wilhelm Conrad Röntgen. "On a new kind of rays". In: *Science* 3.59 (1896), pp. 227–231.
4. Harry Nyquist. "Certain topics in telegraph transmission theory". In: *Transactions of the American Institute of Electrical Engineers* 47.2 (1928), pp. 617–644.
5. Isidor Isaac Rabi. "Space quantization in a gyrating magnetic field". In: *Physical Review* 51.8 (1937), p. 652.
6. Frank Wilcoxon. "Individual Comparisons by Ranking Methods". In: *Biometrics Bulletin* 1.6 (1945), pp. 80–83. ISSN: 00994987. URL: http://www.jstor.org/stable/3001968 (visited on 06/14/2022).
7. Henry B Mann and Donald R Whitney. "On a test of whether one of two random variables is stochastically larger than the other". In: *The annals of mathematical statistics* (1947), pp. 50–60.
8. Claude Elwood Shannon. "Communication in the presence of noise". In: *Proceedings of the IRE* 37.1 (1949), pp. 10–21.
9. Erwin L Hahn. "Spin echoes". In: *Physical review* 80.4 (1950), p. 580.
10. Herbert Robbins and Sutton Monro. "A stochastic approximation method". In: *The annals of mathematical statistics* (1951), pp. 400–407.
11. Frank Rosenblatt. "The perceptron: a probabilistic model for information storage and organization in the brain." In: *Psychological review* 65.6 (1958), p. 386.
12. JR Singer. "Blood flow rates by nuclear magnetic resonance measurements". In: *Science* 130.3389 (1959), pp. 1652–1653.
13. Frank Rosenblatt. "Perceptron simulation experiments". In: *Proceedings of the IRE* 48.3 (1960), pp. 301–309.
14. Raymond Damadian. "Tumor detection by nuclear magnetic resonance". In: *Science* 171.3976 (1971), pp. 1151–1153.

S. Chatterjee, *Deep Learning Innovations in MRI Reconstruction and Analysis*,
https://doi.org/10.1007/978-3-658-50741-1

15. Paul C Lauterbur. “Image formation by induced local interactions: examples employing nuclear magnetic resonance”. In: *nature* 242.5394 (1973), pp. 190–191.
16. Kunihiko Fukushima. “Neural network model for a mechanism of pattern recognition unaffected by shift in position-Neocognitron”. In: *IEICE Technical Report, A* 62.10 (1979), pp. 658–665.
17. Berthold K. P. Horn and Brian G. Schunck. “Determining Optical Flow”. In: *Artif. Intell.* 17 (1981), pp. 185–203.
18. Kunihiko Fukushima and Sei Miyake. “Neocognitron: A self-organizing neural network model for a mechanism of visual pattern recognition”. In: *Competition and cooperation in neural nets*. Springer, 1982, pp. 267–285.
19. David E Rumelhart, Geoffrey E Hinton, and Ronald J Williams. *Learning internal representations by error propagation*. Tech. rep. California Univ San Diego La Jolla Inst for Cognitive Science, 1985.
20. JL Ackerman, MC Offutt, RB Buxton, and TJ Brady. “Rapid 3D tracking of small RF coils”. In: *Proceedings of the 5th annual meeting of SMRM, Montreal, Canada*. 1986, p. 1131.
21. Rina Dechter. “Learning while searching in constraint-satisfaction problems”. In: (1986).
22. E Mark Haacke and John L Patrick. “Reducing motion artifacts in two-dimensional Fourier transform imaging”. In: *Magnetic resonance imaging* 4.4 (1986), pp. 359–376.
23. David E Rumelhart, Geoffrey E Hinton, and Ronald J Williams. “Learning representations by back-propagating errors”. In: *nature* 323.6088 (1986), pp. 533–536.
24. Stephen M Pizer, E Philip Amburn, John D Austin, Robert Cromartie, Ari Geselowitz, Trey Greer, Bart ter Haar Romeny, John B Zimmerman, and Karel Zuiderveld. “Adaptive histogram equalization and its variations”. In: *Computer vision, graphics, and image processing* 39.3 (1987), pp. 355–368.
25. Y-T Zhou, Rama Chellappa, Aseem Vaid, and B Keith Jenkins. “Image restoration using a neural network”. In: *IEEE transactions on acoustics, speech, and signal processing* 36.7 (1988), pp. 1141–1151.
26. Richard L Ehman and Joel P Felmlee. “Adaptive technique for high-definition MR imaging of moving structures.” In: *Radiology* 173.1 (1989), pp. 255–263.
27. Kurt Hornik, Maxwell Stinchcombe, and Halbert White. “Multilayer feedforward networks are universal approximators”. In: *Neural networks* 2.5 (1989), pp. 359–366.
28. Yann LeCun, Bernhard Boser, John Denker, Donnie Henderson, Richard Howard, Wayne Hubbard, and Lawrence Jackel. “Handwritten digit recognition with a back-propagation network”. In: *Advances in neural information processing systems* 2 (1989).
29. Yann LeCun, Bernhard Boser, John S Denker, Donnie Henderson, Richard E Howard, Wayne Hubbard, and Lawrence D Jackel. “Backpropagation applied to handwritten zip code recognition”. In: *Neural computation* 1.4 (1989), pp. 541–551.
30. Pietro Perona and Jitendra Malik. “Scale-space and edge detection using anisotropic diffusion”. In: *IEEE Transactions on pattern analysis and machine intelligence* 12.7 (1990), pp. 629–639.
31. Thomas L Chenevert and James G Pipe. “Effect of bulk tissue motion on quantitative perfusion and diffusion magnetic resonance imaging”. In: *Magnetic resonance in medicine* 19.2 (1991), pp. 261–265.
32. K. Hornik. “Approximation Capabilities of Multilayer Feedforward Networks”. In: *Neural Networks* 4(2) (1991), pp. 131–270.

33. Mark A Kramer. "Nonlinear principal component analysis using autoassociative neural networks". In: *AIChE journal* 37.2 (1991), pp. 233–243.
34. Jane Bromley, James W Bentz, Léon Bottou, Isabelle Guyon, Yann LeCun, Cliff Moore, Eduard Säckinger, and Roopak Shah. "Signature verification using a "siamese" time delay neural network". In: *International Journal of Pattern Recognition and Artificial Intelligence* 7.04 (1993), pp. 669–688.
35. Rafael C. Gonzalez and Richard E. Woods. *Digital Image Processing*. Addison-Wesley, 1993.
36. Moshe Leshno, Vladimir Ya Lin, Allan Pinkus, and Shimon Schocken. "Multilayer feedforward networks with a nonpolynomial activation function can approximate any function". In: *Neural networks* 6.6 (1993), pp. 861–867.
37. Zhuo Wu Fu, Yi Wang, Roger C Grimm, Phillip J Rossman, Joel P Felmlee, Stephen J Riederer, and Richard L Ehman. "Orbital navigator echoes for motion measurements in magnetic resonance imaging". In: *Magnetic resonance in medicine* 34.5 (1995), pp. 746–753.
38. R Jain, R Kasturi, and BG Schunck. "Machine Vision McGRAW-HILL International Editions". In: *New York* (1995).
39. S.-R. Zhao and H. Halling. "A new Fourier method for fan beam reconstruction". In: *1995 IEEE Nuclear Science Symposium and Medical Imaging Conference Record*. Vol. 2. 1995, 1287–1291 vol.2. https://doi.org/10.1109/NSSMIC.1995.510494.
40. Christine C Lee, Clifford R Jack Jr, Roger C Grimm, Phillip J Rossman, Joel P Felmlee, Richard L Ehman, and Stephen J Riederer. "Real-time adaptive motion correction in functional MRI". In: *Magnetic Resonance in Medicine* 36.3 (1996), pp. 436–444.
41. David Atkinson, Derek LG Hill, Peter NR Stoyle, Paul E Summers, and Steven F. Keevil. "Automatic correction of motion artifacts in magnetic resonance images using an entropy focus criterion". In: *IEEE Transactions on Medical imaging* 16.6 (1997), pp. 903–910.
42. J Andrew Derbyshire, Graham A Wright, R Mark Henkelman, and R Scott Hinks. "Dynamic scan-plane tracking using MR position monitoring". In: *Journal of magnetic resonance imaging* 8.4 (1998), pp. 924–932.
43. Alejandro F Frangi, Wiro J Niessen, Koen L Vincken, and Max A Viergever. "Multiscale vessel enhancement filtering". In: *International conference on medical image computing and computer-assisted intervention*. Springer. 1998, pp. 130–137.
44. Colin J Holmes, Rick Hoge, Louis Collins, Roger Woods, Arthur W Toga, and Alan C Evans. "Enhancement of MR images using registration for signal averaging". In: *Journal of computer assisted tomography* 22.2 (1998), pp. 324–333.
45. Wilfried Kubinger, Markus Vincze, and Minu Ayromlou. "The role of gamma correction in colour image processing". In: *9th European Signal Processing Conference (EUSIPCO 1998)*. IEEE. 1998, pp. 1–4.
46. Qing Huo Liu and Nhu Nguyen. "An accurate algorithm for nonuniform fast Fourier transforms (NUFFT's)". In: *IEEE Microwave and guided wave letters* 8.1 (1998), pp. 18–20.
47. Gustav J Braun and Mark D Fairchild. "Image lightness rescaling using sigmoidal contrast enhancement functions". In: *Journal of Electronic Imaging* 8.4 (1999), pp. 380–393.
48. Anders M Dale, Bruce Fischl, and Martin I Sereno. *Cortical Surface-Based Analysis I. Segmentation and Surface Reconstruction*. 1999. URL: http://www.idealibrary.com.

49. Yann LeCun, Patrick Haffner, Léon Bottou, and Yoshua Bengio. "Object recognition with gradient-based learning". In: *Shape, contour and grouping in computer vision*. Springer, 1999, pp. 319–345.
50. Stéphane Mallat. *A wavelet tour of signal processing*. Elsevier, 1999.
51. James G Pipe. "Motion correction with PROPELLER MRI: application to head motion and free-breathing cardiac imaging". In: *Magnetic Resonance in Medicine: An Official Journal of the International Society for Magnetic Resonance in Medicine* 42.5 (1999), pp. 963–969.
52. Klaas P Pruessmann, Markus Weiger, Markus B Scheidegger, and Peter Boesiger. "SENSE: sensitivity encoding for fast MRI". In: *Magnetic Resonance in Medicine: An Official Journal of the International Society for Magnetic Resonance in Medicine* 42.5 (1999), pp. 952–962.
53. Guo W Wei. "Generalized Perona-Malik equation for image restoration". In: *IEEE Signal processing letters* 6.7 (1999), pp. 165–167.
54. Walter F Block, Andrew V Barger, and Charles A Mistretta. "Vastly undersampled isotropic projection imaging". In: *Proc. International Society for Magnetic Resonance in Medicine*. Vol. 8. 2000, p. 161.
55. Takashi Fuse, Eihan Shimizu, and Morito Tsutsumi. "A comparative study on gradient-based approaches for optical flow estimation". In: *International Archives of Photogrammetry and Remote Sensing* 33.B5/1; PART 5 (2000), pp. 269–276.
56. Gary D Knott. *Interpolating cubic splines*. Vol. 18. Springer Science & Business Media, 2000.
57. Kiaran P McGee, Joel P Felmlee, Armando Manduca, Stephen J Riederer, and Richard L Ehman. "Rapid autocorrection using prescan navigator echoes". In: *Magnetic Resonance in Medicine: An Official Journal of the International Society for Magnetic Resonance in Medicine* 43.4 (2000), pp. 583–588.
58. April Fritz, Constance Percy, Andrew Jack, Kanagaratnam Shanmugaratnam, L Sobin, D Maxwell Parkin, S Whelan, World Health Organization, et al. "International classification of diseases for oncology". In: *Geneva: World Health Organization* 3 (2001).
59. Klaas P Pruessmann, Markus Weiger, Peter Börnert, and Peter Boesiger. "Advances in sensitivity encoding with arbitrary k-space trajectories". In: *Magnetic Resonance in Medicine: An Official Journal of the International Society for Magnetic Resonance in Medicine* 46.4 (2001), pp. 638–651.
60. J Vujic, A Marincic, M Ercegovac, and B Milovanovic. "Nikola tesla: 145 years of visionary ideas". In: *5th International Conference on Telecommunications in Modern Satellite, Cable and Broadcasting Service. TELSIKS 2001. Proceedings of Papers (Cat. No. 01EX517)*. Vol. 1. IEEE. 2001, pp. 323–326.
61. Mark Bydder, David J Larkman, and Joseph V Hajnal. "Detection and elimination of motion artifacts by regeneration of k-space". In: *Magnetic Resonance in Medicine: An Official Journal of the International Society for Magnetic Resonance in Medicine* 47.4 (2002), pp. 677–686.
62. Mark A Griswold, Peter M Jakob, Robin M Heidemann, Mathias Nittka, Vladimir Jellus, Jianmin Wang, Berthold Kiefer, and Axel Haase. "Generalized autocalibrating partially parallel acquisitions (GRAPPA)". In: *Magnetic Resonance in Medicine: An Official Journal of the International Society for Magnetic Resonance in Medicine* 47.6 (2002), pp. 1202–1210.

63. Stephen M Smith. "Fast robust automated brain extraction". In: *Human brain mapping* 17.3 (2002), pp. 143–155.
64. Zhou Wang and Alan C Bovik. "A universal image quality index". In: *IEEE signal processing letters* 9.3 (2002), pp. 81–84.
65. Edward Brian Welch, Armando Manduca, Roger C Grimm, Heidi A Ward, and Clifford R Jack Jr. "Spherical navigator echoes for full 3D rigid body motion measurement in MRI". In: *Magnetic resonance in medicine: an official journal of the international society for magnetic resonance in medicine* 47.1 (2002), pp. 32–41.
66. C Canero and Petia Radeva. "Vesselness enhancement diffusion". In: *Pattern Recognition Letters* 24.16 (2003), pp. 3141–3151.
67. Patrice Y Simard, David Steinkraus, John C Platt, et al. "Best practices for convolutional neural networks applied to visual document analysis." In: *Icdar*. Vol. 3. 2003. 2003.
68. Jeffrey Tsao, Peter Boesiger, and Klaas P Pruessmann. "k-t BLAST and k-t SENSE: dynamic MRI with high frame rate exploiting spatiotemporal correlations". In: *Magnetic Resonance in Medicine: An Official Journal of the International Society for Magnetic Resonance in Medicine* 50.5 (2003), pp. 1031–1042.
69. Matt A Bernstein, Kevin F King, and Xiaohong Joe Zhou. *Handbook of MRI pulse sequences*. Elsevier, 2004.
70. Chunlei Liu, Roland Bammer, Dong-hyun Kim, and Michael E Moseley. "Self-navigated interleaved spiral (SNAILS): application to high-resolution diffusion tensor imaging". In: *Magnetic Resonance in Medicine: An Official Journal of the International Society for Magnetic Resonance in Medicine* 52.6 (2004), pp. 1388–1396.
71. Kyoung-Su Oh and Keechul Jung. "GPU implementation of neural networks". In: *Pattern Recognition* 37.6 (2004), pp. 1311–1314.
72. Clifton Phua, Damminda Alahakoon, and Vincent Lee. "Minority report in fraud detection: classification of skewed data". In: *Acm sigkdd explorations newsletter* 6.1 (2004), pp. 50–59.
73. Stephen M Smith, Mark Jenkinson, Mark W Woolrich, Christian F Beckmann, Timothy EJ Behrens, Heidi Johansen-Berg, Peter R Bannister, Marilena De Luca, Ivana Drobnjak, David E Flitney, et al. "Advances in functional and structural MR image analysis and implementation as FSL". In: *Neuroimage* 23 (2004), S208–S219.
74. Zhou Wang, Alan C Bovik, Hamid R Sheikh, and Eero P Simoncelli. "Image quality assessment: from error visibility to structural similarity". In: *IEEE transactions on image processing* 13.4 (2004), pp. 600–612.
75. Dale L Bailey, Michael N Maisey, David W Townsend, and Peter E Valk. *Positron emission tomography*. Vol. 2. Springer, 2005.
76. JA Feekes, SW Hsu, JC Chaloupka, and MD Cassell. "Tertiary microvascular territories define lacunar infarcts in the basal ganglia". In: *Annals ofNeurology* 58.1 (2005), pp. 18–30.
77. Mark Jenkinson, Mickael Pechaud, Stephen Smith, et al. "BET2: MR-based estimation of brain, skull and scalp surfaces". In: *Eleventh annual meeting of the organization for human brain mapping*. Vol. 17. Toronto. 2005, p. 167.
78. Rashindra Manniesing and Wiro Niessen. "Multiscale vessel enhancing diffusion in CT angiography noise filtering". In: *Biennial International Conference on Information Processing in Medical Imaging*. Springer. 2005, pp. 138–149.
79. Stanley Osher, Martin Burger, Donald Goldfarb, Jinjun Xu, and Wotao Yin. "An iterative regularization method for total variation-based image restoration". In: *Multiscale*

Modeling and Simulation 4.2 (2005), pp. 460–489. ISSN: 15403459. https://doi.org/10.1137/040605412.

80. Dave Steinkraus, Ian Buck, and PY Simard. "Using GPUs for machine learning algorithms". In: *Eighth International Conference on Document Analysis and Recognition (ICDAR'05)*. IEEE. 2005, pp. 1115–1120.
81. Marleine Tremblay, Fred Tam, and Simon J Graham. "Retrospective coregistration of functional magnetic resonance imaging data using external monitoring". In: *Magnetic Resonance in Medicine: An Official Journal of the International Society for Magnetic Resonance in Medicine* 53.1 (2005), pp. 141–149.
82. David L Donoho. "Compressed sensing". In: *IEEE Transactions on information theory* 52.4 (2006), pp. 1289–1306.
83. JA Feekes and MD Cassell. "The vascular supply of the functional compartments of the human striatum". In: *Brain : a Journal of Neurology* 129.1 (2006), pp. 2189–201.
84. Tal Geva. "Magnetic resonance imaging: historical perspective". In: *Journal of cardiovascular magnetic resonance* 8.4 (2006), pp. 573–580.
85. Michael Lustig, Juan M Santos, David L Donoho, and John M Pauly. "kt SPARSE: High frame rate dynamic MRI exploiting spatio-temporal sparsity". In: *Proceedings of the 13th annual meeting of ISMRM, Seattle*. Vol. 2420. 2006.
86. Michael K O'Connor and Brad J Kemp. "Single-photon emission computed tomography/computed tomography: basic instrumentation and innovations". In: *Seminars in nuclear medicine*. Vol. 36. 4. Elsevier. 2006, pp. 258–266.
87. Oliver Speck, J Hennig, and M Zaitsev. "Prospective real-time slice-by-slice motion correction for fMRI in freely moving subjects". In: *Magnetic Resonance Materials in Physics, Biology and Medicine* 19.2 (2006), pp. 55–61.
88. Trong-Kha Truong, Donald W Chakeres, David Q Beversdorf, Douglas W Scharre, and Petra Schmalbrock. "Effects of static and radiofrequency magnetic field inhomogeneity in ultra-high field magnetic resonance imaging". In: *Magnetic resonance imaging* 24.2 (2006), pp. 103–112.
89. André JW Van der Kouwe, Thomas Benner, and Anders M Dale. "Real-time rigid body motion correction and shimming using cloverleaf navigators". In: *Magnetic resonance in medicine: an official journal of the international society for magnetic resonance in medicine* 56.5 (2006), pp. 1019–1032.
90. Maxim Zaitsev, Christian Dold, Georgios Sakas, Jürgen Hennig, and Oliver Speck. "Magnetic resonance imaging of freely moving objects: prospective real-time motion correction using an external optical motion tracking system". In: *Neuroimage* 31.3 (2006), pp. 1038–1050.
91. Roland Bammer, Murat Aksoy, and Chunlei Liu. "Augmented generalized SENSE reconstruction to correct for rigid body motion". In: *Magnetic Resonance in Medicine: An Official Journal of the International Society for Magnetic Resonance in Medicine* 57.1 (2007), pp. 90–102.
92. Kai Tobias Block, Martin Uecker, and Jens Frahm. "Undersampled radial MRI with multiple coils. Iterative image reconstruction using a total variation constraint". In: *Magnetic Resonance in Medicine: An Official Journal of the International Society for Magnetic Resonance in Medicine* 57.6 (2007), pp. 1086–1098.
93. David A Clifton, Peter R Bannister, and Lionel Tarassenko. "A framework for novelty detection in jet engine vibration data". In: *Key engineering materials*. Vol. 347. Trans Tech Publ. 2007, pp. 305–310.

94. Jeffrey A Fessler. "On NUFFT-based gridding for non-Cartesian MRI". In: *Journal of magnetic resonance* 188.2 (2007), pp. 191–195.
95. Michael Lustig, David Donoho, and John M. Pauly. "Sparse MRI: The application of compressed sensing for rapid MR imaging". In: *Magnetic resonance in medicine* 58.6 (2007), pp. 1182–1195. ISSN: 0740-3194. https://doi.org/10.1002/mrm.21391.
96. Daniel S. Marcus, Tracy H. Wang, Jamie Parker, John G. Csernansky, John C. Morris, and Randy L. Buckner. "Open Access Series of Imaging Studies (OASIS): cross-sectional MRI data in young, middle aged, nondemented, and demented older adults". In: *Journal of cognitive neuroscience* 19.9 (2007), pp. 1498–1507. ISSN: 0898-929X. https://doi.org/10.1162/jocn.2007.19.9.1498.
97. Brian B Avants, Charles L Epstein, Murray Grossman, and James C Gee. "Symmetric diffeomorphic image registration with cross-correlation: evaluating automated labeling of elderly and neurodegenerative brain". In: *Medical image analysis* 12.1 (2008), pp. 26–41.
98. Jeroen Hendrikse, Jaco J. Zwanenburg, Fredy Visser, Taro Takahara, and Peter Luijten. "Noninvasive depiction of the lenticulostriate arteries with time-of-flight MR angiography at 7.0 T". In: *Cerebrovascular Diseases* 26.6 (2008), pp. 624–629. ISSN: 10159770. https://doi.org/10.1159/000166838.
99. Alireza Nasiri Avanaki. "Exact global histogram specification optimized for structural similarity". In: *Optical review* 16.6 (2009), pp. 613–621.
100. Brian B Avants, Nick Tustison, Gang Song, et al. "Advanced normalization tools (ANTS)". In: *Insight j* 2.365 (2009), pp. 1–35.
101. Yoshua Bengio. *Learning deep architectures for AI*. Now Publishers Inc, 2009.
102. Candice A Bookwalter, Mark A Griswold, and Jeffrey L Duerk. "Multiple overlapping k-space junctions for investigating translating objects (MOJITO)". In: *IEEE transactions on medical imaging* 29.2 (2009), pp. 339–349.
103. Felix A Breuer, Stephan AR Kannengiesser, Martin Blaimer, Nicole Seiberlich, Peter M Jakob, and Mark A Griswold. "General formulation for quantitative G-factor calculation in GRAPPA reconstructions". In: *Magnetic Resonance in Medicine: An Official Journal of the International Society for Magnetic Resonance in Medicine* 62.3 (2009), pp. 739–746.
104. Jia Deng, Wei Dong, Richard Socher, Li-Jia Li, Kai Li, and Li Fei-Fei. "Imagenet: A large-scale hierarchical image database". In: *2009 IEEE conference on computer vision and pattern recognition*. Ieee. 2009, pp. 248–255.
105. Gabor T Herman. *Fundamentals of computerized tomography: image reconstruction from projections*. Springer Science & Business Media, 2009.
106. Geoffrey E Hinton. "Deep belief networks". In: *Scholarpedia* 4.5 (2009), p. 5947.
107. Peter Jezzard. "Chapter 29—The Physical Basis of Spatial Distortions in Magnetic Resonance Images". In: *Handbook of Medical Image Processing and Analysis (Second Edition)*. Ed. by ISAAC N. BANKMAN. Second Edition. Burlington: Academic Press, 2009, pp. 499–514. ISSN: 978-0-12-373904-9. https://doi.org/10.1016/B978-012373904-9.50039-8. URL: https://www.sciencedirect.com/science/article/pii/B9780123739049500398.
108. Hong Jung, Kyunghyun Sung, Krishna S Nayak, Eung Yeop Kim, and Jong Chul Ye. "k-t FOCUSS: a general compressed sensing framework for high resolution dynamic MRI". In: *Magnetic Resonance in Medicine: An Official Journal of the International Society for Magnetic Resonance in Medicine* 61.1 (2009), pp. 103–116.

109. Chang Ki Kang, Cheol Wan Park, Jae Yong Han, Sang Hoon Kim, Chan A. Park, Kyoung Nam Kim, Suk Min Hong, Young Bo Kim, Kendall H. Lee, and Zang Hee Cho. "Imaging and analysis of lenticulostriate arteries using 7.0-Tesla magnetic resonance angiography". In: *Magnetic Resonance in Medicine* 61.1 (2009), pp. 136–144. ISSN: 07403194. https://doi.org/10.1002/mrm.21786.
110. Andreas H Mahnken, Jens Ricke, and Kai E Wilhelm. *CT-and MR-guided Interventions in Radiology*. Vol. 22. Springer, 2009.
111. Melvyn B Ooi, Sascha Krueger, William J Thomas, Srirama V Swaminathan, and Truman R Brown. "Prospective real-time correction for arbitrary head motion using active markers". In: *Magnetic Resonance in Medicine: An Official Journal of the International Society for Magnetic Resonance in Medicine* 62.4 (2009), pp. 943–954.
112. Lei Qin, Peter van Gelderen, John Andrew Derbyshire, Fenghua Jin, Jongho Lee, Jacco A de Zwart, Yang Tao, and Jeff H Duyn. "Prospective head-movement correction for high-resolution MRI using an in-bore optical tracking system". In: *Magnetic Resonance in Medicine: An Official Journal of the International Society for Magnetic Resonance in Medicine* 62.4 (2009), pp. 924–934.
113. Mark W Woolrich, Saad Jbabdi, Brian Patenaude, Michael Chappell, Salima Makni, Timothy Behrens, Christian Beckmann, Mark Jenkinson, and Stephen M Smith. "Bayesian analysis of neuroimaging data in FSL". In: *Neuroimage* 45.1 (2009), S173–S186.
114. Wei Lin, Feng Huang, Peter Börnert, Yu Li, and Arne Reykowski. "Motion correction using an enhanced floating navigator and GRAPPA operations". In: *Magnetic Resonance in Medicine: An Official Journal of the International Society for Magnetic Resonance in Medicine* 63.2 (2010), pp. 339–348.
115. Jing Liu, Pascal Spincemaille, Noel CF Codella, Thanh D Nguyen, Martin R Prince, and Yi Wang. "Respiratory and cardiac self-gated free-breathing cardiac CINE imaging with multiecho 3D hybrid radial SSFP acquisition". In: *Magnetic resonance in medicine* 63.5 (2010), pp. 1230–1237.
116. Michael Lustig and John M Pauly. "SPIRiT: iterative self-consistent parallel imaging reconstruction from arbitrary k-space". In: *Magnetic resonance in medicine* 64.2 (2010), pp. 457–471.
117. Vinod Nair and Geoffrey E Hinton. "Rectified linear units improve restricted boltzmann machines". In: *Icml*. 2010.
118. Lisa Torrey and Jude Shavlik. "Transfer learning". In: *Handbook of research on machine learning applications and trends: algorithms, methods, and techniques*. 2010, pp. 242–264.
119. Nicholas J Tustison, Brian B Avants, Philip A Cook, Yuanjie Zheng, Alexander Egan, Paul A Yushkevich, and James C Gee. "N4ITK: improved N3 bias correction". In: *IEEE transactions on medical imaging* 29.6 (2010), pp. 1310–1320.
120. Martin Uecker, Shuo Zhang, and Jens Frahm. "Nonlinear inverse reconstruction for real-time MRI of the human heart using undersampled radial FLASH". In: *Magnetic Resonance in Medicine: An Official Journal of the International Society for Magnetic Resonance in Medicine* 63.6 (2010), pp. 1456–1462.
121. Brian C Andrews-Shigaki, Brian SR Armstrong, Maxim Zaitsev, and Thomas Ernst. "Prospective motion correction for magnetic resonance spectroscopy using single camera retro-grate reflector optical tracking". In: *Journal of Magnetic Resonance Imaging* 33.2 (2011), pp. 498–504.

122. Brian B Avants, Nicholas J Tustison, Jue Wu, Philip A Cook, and James C Gee. "An open source multivariate framework for n-tissue segmentation with evaluation on public data". In: *Neuroinformatics* 9.4 (2011), pp. 381–400.
123. Christoph Forman, Murat Aksoy, Joachim Hornegger, and Roland Bammer. "Self-encoded marker for optical prospective head motion correction in MRI". In: *Medical image analysis* 15.5 (2011), pp. 708–719.
124. Rafael C. Gonzalez and Richard E. Woods. *Digital Image Processing, Third Edition*. Pearson, 2011.
125. Florian Knoll, Kristian Bredies, Thomas Pock, and Rudolf Stollberger. "Second order total generalized variation (TGV) for MRI". In: *Magnetic resonance in medicine* 65.2 (2011), pp. 480–491.
126. Tobias Kober, José P Marques, Rolf Gruetter, and Gunnar Krueger. "Head motion detection using FID navigators". In: *Magnetic resonance in medicine* 66.1 (2011), pp. 135–143.
127. Andrew Ng et al. "Sparse autoencoder". In: *CS294A Lecture notes* 72.2011 (2011), pp. 1–19.
128. Davide Piccini, Arne Littmann, Sonia Nielles-Vallespin, and Michael O Zenge. "Spiral phyllotaxis: the natural way to construct a 3D radial trajectory in MRI". In: *Magnetic resonance in medicine* 66.4 (2011), pp. 1049–1056.
129. Nicole Seiberlich, Philipp Ehses, Jeff Duerk, Robert Gilkeson, and Mark Griswold. "Improved radial GRAPPA calibration for real-time free-breathing cardiac imaging". In: *Magnetic resonance in medicine* 65.2 (2011), pp. 492–505.
130. Nicole Seiberlich, Gregory Lee, Philipp Ehses, Jeffrey L Duerk, Robert Gilkeson, and Mark Griswold. "Improved temporal resolution in cardiac imaging using through-time spiral GRAPPA". In: *Magnetic resonance in medicine* 66.6 (2011), pp. 1682–1688.
131. Christoph Sommer, Christoph Straehle, Ullrich Kothe, and Fred A. Hamprecht. "Ilastik: Interactive learning and segmentation toolkit". In: *Proceedings—International Symposium on Biomedical Imaging* September 2014 (2011), pp. 230–233. ISSN: 19457928. https://doi.org/10.1109/ISBI.2011.5872394.
132. Matthew D Zeiler, Graham W Taylor, and Rob Fergus. "Adaptive deconvolutional networks for mid and high level feature learning". In: *2011 International Conference on Computer Vision*. IEEE. 2011, pp. 2018–2025.
133. Joseph Y Cheng, Marcus T Alley, Charles H Cunningham, Shreyas S Vasanawala, John M Pauly, and Michael Lustig. "Nonrigid motion correction in 3D using autofocusing withlocalized linear translations". In: *Magnetic resonance in medicine* 68.6 (2012), pp. 1785–1797.
134. Anagha Deshmane, Vikas Gulani, Mark A. Griswold, and Nicole Seiberlich. "Parallel MR imaging". In: *Journal of Magnetic Resonance Imaging* 36.1 (2012), pp. 55–72. ISSN: 10531807. https://doi.org/10.1002/jmri.23639.
135. Ruud B van Heeswijk, Gabriele Bonanno, Simone Coppo, Andrew Coristine, Tobias Kober, and Matthias Stuber. "Motion compensation strategies in magnetic resonance imaging". In: *Critical Reviews in Biomedical Engineering* 40.2 (2012).
136. Geoffrey Hinton, Nitish Srivastava, and Kevin Swersky. "Neural networks for machine learning lecture 6a overview of mini-batch gradient descent". In: 14.8 (2012), p. 2. URL: http://www.cs.toronto.edu/~hinton/coursera/lecture6/lec6.pdf.
137. Geoffrey E Hinton, Nitish Srivastava, Alex Krizhevsky, Ilya Sutskever, and Ruslan R Salakhutdinov. "Improving neural networks by preventing co-adaptation of feature detectors". In: *arXiv preprint* arXiv:1207.0580 (2012).

138. Mark Jenkinson, Christian F Beckmann, Timothy EJ Behrens, Mark W Woolrich, and Stephen M Smith. “Fsl”. In: *Neuroimage* 62.2 (2012), pp. 782–790.
139. Dongil Kim, Pilsung Kang, Sungzoon Cho, Hyoung-joo Lee, and Seungyong Doh. “Machine learning-based novelty detection for faulty wafer detection in semiconductor manufacturing”. In: *Expert Systems with Applications* 39.4 (2012), pp. 4075–4083.
140. Alex Krizhevsky, Ilya Sutskever, and Geoffrey E Hinton. “Imagenet classification with deep convolutional neural networks”. In: *Advances in neural information processing systems* 25 (2012), pp. 1097–1105.
141. Julian Maclaren, Brian SR Armstrong, Robert T Barrows, KA Danishad, Thomas Ernst, Colin L Foster, Kazim Gumus, Michael Herbst, Ilja Y Kadashevich, Todd P Kusik, et al. “Measurement and correction of microscopic head motion during magnetic resonance imaging of the brain”. In: *PloS one* 7.11 (2012), e48088.
142. Johan Pallud, Luc Taillandier, Laurent Capelle, Denys Fontaine, Matthieu Peyre, François Ducray, Hugues Duffau, and Emmanuel Mandonnet. “Quantitative morphological magnetic resonance imaging follow-up of low-grade glioma: a plea for systematic measurement of growth rates”. In: *Neurosurgery* 71.3 (2012), pp. 729–740.
143. Esben Plenge, Dirk HJ Poot, Monique Bernsen, Gyula Kotek, Gavin Houston, Piotr Wielopolski, Louise van der Weerd, Wiro J Niessen, and Erik Meijering. “Super-resolution methods in MRI: can they improve the trade-off between resolution, signal-to-noise ratio, and acquisition time?” In: *Magnetic resonance in medicine* 68.6 (2012), pp. 1983–1993.
144. Jessica Schulz, Thomas Siegert, Enrico Reimer, Christian Labadie, Julian Maclaren, Michael Herbst, Maxim Zaitsev, and Robert Turner. “An embedded optical tracking system for motion-corrected magnetic resonance imaging at 7T”. In: *Magnetic Resonance Materials in Physics, Biology and Medicine* 25.6 (2012), pp. 443–453.
145. M Dylan Tisdall, Aaron T Hess, Martin Reuter, Ernesta M Meintjes, Bruce Fischl, and André JW van der Kouwe. “Volumetric navigators for prospective motion correction and selective reacquisition in neuroanatomical MRI”. In: *Magnetic resonance in medicine* 68.2 (2012), pp. 389–399.
146. Eric Van Reeth, Ivan WK Tham, Cher Heng Tan, and Chueh Loo Poh. “Super-resolution in magnetic resonance imaging: a review”. In: *Concepts in Magnetic Resonance Part A* 40.6 (2012), pp. 306–325.
147. Kenneth Clark, Bruce Vendt, Kirk Smith, John Freymann, Justin Kirby, Paul Koppel, Stephen Moore, Stanley Phillips, David Maffitt, Michael Pringle, Lawrence Tarbox, and Fred Prior. “The Cancer Imaging Archive (TCIA): Maintaining and Operating a Public Information Repository”. In: *Journal of Digital Imaging* 26.6 (Dec. 2013), pp. 1045–1057. ISSN: 1618-727X. https://doi.org/10.1007/s10278-013-9622-7. URL: https://doi.org/10.1007/s10278-013-9622-7.
148. Alex Graves. “Generating sequences with recurrent neural networks”. In: *arXiv preprint* arXiv:1308.0850 (2013).
149. Diederik P Kingma and Max Welling. “Auto-encoding variational bayes”. In: *arXiv preprint* arXiv:1312.6114 (2013).
150. Alexander Loktyushin, Hannes Nickisch, Rolf Pohmann, and Bernhard Schölkopf. “Blind retrospective motion correction of MR images”. In: *Magnetic resonance in medicine* 70.6 (2013), pp. 1608–1618.
151. Bradley Christopher Lowekamp, David T Chen, Luis Ibáñez, and Daniel Blezek. “The design of SimpleITK”. In: *Frontiers in neuroinformatics* 7 (2013), p. 45.

152. Andrew L Maas, Awni Y Hannun, Andrew Y Ng, et al. "Rectifier nonlinearities improve neural network acoustic models". In: *Proc. icml.* Vol. 30. 1. Citeseer. 2013, p. 3.
153. Julian Maclaren, Michael Herbst, Oliver Speck, and Maxim Zaitsev. "Prospective motion correction in brain imaging: A review". In: *Magnetic Resonance in Medicine* 69.3 (2013), pp. 621–636. ISSN: 1522-2594. https://doi.org/10.1002/mrm.24314. URL: http://dx.doi.org/10.1002/mrm.24314.
154. Melvyn B Ooi, Murat Aksoy, Julian Maclaren, Ronald D Watkins, and Roland Bammer. "Prospective motion correction using inductively coupled wireless RF coils". In: *Magnetic resonance in medicine* 70.3 (2013), pp. 639–647.
155. Lei Qin, Ehud J Schmidt, Zion Tsz Ho Tse, Juan Santos, William S Hoge, Clare Tempany-Afdhal, Kim Butts-Pauly, and Charles L Dumoulin. "Prospective motion correction using tracking coils". In: *Magnetic resonance in medicine* 69.3 (2013), pp. 749–759.
156. Ilya Sutskever, James Martens, George Dahl, and Geoffrey Hinton. "On the importance of initialization and momentum in deep learning". In: *International conference on machine learning*. PMLR. 2013, pp. 1139–1147.
157. Ghislain Vaillant, Claudia Prieto, Christoph Kolbitsch, Graeme Penney, and Tobias Schaeffter. "Retrospective rigid motion correction in k-space for segmented radial MRI". In: *IEEE transactions on medical imaging* 33.1 (2013), pp. 1–10.
158. Joanna M Wardlaw, Eric E Smith, Geert J Biessels, Charlotte Cordonnier, Franz Fazekas, Richard Frayne, Richard I Lindley, John T O'Brien, Frederik Barkhof, Oscar R Benavente, et al. "Neuroimaging standards for research into small vessel disease and its contribution to ageing and neurodegeneration". In: *The Lancet Neurology* 12.8 (2013), pp. 822–838.
159. Joanna M. Wardlaw, Colin Smith, and Martin Dichgans. "Mechanisms of sporadic cerebral small vessel disease: Insights from neuroimaging". In: *The Lancet Neurology* 12.5 (2013), pp. 483–497. ISSN: 14744422. https://doi.org/10.1016/S1474-4422(13)70060-7. URL: http://dx.doi.org/10.1016/S1474-4422(13)70060-7.
160. Harris A Ahmad, Hui Jing Yu, and Colin G Miller. "Medical imaging modalities". In: *Medical Imaging in Clinical Trials*. Springer, 2014, pp. 3–26.
161. Guillaume Alain and Yoshua Bengio. "What regularized auto-encoders learn from the data-generating distribution". In: *The Journal of Machine Learning Research* 15.1 (2014), pp. 3563–3593.
162. Kai Tobias Block, Hersh Chandarana, Sarah Milla, Mary Bruno, Tom Mulholland, Girish Fatterpekar, Mari Hagiwara, Robert Grimm, Christian Geppert, Berthold Kiefer, and Daniel K. Sodickson. "Towards Routine Clinical Use of Radial Stack-of-Stars 3D Gradient-Echo Sequences for Reducing Motion Sensitivity". In: *Journal of the Korean Society of Magnetic Resonance in Medicine* 18.2 (2014), p. 87. ISSN: 1226-9751. https://doi.org/10.13104/jksmrm.2014.18.2.87.
163. Li Feng, Robert Grimm, Kai Tobias Block, Hersh Chandarana, Sungheon Kim, Jian Xu, Leon Axel, Daniel K. Sodickson, and Ricardo Otazo. "Golden-angle radial sparse parallel MRI: combination of compressed sensing, parallel imaging, and golden-angle radial sampling for fast and flexible dynamic volumetric MRI". In: *Magnetic resonance in medicine* 72.3 (2014), pp. 707–717. ISSN: 0740-3194. https://doi.org/10.1002/mrm.24980.
164. Eleftherios Garyfallidis, Matthew Brett, Bagrat Amirbekian, Ariel Rokem, Stefan Van Der Walt, Maxime Descoteaux, and Ian Nimmo-Smith. "Dipy, a library for the analysis of diffusion MRI data". In: *Frontiers in neuroinformatics* 8 (2014), p. 8.

165. Ian Goodfellow, Jean Pouget-Abadie, Mehdi Mirza, Bing Xu, David Warde-Farley, Sherjil Ozair, Aaron Courville, and Yoshua Bengio. "Generative adversarial nets". In: *Advances in neural information processing systems* 27 (2014).
166. Michael Herbst, Julian Maclaren, Cris Lovell-Smith, Rebecca Sostheim, Karl Egger, Andreas Harloff, Jan Korvink, Juergen Hennig, and Maxim Zaitsev. "Reproduction of Motion Artifacts for Performance Analysis of Prospective Motion Correction in MRI". In: *Magnetic resonance in medicine : official journal of the Society of Magnetic Resonance in Medicine / Society of Magnetic Resonance in Medicine* 71.1 (Jan. 2014), 10.1002/mrm.24645. https://doi.org/10.1002/mrm.24645. URL: http://www.ncbi.nlm.nih.gov/pmc/articles/PMC3674114/.
167. Boseul Jeong, Dae Seob Choi, Hwa Seon Shin, Hye Young Choi, Mi Jung Park, Kyung Nyeo Jeon, Jae Beom Na, and Sung Hoon Chung. "T1-weighted FLAIR MR imaging for the evaluation of enhancing brain tumors: comparison with spin echo imaging". In: *Journal of the Korean Society of Magnetic Resonance in Medicine* 18.2 (2014), pp. 151–156.
168. Diederik P Kingma and Jimmy Ba. "Adam: A method for stochastic optimization". In: *arXiv preprint* arXiv:1412.6980 (2014).
169. Bjoern H Menze, Andras Jakab, Stefan Bauer, Jayashree Kalpathy-Cramer, Keyvan Farahani, Justin Kirby, Yuliya Burren, Nicole Porz, Johannes Slotboom, Roland Wiest, et al. "The multimodal brain tumor image segmentation benchmark (BRATS)". In: *IEEE transactions on medical imaging* 34.10 (2014), pp. 1993–2024.
170. Marco AF Pimentel, David A Clifton, Lei Clifton, and Lionel Tarassenko. "A review of novelty detection". In: *Signal Processing* 99 (2014), pp. 215–249.
171. Saikat Sengupta, Sasidhar Tadanki, John C Gore, and E Brian Welch. "Prospective real-time head motion correction using inductively coupled wireless NMR probes". In: *Magnetic resonance in medicine* 72.4 (2014), pp. 971–985.
172. Karen Simonyan and Andrew Zisserman. "Very deep convolutional networks for large-scale image recognition". In: *arXiv preprint* arXiv:1409.1556 (2014).
173. Martin Uecker, Peng Lai, Mark J Murphy, Patrick Virtue, Michael Elad, John M Pauly, Shreyas S Vasanawala, and Michael Lustig. "ESPIRiT—an eigenvalue approach to autocalibrating parallel MRI: where SENSE meets GRAPPA". In: *Magnetic resonance in medicine* 71.3 (2014), pp. 990–1001.
174. Matthew D Zeiler and Rob Fergus. "Visualizing and understanding convolutional networks". In: *European conference on computer vision*. Springer. 2014, pp. 818–833.
175. Jinwon An and Sungzoon Cho. "Variational autoencoder based anomaly detection using reconstruction probability". In: *Special Lecture on IE* 2.1 (2015), pp. 1–18.
176. Michael A Bruno, Eric A Walker, and Hani H Abujudeh. "Understanding and confronting our mistakes: the epidemiology of error in radiology and strategies for error reduction". In: *Radiographics* 35.6 (2015), pp. 1668–1676.
177. Elizabeth B Claus, Kyle M Walsh, John K Wiencke, Annette M Molinaro, Joseph L Wiemels, Joellen M Schildkraut, Melissa L Bondy, Mitchel Berger, Robert Jenkins, and Margaret Wrensch. "Survival and low-grade glioma: the emergence of genetic information". In: *Neurosurgical focus* 38.1 (2015), E6.
178. Djork-Arné Clevert, Thomas Unterthiner, and Sepp Hochreiter. "Fast and accurate deep network learning by exponential linear units (elus)". In: *arXiv preprint* arXiv:1511.07289 (2015).

179. Kaiming He, Xiangyu Zhang, Shaoqing Ren, and Jian Sun. "Delving deep into rectifiers: Surpassing human-level performance on imagenet classification". In: *Proceedings of the IEEE international conference on computer vision*. 2015, pp. 1026–1034.
180. Jithin Saji Isaac and Ramesh Kulkarni. "Super resolution techniques for medical image processing". In: *2015 International Conference on Technologies for Sustainable Development (ICTSD)*. IEEE. 2015, pp. 1–6.
181. Max Jaderberg, Karen Simonyan, Andrew Zisserman, and Koray Kavukcuoglu. "Spatial Transformer Networks". In: *NIPS*. 2015.
182. Oren N Jaspan, Roman Fleysher, and Michael L Lipton. "Compressed sensing MRI: a review of the clinical literature". In: *The British journal of radiology* 88.1056 (2015), p. 20150487.
183. Tim Jerman, Franjo Pernuš, Boštjan Likar, and Žiga Špiclin. "Beyond Frangi: an improved multiscale vesselness filter". In: *Medical Imaging 2015: Image Processing*. Vol. 9413. International Society for Optics and Photonics. 2015, 94132A.
184. Yann LeCun, Yoshua Bengio, and Geoffrey Hinton. "Deep learning". In: *nature* 521.7553 (2015), pp. 436–444.
185. Volodymyr Mnih, Koray Kavukcuoglu, David Silver, Andrei A Rusu, Joel Veness, Marc G Bellemare, Alex Graves, Martin Riedmiller, Andreas K Fidjeland, Georg Ostrovski, et al. "Human-level control through deep reinforcement learning". In: *nature* 518.7540 (2015), pp. 529–533.
186. Olaf Ronneberger, Philipp Fischer, and Thomas Brox. "U-net: Convolutional networks for biomedical image segmentation". In: *International Conference on Medical image computing and computer-assisted intervention*. Springer. 2015, pp. 234–241.
187. Holger R Roth, Le Lu, Ari Seff, Kevin M Cherry, Joanne Hoffman, Shijun Wang, Jiamin Liu, Evrim Turkbey, and Ronald M Summers. *A new 2.5 D representation for lymph node detection in CT. The Cancer Imaging Archive*. 2015. https://doi.org/10.7937/K9/TCIA.2015.AQIIDCNM.
188. Jürgen Schmidhuber. "Deep learning in neural networks: An overview". In: *Neural networks* 61 (2015), pp. 85–117.
189. Aditya Singh, Benjamin Zahneisen, Brian Keating, Michael Herbst, Linda Chang, Maxim Zaitsev, and Thomas Ernst. "Optical Tracking With Two Markers for Robust Prospective Motion Correction for Brain Imaging". In: *Magma (New York, N.Y.)* 28.6 (Dec. 2015), pp. 523–534. https://doi.org/10.1007/s10334-015-0493-4. URL: http://www.ncbi.nlm.nih.gov/pmc/articles/PMC5259555/.
190. Daniel Stucht, K. Appu Danishad, Peter Schulze, Frank Godenschweger, Maxim Zaitsev, and Oliver Speck. "Highest Resolution In Vivo Human Brain MRI Using Prospective Motion Correction". In: *PLOS ONE* 10.7 (July 2015), e0133921–. URL: https://doi.org/10.1371/journal.pone.0133921.
191. Maxim Zaitsev, Julian Maclaren, and Michael Herbst. "Motion Artefacts in MRI: a Complex Problem with Many Partial Solutions". In: *Journal of magnetic resonance imaging : JMRI* 42.4 (Oct. 2015), pp. 887–901. https://doi.org/10.1002/jmri.24850. URL: http://www.ncbi.nlm.nih.gov/pmc/articles/PMC4517972/.
192. Martin Arjovsky, Amar Shah, and Yoshua Bengio. "Unitary evolution recurrent neural networks". In: *International Conference on Machine Learning*. PMLR. 2016, pp. 1120–1128.
193. Vinit Baliyan, Chandan J Das, Raju Sharma, and Arun Kumar Gupta. "Diffusion weighted imaging: technique and applications". In: *World journal of radiology* 8.9 (2016), p. 785.

194. Özgün Çiçek, Ahmed Abdulkadir, Soeren S Lienkamp, Thomas Brox, and Olaf Ronneberger. "3D U-Net: learning dense volumetric segmentation from sparse annotation". In: *International conference on medical image computing and computer-assisted intervention*. Springer. 2016, pp. 424–432.
195. Nat Dilokthanakul, Pedro AM Mediano, Marta Garnelo, Matthew CH Lee, Hugh Salimbeni, Kai Arulkumaran, and Murray Shanahan. "Deep unsupervised clustering with gaussian mixture variational autoencoders". In: *arXiv preprint* arXiv:1611.02648 (2016).
196. Leon A Gatys, Alexander S Ecker, and Matthias Bethge. "Image style transfer using convolutional neural networks". In: *Proceedings of the IEEE conference on computer vision and pattern recognition*. 2016, pp. 2414–2423.
197. F Godenschweger, U Kägebein, D Stucht, U Yarach, A Sciarra, R Yakupov, F Lüsebrink, P Schulze, and O Speck. "Motion correction in MRI of the brain". In: *Physics in Medicine & Biology* 61.5 (2016), R32. URL: http://stacks.iop.org/0031-9155/61/i=5/a=R32.
198. Ian Goodfellow, Yoshua Bengio, and Aaron Courville. *Deep Learning*. http://www.deeplearningbook.org. MIT Press, 2016.
199. Nitzan Guberman. "On complex valued convolutional neural networks". In: *arXiv preprint* arXiv:1602.09046 (2016).
200. Kaiming He, Xiangyu Zhang, Shaoqing Ren, and Jian Sun. "Deep Residual Learning for Image Recognition". In: *IEEE Conference on Computer Vision and Pattern Recognition (CVPR)* (2016), pp. 770–778. https://doi.org/10.1109/CVPR.2016.90.
201. Dan Hendrycks and Kevin Gimpel. "Gaussian error linear units (gelus)". In: *arXiv preprint* arXiv:1606.08415 (2016).
202. Max Jaderberg, Karen Simonyan, Andrew Zisserman, and Koray Kavukcuoglu. "Spatial Transformer Networks". In: arXiv:1506.02025 *[cs]* (Feb. 2016). arXiv: 1506.02025. URL: http://arxiv.org/abs/1506.02025.
203. Justin Johnson, Alexandre Alahi, and Li Fei-Fei. "Perceptual losses for real-time style transfer and super-resolution". In: *European conference on computer vision*. Springer. 2016, pp. 694–711.
204. Yudong Liang, Jinjun Wang, Sanping Zhou, Yihong Gong, and Nanning Zheng. "Incorporating image priors with deep convolutional neural networks for image super-resolution". In: *Neurocomputing* 194 (2016), pp. 340–347.
205. Wei Liao, Karl Rohr, Chang Ki Kang, Zang Hee Cho, Life Fellow, and Stefan Wörz. "Automatic 3D segmentation and quantification of lenticulostriate arteries from high-resolution 7 tesla MRA images". In: *IEEE Transactions on Image Processing* 25.1 (2016), pp. 400–413. ISSN: 10577149. https://doi.org/10.1109/TIP.2015.2499085.
206. Alireza Makhzani, Jonathon Shlens, Navdeep Jaitly, Ian Goodfellow, and Brendan Frey. *Adversarial Autoencoders*. 2016. arXiv: https://arxiv.org/abs/1511.05644 [cs.LG].
207. Augustus Odena, Vincent Dumoulin, and Chris Olah. "Deconvolution and checkerboard artifacts". In: *Distill* 1.10 (2016), e3.
208. Sérgio Pereira, Adriano Pinto, Victor Alves, and Carlos A. Silva. "Brain tumor segmentation using convolutional neural networks in MRI images". In: *IEEE transactions on medical imaging* 35.5 (2016), pp. 1240–1251.
209. David Silver, Aja Huang, Chris J Maddison, Arthur Guez, Laurent Sifre, George Van Den Driessche, Julian Schrittwieser, Ioannis Antonoglou, Veda Panneershelvam, Marc Lanctot, et al. "Mastering the game of Go with deep neural networks and tree search". In: *nature* 529.7587 (2016), pp. 484–489.

210. Xuchen Song, Xue Gao, Yanfang Ding, and Zhixin Wang. "A handwritten Chinese characters recognition method based on sample set expansion and CNN". In: *2016 3rd International Conference on Systems and Informatics (ICSAI)*. IEEE. 2016, pp. 843–849.
211. Christian Szegedy, Vincent Vanhoucke, Sergey Ioffe, Jon Shlens, and Zbigniew Wojna. "Rethinking the inception architecture for computer vision". In: *Proceedings of the IEEE conference on computer vision and pattern recognition*. 2016, pp. 2818–2826.
212. Shanshan Wang, Zhenghang Su, Leslie Ying, Xi Peng, Shun Zhu, Feng Liang, Dagan Feng, and Dong Liang. "Accelerating magnetic resonance imaging via deep learning". In: *2016 IEEE 13th International Symposium on Biomedical Imaging (ISBI)*. IEEE. 2016, pp. 514–517.
213. Sebastien C Wong, Adam Gatt, Victor Stamatescu, and Mark D McDonnell. "Understanding data augmentation for classification: when to warp?" In: *2016 international conference on digital image computing: techniques and applications (DICTA)*. IEEE. 2016, pp. 1–6.
214. Tinghui Zhou, Philipp Krähenbühl, Mathieu Aubry, Qixing Huang, and Alexei A. Efros. "Learning Dense Correspondence via 3D-Guided Cycle Consistency". In: *2016 IEEE Conference on Computer Vision and Pattern Recognition (CVPR)* (2016), pp. 117–126.
215. Andrew Aitken, Christian Ledig, Lucas Theis, Jose Caballero, Zehan Wang, and Wenzhe Shi. "Checkerboard artifact free sub-pixel convolution: A note on sub-pixel convolution, resize convolution and convolution resize". In: *arXiv preprint* arXiv:1707.02937 (2017).
216. Wenjia Bai, Matthew Sinclair, Giacomo Tarroni, Ozan Oktay, Martin Rajchl, Ghislain Vaillant, Aaron M Lee, Nay Aung, Elena Lukaschuk, Mihir M Sanghvi, et al. "Human-level CMR image analysis with deep fully convolutional networks". In: (2017).
217. Spyridon Bakas, Hamed Akbari, Aristeidis Sotiras, Michel Bilello, Martin Rozycki, Justin S Kirby, John B Freymann, Keyvan Farahani, and Christos Davatzikos. "Advancing the cancer genome atlas glioma MRI collections with expert segmentation labels and radiomic features". In: *Scientific data* 4.1 (2017), pp. 1–13.
218. Joerg Barkhausen, Thomas Kahn, Gabriele A Krombach, Christiane K Kuhl, Joachim Lotz, David Maintz, Jens Ricke, Stefan O Schoenberg, Thomas J Vogl, Frank K Wacker, et al. "White paper: Interventional MRI: Current status and potential for development considering economic perspectives, part 1: General application". In: *RoeFo-Fortschritte auf dem Gebiet der Roentgenstrahlen und der bildgebenden Verfahren*. Vol. 189. 07. Georg Thieme Verlag KG. 2017, pp. 611–623.
219. Gerda Bortsova, Gijs van Tulder, Florian Dubost, Tingying Peng, Nassir Navab, Aad van der Lugt, Daniel Bos, and Marleen De Bruijne. "Segmentation of intracranial arterial calcification with deeply supervised residual dropout networks". In: *Lecture Notes in Computer Science (including subseries Lecture Notes in Artificial Intelligence and Lecture Notes in Bioinformatics)*. 2017. ISSN: 9783319661780. https://doi.org/10.1007/978-3-319-66179-7_41. arXiv: https://arxiv.org/abs/1706.01148.
220. Adrian P Brady. "Error and discrepancy in radiology: inevitable or avoidable?" In: *Insights into imaging* 8.1 (2017), pp. 171–182.
221. Supriyo Chakraborty, Richard Tomsett, Ramya Raghavendra, Daniel Harborne, Moustafa Alzantot, Federico Cerutti, Mani Srivastava, Alun Preece, Simon Julier, Raghuveer M Rao, et al. "Interpretability of deep learning models: a survey of results". In: *2017 IEEE smartworld, ubiquitous intelligence & computing, advanced & trusted computed, scalable computing & communications, cloud*

& big data computing, Internet of people and smart city innovation (smartworld/SCALCOM/UIC/ATC/CBDcom/IOP/SCI). IEEE. 2017, pp. 1–6.
222. Jun Cheng. *Brain tumor dataset. figshare. dataset*. 2017. https://doi.org/10.6084/m9.figshare.1512427.v5.
223. Hwan-ho Cho and Hyunjin Park. "Classification of low-grade and high-grade glioma using multi-modal image radiomics features". In: *2017 39th Annual International Conference of the IEEE Engineering in Medicine and Biology Society (EMBC)*. IEEE. 2017, pp. 3081–3084.
224. Florian Dubost, Gerda Bortsova, Hieab Adams, Arfan Ikram, Wiro J Niessen, Meike Vernooij, and Marleen De Bruijne. "Gp-unet: Lesion detection from weak labels with a 3d regression network". In: *International Conference on Medical Image Computing and Computer-Assisted Intervention*. Springer. 2017, pp. 214–221.
225. Jesse Hamilton, Dominique Franson, and Nicole Seiberlich. "Recent advances in parallel imaging for MRI". In: *Progress in nuclear magnetic resonance spectroscopy* 101 (2017), pp. 71–95.
226. Chih Yang Hsu, Mahsa Ghaffari, Ali Alaraj, Michael Flannery, Xiaohong Joe Zhou, and Andreas Linninger. "Gap-free segmentation of vascular networks with automatic image processing pipeline". In: *Computers in Biology and Medicine* 82.January (2017), pp. 29–39. ISSN: 18790534. https://doi.org/10.1016/j.compbiomed.2017.01.012.
227. Kyong Hwan Jin, Michael T McCann, Emmanuel Froustey, and Michael Unser. "Deep convolutional neural network for inverse problems in imaging". In: *IEEE Transactions on Image Processing* 26.9 (2017), pp. 4509–4522.
228. Saed Khawaldeh, Usama Pervaiz, Azhar Rafiq, and Rami S Alkhawaldeh. "Noninvasive grading of glioma tumor using magnetic resonance imaging with convolutional neural networks". In: *Applied Sciences* 8.1 (2017), p. 27.
229. Günter Klambauer, Thomas Unterthiner, Andreas Mayr, and Sepp Hochreiter. "Self-normalizing neural networks". In: *Proceedings of the 31st international conference on neural information processing systems*. 2017, pp. 972–981.
230. Ghazanfar Latif, M Mohsin Butt, Adil H Khan, Omair Butt, and DNF Awang Iskandar. "Multiclass brain Glioma tumor classification using block-based 3D Wavelet features of MR images". In: *2017 4th International Conference on Electrical and Electronic Engineering (ICEEE)*. IEEE. 2017, pp. 333–337.
231. Geert Litjens, Thijs Kooi, Babak Ehteshami Bejnordi, Arnaud Arindra Adiyoso Setio, Francesco Ciompi, Mohsen Ghafoorian, Jeroen A.W.M. van der Laak, Bram van Ginneken, and Clara I. Sánchez. "A survey on deep learning in medical image analysis". In: *Medical Image Analysis* 42 (2017), pp. 60–88. ISSN: 1361-8415. https://doi.org/10.1016/j.media.2017.07.005. URL: http://dx.doi.org/10.1016/j.media.2017.07.005.
232. Christopher T Lloyd, Alessandro Sorichetta, and Andrew J Tatem. "High resolution global gridded data for use in population studies". In: *Scientific data* 4.1 (2017), pp. 1–17.
233. Ilya Loshchilov and Frank Hutter. "Decoupled weight decay regularization". In: *arXiv preprint* arXiv:1711.05101 (2017).
234. Donald W McRobbie, Elizabeth A Moore, Martin J Graves, and Martin R Prince. *MRI from Picture to Proton*. Cambridge university press, 2017.
235. Craig H. Meyer. *Cartesian & Non-Cartesian Sampling Schemes—Advantages & Disadvantages*. 2017. URL: http://indexsmart.mirasmart.com/ISMRM2017/PDFfiles/8107.html (visited on 02/11/2021).

236. Paulius Micikevicius, Sharan Narang, Jonah Alben, Gregory Diamos, Erich Elsen, David Garcia, Boris Ginsburg, Michael Houston, Oleksii Kuchaiev, Ganesh Venkatesh, et al. "Mixed precision training". In: *arXiv preprint* arXiv:1710.03740 (2017).
237. Aaron van den Oord, Oriol Vinyals, and Koray Kavukcuoglu. "Neural discrete representation learning". In: *arXiv preprint* arXiv:1711.00937 (2017).
238. Chi-Hieu Pham, Aurélien Ducournau, Ronan Fablet, and François Rousseau. "Brain MRI super-resolution using deep 3D convolutional networks". In: *2017 IEEE 14th International Symposium on Biomedical Imaging (ISBI 2017)*. IEEE. 2017, pp. 197–200.
239. Shan E Ahmed Raza, Linda Cheung, David Epstein, Stella Pelengaris, Michael Khan, and Nasir M Rajpoot B. "MIMONet : Gland Segmentation Using Neural Network". In: *Computer Methods and Programs in Biomedicine* 1.d (2017), pp. 698–706. ISSN: 18727565. https://doi.org/10.1007/978-3-319-60964-5. arXiv: arXiv:1705.03820v3.
240. Gabriel Prieto Renieblas, Agustin Turrero Nogues, Alberto Munoz Gonzalez, Nieves Gomez-Leon, and Eduardo Guibelalde Del Castillo. "Structural similarity index family for image quality assessment in radiological images". In: *Journal of medical imaging* 4.3 (2017), p. 035501. https://doi.org/10.1117/1.JMI.4.3.035501.
241. Giorgio Roffo. "Ranking to learn and learning to rank: On the role of ranking in pattern recognition applications". In: *arXiv preprint* arXiv:1706.05933 (2017).
242. Mrinmoy Sandilya and SR Nirmala. "Compressed sensing trends in magnetic resonance imaging". In: *Engineering science and technology, an international journal* 20.4 (2017), pp. 1342–1352.
243. Nadia Solowij, Andrew Zalesky, Valentina Lorenzetti, and M Yücel. "Chronic Cannabis Use and Axonal Fiber Connectivity". In: *Handbook of Cannabis and Related Pathologies*. Elsevier, 2017, pp. 391–400.
244. Ludovic Trottier, Philippe Giguere, and Brahim Chaib-Draa. "Parametric exponential linear unit for deep convolutional neural networks". In: *2017 16th IEEE International Conference on Machine Learning and Applications (ICMLA)*. IEEE. 2017, pp. 207–214.
245. Saining Xie, Ross Girshick, Piotr Dollár, Zhuowen Tu, and Kaiming He. "Aggregated residual transformations for deep neural networks". In: *Proceedings of the IEEE conference on computer vision and pattern recognition*. 2017, pp. 1492–1500.
246. Guang Yang, Simiao Yu, Hao Dong, Greg Slabaugh, Pier Luigi Dragotti, Xujiong Ye, Fangde Liu, Simon Arridge, Jennifer Keegan, Yike Guo, et al. "DAGAN: deep de-aliasing generative adversarial networks for fast compressed sensing MRI reconstruction". In: *IEEE transactions on medical imaging* 37.6 (2017), pp. 1310–1321.
247. Guodong Zeng, Xin Yang, Jing Li, Lequan Yu, Pheng-Ann Heng, and Guoyan Zheng. "3D U-net with multi-level deep supervision: fully automatic segmentation of proximal femur in 3D MR images". In: *International workshop on machine learning in medical imaging*. Springer. 2017, pp. 274–282.
248. Jaco J.M. Zwanenburg and Matthias J.P. Van Osch. "Targeting cerebral small vessel disease with MRI". In: *Stroke* 48.11 (2017), pp. 3175–3182. ISSN: 15244628. https://doi.org/10.1161/STROKEAHA.117.016996.
249. Jonas Adler and Ozan Öktem. "Learned primal-dual reconstruction". In: *IEEE transactions on medical imaging* 37.6 (2018), pp. 1322–1332.
250. Samet Akcay, Amir Atapour-Abarghouei, and Toby P Breckon. "Ganomaly: Semi-supervised anomaly detection via adversarial training". In: *Asian conference on computer vision*. Springer. 2018, pp. 622–637.

251. Wenjia Bai, Matthew Sinclair, Giacomo Tarroni, Ozan Oktay, Martin Rajchl, Ghislain Vaillant, Aaron M Lee, Nay Aung, Elena Lukaschuk, Mihir M Sanghvi, et al. "Automated cardiovascular magnetic resonance image analysis with fully convolutional networks". In: *Journal of Cardiovascular Magnetic Resonance* 20.1 (2018), pp. 1–12.
252. Spyridon Bakas, Mauricio Reyes, Andras Jakab, Stefan Bauer, Markus Rempfler, Alessandro Crimi, Russell Takeshi Shinohara, Christoph Berger, Sung Min Ha, Martin Rozycki, et al. "Identifying the best machine learning algorithms for brain tumor segmentation, progression assessment, and overall survival prediction in the BRATS challenge". In: *arXiv preprint* arXiv:1811.02629 (2018).
253. Nicolas Basty and Vicente Grau. "Super resolution of cardiac cine MRI sequences using deep learning". In: *Image Analysis for Moving Organ, Breast, and Thoracic Images*. Springer, 2018, pp. 23–31.
254. Michaël Bernier, Stephen C. Cunnane, and Kevin Whittingstall. "The morphology of the human cerebrovascular system". In: *Human Brain Mapping* 39.12 (2018), pp. 4962–4975. ISSN: 10970193. https://doi.org/10.1002/hbm.24337.
255. Paul Blanc-Durand, Axel Van Der Gucht, Niklaus Schaefer, Emmanuel Itti, and John O. Prior. "Automatic lesion detection and segmentation of18F-FET PET in gliomas: A full 3D U-Net convolutional neural network study". In: *PLoS ONE* 13.4 (2018), pp. 1–11. ISSN: 19326203. https://doi.org/10.1371/journal.pone.0195798.
256. Tomas Budrys, Vincentas Veikutis, Saulius Lukosevicius, Rymante Gleizniene, Egle Monastyreckiene, and Ilona Kulakiene. "Artifacts in magnetic resonance imaging: how it can really affect diagnostic image quality and confuse clinical diagnosis?" In: *Journal of Vibroengineering* 20.2 (2018), pp. 1202–1213.
257. Eduardo Castro, Jaime S Cardoso, and Jose Costa Pereira. "Elastic deformations for data augmentation in breast cancer mass detection". In: *2018 IEEE EMBS International Conference on Biomedical & Health Informatics (BHI)*. IEEE. 2018, pp. 230–234.
258. Li Chen, Mahmud Mossa-Basha, Niranjan Balu, Gador Canton, Jie Sun, Kristi Pimentel, Thomas S. Hatsukami, Jenq Neng Hwang, and Chun Yuan. "Development of a quantitative intracranial vascular features extraction tool on 3D MRA using semiautomated open-curve active contour vessel tracing". In: *Magnetic Resonance in Medicine* 79.6 (2018), pp. 3229–3238. ISSN: 15222594. https://doi.org/10.1002/mrm.26961.
259. Hyeong-Seok Choi, Jang-Hyun Kim, Jaesung Huh, Adrian Kim, Jung-Woo Ha, and Kyogu Lee. "Phase-aware speech enhancement with deep complex u-net". In: *International Conference on Learning Representations*. 2018.
260. Ben A Duffy, Wenlu Zhang, Haoteng Tang, Lu Zhao, Meng Law, Arthur W Toga, and Hosung Kim. "Retrospective correction of motion artifact affected structural MRI images using deep learning of simulated motion". In: (2018).
261. Stefan Elfwing, Eiji Uchibe, and Kenji Doya. "Sigmoid-weighted linear units for neural network function approximation in reinforcement learning". In: *Neural Networks* 107 (2018), pp. 3–11.
262. Taejoon Eo, Yohan Jun, Taeseong Kim, Jinseong Jang, Ho-Joon Lee, and Dosik Hwang. "KIKI-net: cross-domain convolutional neural networks for reconstructing undersampled magnetic resonance images". In: *Magnetic resonance in medicine* 80.5 (2018), pp. 2188–2201.
263. Anna Fabijańska. "Segmentation of corneal endothelium images using a U-Net-based convolutional neural network". In: *Artificial Intelligence in Medicine* 88 (2018), pp. 1–13. ISSN: 18732860. https://doi.org/10.1016/j.artmed.2018.04.004.

264. Chenjie Ge, Irene Yu-Hua Gu, Asgeir Store Jakola, and Jie Yang. "Deep learning and multi-sensor fusion for glioma classification using multistream 2D convolutional networks". In: *2018 40th Annual International Conference of the IEEE Engineering in Medicine and Biology Society (EMBC)*. IEEE. 2018, pp. 5894–5897.
265. R Guerrero, C Qin, O Oktay, C Bowles, L Chen, R Joules, R Wolz, M del C Valdés-Hernández, David Alexander Dickie, Joanna Wardlaw, et al. "White matter hyperintensity and stroke lesion segmentation and differentiation using convolutional neural networks". In: *NeuroImage: Clinical* 17 (2018), pp. 918–934.
266. Yifan Guo, Weixian Liao, Qianlong Wang, Lixing Yu, Tianxi Ji, and Pan Li. "Multi-dimensional time series anomaly detection: A gru-based gaussian mixture variational autoencoder approach". In: *Asian Conference on Machine Learning*. PMLR. 2018, pp. 97–112.
267. Kerstin Hammernik, Teresa Klatzer, Erich Kobler, Michael P Recht, Daniel K Sodickson, Thomas Pock, and Florian Knoll. "Learning a variational network for reconstruction of accelerated MRI data". In: *Magnetic resonance in medicine* 79.6 (2018), pp. 3055–3071.
268. Nicholas Heller, Joshua Dean, and Nikolaos Papanikolopoulos. "Imperfect Segmentation Labels: How Much Do They Matter?" In: *Intravascular Imaging and Computer Assisted Stenting and Large-Scale Annotation of Biomedical Data and Expert Label Synthesis*. Ed. by Danail Stoyanov, Zeike Taylor, Simone Balocco, Raphael Sznitman, Anne Martel, Lena Maier-Hein, Luc Duong, Guillaume Zahnd, Stefanie Demirci, Shadi Albarqouni, Su-Lin Lee, Stefano Moriconi, Veronika Cheplygina, Diana Mateus, Emanuele Trucco, Eric Granger, and Pierre Jannin. Cham: Springer International Publishing, 2018, pp. 112–120. ISSN: 978-3-030-01364-6.
269. Chang Min Hyun, Hwa Pyung Kim, Sung Min Lee, Sungchul Lee, and Jin Keun Seo. "Deep learning for undersampled MRI reconstruction". In: *Physics in Medicine & Biology* 63.13 (2018), p. 135007.
270. Fabian Isensee, Jens Petersen, Andre Klein, David Zimmerer, Paul F Jaeger, Simon Kohl, Jakob Wasserthal, Gregor Koehler, Tobias Norajitra, Sebastian Wirkert, et al. "nnu-net: Self-adapting framework for u-net-based medical image segmentation". In: *arXiv preprint* arXiv:1809.10486 (2018).
271. Hoyeon Lee, Jongha Lee, Hyeongseok Kim, Byungchul Cho, and Seungryong Cho. "Deep-neural-network-based sinogram synthesis for sparse-view CT image reconstruction". In: *IEEE Transactions on Radiation and Plasma Medical Sciences* 3.2 (2018), pp. 109–119.
272. Xiaomeng Li, Lequan Yu, Hao Chen, Chi-Wing Fu, and Pheng-Ann Heng. "Semi-supervised skin lesion segmentation via transformation consistent self-ensembling model". In: *arXiv preprint* arXiv:1808.03887 (2018).
273. Chen-Hsuan Lin, Ersin Yumer, Oliver Wang, Eli Shechtman, and Simon Lucey. "ST-GAN: Spatial Transformer Generative Adversarial Networks for Image Compositing". In: *2018 IEEE/CVF Conference on Computer Vision and Pattern Recognition*. Salt Lake City, UT: IEEE, June 2018, pp. 9455–9464. ISSN: 978-1-5386-6420-9. https://doi.org/10.1109/CVPR.2018.00985. URL: https://ieeexplore.ieee.org/document/8579083/.
274. Jyh-Miin Lin. "Python non-uniform fast Fourier transform (PyNUFFT): An accelerated non-Cartesian MRI package on a heterogeneous platform (CPU/GPU)". In: *Journal of Imaging* 4.3 (2018), p. 51.

275. Pengju Liu, Hongzhi Zhang, Kai Zhang, Liang Lin, and Wangmeng Zuo. “Multi-level wavelet-CNN for image restoration”. In: *Proceedings of the IEEE conference on computer vision and pattern recognition workshops*. 2018, pp. 773–782.
276. Hendrik Mattern, Alessandro Sciarra, Frank Godenschweger, Daniel Stucht, Falk Lüsebrink, Georg Rose, and Oliver Speck. “Prospective motion correction enables highest resolution time-of-flight angiography at 7T”. In: *Magnetic Resonance in Medicine* 80.1 (2018), pp. 248–258. ISSN: 15222594. https://doi.org/10.1002/mrm.27033.
277. Ozan Oktay, Jo Schlemper, Loic Le Folgoc, Matthew Lee, Mattias Heinrich, Kazunari Misawa, Kensaku Mori, Steven McDonagh, Nils Y Hammerla, Bernhard Kainz, et al. “Attention u-net: Learning where to look for the pancreas”. In: *MIDL 2018*. 2018.
278. Kamlesh Pawar, Zhaolin Chen, N Jon Shah, and Gary F Egan. “Motion correction in MRI using deep convolutional neural network”. In: *Proceedings of the ISMRM Scientific Meeting & Exhibition, Paris*. Vol. 1174. 2018.
279. Chen Qin, Jo Schlemper, Jose Caballero, Anthony N Price, Joseph V Hajnal, and Daniel Rueckert. “Convolutional recurrent neural networks for dynamic MR image reconstruction”. In: *IEEE transactions on medical imaging* 38.1 (2018), pp. 280–290.
280. Muhammad Imran Razzak, Muhammad Imran, and Guandong Xu. “Efficient brain tumor segmentation with multiscale two-pathway-group conventional neural networks”. In: *IEEE journal of biomedical and health informatics* 23.5 (2018), pp. 1911–1919.
281. Manassés Ribeiro, André Eugênio Lazzaretti, and Heitor Silvério Lopes. “A study of deep convolutional auto-encoders for anomaly detection in videos”. In: *Pattern Recognition Letters* 105 (2018), pp. 13–22.
282. Seyed Sadegh Mohseni Salehi, Seyed Raein Hashemi, Clemente Velasco-Annis, Abdelhakim Ouaalam, Judy A Estroff, Deniz Erdogmus, Simon K Warfield, and Ali Gholipour. “Real-time automatic fetal brain extraction in fetal MRI by deep learning”. In: *2018 IEEE 15th International Symposium on Biomedical Imaging (ISBI 2018)*. IEEE. 2018, pp. 720–724.
283. Robin Sandkühler, Christoph Jud, Simon Andermatt, and Philippe C. Cattin. “AirLab: Autograd Image Registration Laboratory”. In: *CoRR* abs / 1806.09907 (2018). arXiv: 1806.09907. URL: http://arxiv.org/abs/1806.09907.
284. Simone Scardapane, Steven Van Vaerenbergh, Amir Hussain, and Aurelio Uncini. “Complex-valued neural networks with nonparametric activation functions”. In: *IEEE Transactions on Emerging Topics in Computational Intelligence* 4.2 (2018), pp. 140–150.
285. Iram Shahzadi, Tong Boon Tang, Fabrice Meriadeau, and Abdul Quyyum. “CNN-LSTM: Cascaded framework for brain tumour classification”. In: *2018 IEEE-EMBS Conference on Biomedical Engineering and Sciences (IECBES)*. IEEE. 2018, pp. 633–637.
286. Chiheb Trabelsi, Olexa Bilaniuk, Ying Zhang, Dmitriy Serdyuk, Sandeep Subramanian, Joao Felipe Santos, Soroush Mehri, Negar Rostamzadeh, Yoshua Bengio, and Christopher J Pal. “Deep Complex Networks”. In: *International Conference on Learning Representations*. 2018.
287. Du Tran, Heng Wang, Lorenzo Torresani, Jamie Ray, Yann LeCun, and Manohar Paluri. “A closer look at spatiotemporal convolutions for action recognition”. In: *Proceedings of the IEEE conference on Computer Vision and Pattern Recognition*. 2018, pp. 6450–6459.

288. Dmitry Ulyanov, Andrea Vedaldi, and Victor Lempitsky. "Deep image prior". In: *Proceedings of the IEEE conference on computer vision and pattern recognition*. 2018, pp. 9446–9454.
289. Bob D. de Vos, Floris F. Berendsen, Max A. Viergever, Hessam Sokooti, Marius Staring, and Ivana Isgum. "A Deep Learning Framework for Unsupervised Affine and Deformable Image Registration". In: arXiv:1809.06130 *[cs]* (Dec. 2018). arXiv: 1809.06130. URL: http://arxiv.org/abs/1809.06130.
290. Mei Wang and Weihong Deng. "Deep visual domain adaptation: A survey". In: *Neurocomputing* 312 (2018), pp. 135–153.
291. Haowen Xu, Wenxiao Chen, Nengwen Zhao, Zeyan Li, Jiahao Bu, Zhihan Li, Ying Liu, Youjian Zhao, Dan Pei, Yang Feng, et al. "Unsupervised anomaly detection via variational auto-encoder for seasonal kpis in web applications". In: *Proceedings of the 2018 World Wide Web Conference*. 2018, pp. 187–196.
292. Yang Yang, Lin-Feng Yan, Xin Zhang, Yu Han, Hai-Yan Nan, Yu-Chuan Hu, Bo Hu, Song-Lin Yan, Jin Zhang, Dong-Liang Cheng, et al. "Glioma grading on conventional MR images: a deep learning study with transfer learning". In: *Frontiers in neuroscience* (2018), p. 804.
293. Jure Zbontar, Florian Knoll, Anuroop Sriram, Tullie Murrell, Zhengnan Huang, Matthew J Muckley, Aaron Defazio, Ruben Stern, Patricia Johnson, Mary Bruno, et al. "fastMRI: An open dataset and benchmarks for accelerated MRI". In: *arXiv preprint* arXiv:1811.08839 (2018).
294. Kun Zeng, Hong Zheng, Congbo Cai, Yu Yang, Kaihua Zhang, and Zhong Chen. "Simultaneous single-and multi-contrast super-resolution for brain MRI images based on a convolutional neural network". In: *Computers in biology and medicine* 99 (2018), pp. 133–141.
295. Jun Zhang. "Inverse-Consistent Deep Networks for Unsupervised Deformable Image Registration". In: arXiv:1809.03443 *[cs]* (Sept. 2018). arXiv: 1809.03443. URL: http://arxiv.org/abs/1809.03443.
296. Richard Zhang, Phillip Isola, Alexei A Efros, Eli Shechtman, and Oliver Wang. "The unreasonable effectiveness of deep features as a perceptual metric". In: *Proceedings of the IEEE conference on computer vision and pattern recognition*. 2018, pp. 586–595.
297. Bo Zhu, Jeremiah Z Liu, Stephen F Cauley, Bruce R Rosen, and Matthew S Rosen. "Image reconstruction by domain-transform manifold learning". In: *Nature* 555.7697 (2018), pp. 487–492.
298. David Zimmerer, Simon AA Kohl, Jens Petersen, Fabian Isensee, and Klaus H Maier-Hein. "Context-encoding variational autoencoder for unsupervised anomaly detection". In: *arXiv preprint* arXiv:1812.05941 (2018).
299. Nyoman Abiwinanda, Muhammad Hanif, S Tafwida Hesaputra, Astri Handayani, and Tati Rajab Mengko. "Brain tumor classification using convolutional neural network". In: *World congress on medical physics and biomedical engineering 2018*. Springer. 2019, pp. 183–189.
300. N. Abraham and N. M. Khan. "A Novel Focal Tversky Loss Function With Improved Attention U-Net for Lesion Segmentation". In: *2019 IEEE 16th International Symposium on Biomedical Imaging (ISBI 2019)*. 2019, pp. 683–687. https://doi.org/10.1109/ISBI.2019.8759329.
301. Mehmet Akçakaya, Steen Moeller, Sebastian Weingärtner, and Kâ;mil Uğurbil. "Scan–neural-networks for k-space interpolation (RAKI) reconstruction: Database-free deep

learning for fast imaging". In: *Magnetic resonance in medicine* 81.1 (2019), pp. 439–453.

302. R Alaguselvi and Kalpana Murugan. "Image Enhancement Using Convolutional Neural Networks". In: *2019 IEEE International Conference on Clean Energy and Energy Efficient Electronics Circuit for Sustainable Development (INCCES)*. IEEE. 2019, pp. 1–5.
303. Guha Balakrishnan, Amy Zhao, Mert R. Sabuncu, John Guttag, and Adrian V. Dalca. "VoxelMorph: A Learning Framework for Deformable Medical Image Registration". In: *IEEE Transactions on Medical Imaging* 38.8 (Aug. 2019). arXiv: 1809.05231, pp. 1788–1800. ISSN: 0278-0062, 1558-254X. https://doi.org/10.1109/TMI.2019.2897538. URL: http://arxiv.org/abs/1809.05231 (visited on 09/08/2020).
304. Laura Beggel, Michael Pfeiffer, and Bernd Bischl. "Robust anomaly detection in images using adversarial autoencoders". In: *arXiv preprint* arXiv:1901.06355 (2019).
305. Stuart Berg, Dominik Kutra, Thorben Kroeger, Christoph N. Straehle, Bernhard X. Kausler, Carsten Haubold, Martin Schiegg, Janez Ales, Thorsten Beier, Markus Rudy, Kemal Eren, Jaime I. Cervantes, Buote Xu, Fynn Beuttenmueller, Adrian Wolny, Chong Zhang, Ullrich Koethe, Fred A. Hamprecht, and Anna Kreshuk. "ilastik: interactive machine learning for (bio)image analysis". In: *Nature Methods* 16.12 (2019), pp. 1226–1232. ISSN: 1548-7105. https://doi.org/10.1038/s41592-019-0582-9. URL: https://doi.org/10.1038/s41592-019-0582-9.
306. Gerda Bortsova, Florian Dubost, Laurens Hogeweg, Ioannis Katramados, and Marleen de Bruijne. "Semi-supervised Medical Image Segmentation via Learning Consistency Under Transformations". In: *Lecture Notes in Computer Science (including subseries Lecture Notes in Artificial Intelligence and Lecture Notes in Bioinformatics)* 11769 LNCS (2019), pp. 810–818. ISSN: 16113349. DOI: https://doi.org/10.1007/978-3-030-32226-7
307. S Deepak and PM Ameer. "Brain tumor classification using deep CNN features via transfer learning". In: *Computers in biology and medicine* 111 (2019), p. 103345.
308. Blake E. Dewey, Can Zhao, Jacob C. Reinhold, Aaron Carass, Kathryn C. Fitzgerald, Elias S. Sotirchos, Shiv Saidha, Jiwon Oh, Dzung L. Pham, Peter A. Calabresi, Peter C. M. van Zijl, and J. Prince. "DeepHarmony: A deep learning approach to contrast harmonization across scanner changes." In: *Magnetic resonance imaging* (2019).
309. Martin Eschelbach, Ali Aghaeifar, Jonas Bause, Jonas Handwerker, Jens Anders, Eva-Maria Engel, Axel Thielscher, and Klaus Scheffer. "Comparison of prospective head motion correction with NMR field probes and an optical tracking system". In: *Magnetic Resonance in Medicine* 81.1 (2019), pp. 719–729. https://doi.org/10.1002/mrm.27343. eprint: https://onlinelibrary.wiley.com/doi/pdf/10.1002/mrm.27343. URL: https://onlinelibrary.wiley.com/doi/abs/10.1002/mrm.27343.
310. Soundararajan Ezekiel, Larry Pearlstein, Abdullah Ali Alshehri, Adam Lutz, Jackson Zaunegger, and Waleed Farag. "Investigating GAN and VAE to train DCNN". In: *International Journal of Machine Learning and Computing* 9.6 (2019), pp. 774–781.
311. et al. Falcon WA. "PyTorch Lightning". In: *GitHub. Note:* https://github.com/PyTorchLightning/pytorch-lightning 3 (2019).
312. "FIRE: Unsupervised bi-directional inter-modality registration using deep networks". In: (July 2019). URL: http://arxiv.org/abs/1907.05062.
313. Vahid Ghodrati, Jiaxin Shao, Mark Bydder, Ziwu Zhou, Wotao Yin, Kim-Lien Nguyen, Yingli Yang, and Peng Hu. "MR image reconstruction using deep learning: evaluation of

network structure and loss functions". In: *Quantitative imaging in medicine and surgery* 9.9 (2019), pp. 1516–1527.
314. Fantin Girard, Conrad Kavalec, and Farida Cheriet. "Joint segmentation and classification of retinal arteries/veins from fundus images". In: *Artificial intelligence in medicine* 94 (2019), pp. 96–109.
315. Marloes HJ Hagens, Jessica Burggraaff, Iris D Kilsdonk, Serena Ruggieri, Sara Collorone, Rosa Cortese, Niamh Cawley, Emilia Sbardella, Michaela Andelova, Michael Amann, et al. "Impact of 3 Tesla MRI on interobserver agreement in clinically isolated syndrome: a MAGNIMS multicentre study". In: *Multiple Sclerosis Journal* 25.3 (2019), pp. 352–360.
316. Yoseo Han, Leonard Sunwoo, and Jong Chul Ye. "k-space deep learning for accelerated MRI". In: *IEEE transactions on medical imaging* 39.2 (2019), pp. 377–386.
317. Wenhao Jiang, Zhiyu Liu, Kit-Hang Lee, Shihui Chen, Yui-Lun Ng, Qi Dou, Hing-Chiu Chang, and Ka-Wai Kwok. "Respiratory motion correction in abdominal MRI using a densely connected U-Net with GAN-guided training". In: *arXiv preprint* arXiv:1906.09745 (2019).
318. Justin M Johnson and Taghi M Khoshgoftaar. "Survey on deep learning with class imbalance". In: *Journal of Big Data* 6.1 (2019), pp. 1–54.
319. Tae Hyung Kim, Pratyush Garg, and Justin P Haldar. "LORAKI: Autocalibrated recurrent neural networks for autoregressive MRI reconstruction in k-space". In: *arXiv preprint* arXiv:1904.09390 (2019).
320. Thomas Küstner, Karim Armanious, Jiahuan Yang, Bin Yang, Fritz Schick, and Sergios Gatidis. "Retrospective correction of motion-affected MR images using deep learning frameworks". In: *Magnetic resonance in medicine* 82.4 (2019), pp. 1527–1540.
321. Haichun Li, Ao Li, and Minghui Wang. "A novel end-to-end brain tumor segmentation method using improved fully convolutional networks". In: *Computers in biology and medicine* 108 (2019), pp. 150–160.
322. Nanjun Li and Faliang Chang. "Video anomaly detection and localization via multivariate gaussian fully convolution adversarial autoencoder". In: *Neurocomputing* 369 (2019), pp. 92–105.
323. Michelle Livne, Jana Rieger, Orhun Utku Aydin, Abdel Aziz Taha, Ela Marie Akay, Tabea Kossen, Jan Sobesky, John D Kelleher, Kristian Hildebrand, Dietmar Frey, et al. "A U-Net deep learning framework for high performance vessel segmentation in patients with cerebrovascular disease". In: *Frontiers in neuroscience* 13 (2019), p. 97.
324. Dwarikanath Mahapatra. "GAN Based Medical Image Registration". In: arXiv:1805.02369 *[cs]* (Sept. 2019). arXiv: 1805.02369. URL: http://arxiv.org/abs/1805.02369.
325. Hendrik Mattern, Alessandro Sciarra, Falk Lüsebrink, Julio Acosta-Cabronero, and Oliver Speck. "Prospective motion correction improves high-resolution quantitative susceptibility mapping at 7T". In: *Magnetic Resonance in Medicine* 81.3 (2019), pp. 1605–1619. https://doi.org/10.1002/mrm.27509. eprint: https://onlinelibrary.wiley.com/doi/pdf/10.1002/mrm.27509. URL: https://onlinelibrary.wiley.com/doi/abs/10.1002/mrm.27509.
326. Frank Ong and Michael Lustig. "SigPy: a python package for high performance iterative reconstruction". In: *Proc. ISMRM*. 2019.
327. Adam Paszke, Sam Gross, Francisco Massa, Adam Lerer, James Bradbury, Gregory Chanan, Trevor Killeen, Zeming Lin, Natalia Gimelshein, Luca Antiga, et al. "Pytorch:

An imperative style, high-performance deep learning library". In: *Advances in neural information processing systems* 32 (2019).
328. Kamlesh Pawar, Zhaolin Chen, N Jon Shah, and Gary F Egan. "Suppressing motion artefacts in MRI using an Inception-ResNet network with motion simulation augmentation". In: *NMR in Biomedicine* (2019), e4225.
329. Linmin Pei, Lasitha Vidyaratne, Wei-Wen Hsu, Md Monibor Rahman, Khan M Iftekharuddin, Linmin Pei, Lasitha Vidyaratne, Wei-Wen Hsu, Md Monibor Rahman, and Khan M Iftekharuddin. "Brain tumor classification using 3D convolutional neural network". In: *International MICCAI brainlesion workshop*. 2019, pp. 335–342.
330. Chi-Hieu Pham, Carlos Tor-Díez, Hélène Meunier, Nathalie Bednarek, Ronan Fablet, Nicolas Passat, and François Rousseau. "Multiscale brain MRI super-resolution using deep 3D convolutional networks". In: *Computerized Medical Imaging and Graphics* 77 (2019), p. 101647.
331. Jo Schlemper, Ilkay Oksuz, James R Clough, Jinming Duan, Andrew P King, Julia A Schnabel, Joseph V Hajnal, and Daniel Rueckert. "dAUTOMAP: Decomposing AUTOMAP to achieve scalability and enhance performance". In: *arXiv preprint* arXiv:1909.10995 (2019).
332. Hossam H Sultan, Nancy M Salem, and Walid Al-Atabany. "Multi-classification of brain tumor images using deep neural network". In: *IEEE Access* 7 (2019), pp. 69215–69225.
333. Yuanyuan Sun, Lele Xu, Lili Guo, Ye Li, and Yongming Wang. "A Comparison Study of VAE and GAN for Software Fault Prediction". In: *International Conference on Algorithms and Architectures for Parallel Processing*. Springer. 2019, pp. 82–96.
334. Su Yang, Jihoon Kweon, Jae-Hyung Roh, Jae-Hwan Lee, Heejun Kang, Lae-Jeong Park, Dong Jun Kim, Hyeonkyeong Yang, Jaehee Hur, Do-Yoon Kang, et al. "Deep learning segmentation of major vessels in X-ray coronary angiography". In: *Scientific reports* 9.1 (2019), pp. 1–11.
335. Hantao Yao, Feng Dai, Shiliang Zhang, Yongdong Zhang, Qi Tian, and Changsheng Xu. "Dr2-net: Deep residual reconstruction network for image compressive sensing". In: *Neurocomputing* 359 (2019), pp. 483–493.
336. Jong Chul Ye. "Compressed sensing MRI: a review from signal processing perspective". In: *BMC Biomedical Engineering* 1.1 (2019), pp. 1–17.
337. Chi Zhang, Seyed Amir Hossein Hosseini, Steen Moeller, Sebastian Weingärtner, Kamil Ugurbil, and Mehmet Akcakaya. "Scan-specific residual convolutional neural networks for fast MRI using residual RAKI". In: *2019 53rd Asilomar Conference on Signals, Systems, and Computers*. IEEE. 2019, pp. 1476–1480.
338. Wenshuai Zhao and Zengfeng Zeng. "Multi Scale Supervised 3D U-Net for Kidney and Tumor Segmentation". In: 2019. https://doi.org/10.24926/548719.007. arXiv: https://arxiv.org/abs/1908.03204.
339. Alejandro Barredo Arrieta, Natalia Díaz-Rodríguez, Javier Del Ser, Adrien Bennetot, Siham Tabik, Alberto Barbado, Salvador García, Sergio Gil-López, Daniel Molina, Richard Benjamins, et al. "Explainable Artificial Intelligence (XAI): Concepts, taxonomies, opportunities and challenges toward responsible AI". In: *Information Fusion* 58 (2020), pp. 82–115.
340. Milica M Badža and Marko Č Barjaktarović. "Classification of brain tumors from MRI images using a convolutional neural network". In: *Applied Sciences* 10.6 (2020), p. 1999.
341. Christoph Baur, Benedikt Wiestler, Shadi Albarqouni, and Nassir Navab. "Bayesian Skip-Autoencoders for Unsupervised Hyperintense Anomaly Detection in High Reso-

lution Brain Mri". In: *2020 IEEE 17th International Symposium on Biomedical Imaging (ISBI)*. 2020, pp. 1905–1909. https://doi.org/10.1109/ISBI45749.2020.9098686.
342. Christoph Baur, Benedikt Wiestler, Shadi Albarqouni, and Nassir Navab. *Scale-Space Autoencoders for Unsupervised Anomaly Segmentation in Brain MRI*. 2020. arXiv: https://arxiv.org/abs/2006.12852.
343. Emma Beede, Elizabeth Baylor, Fred Hersch, Anna Iurchenko, Lauren Wilcox, Paisan Ruamviboonsuk, and Laura M Vardoulakis. "A human-centered evaluation of a deep learning system deployed in clinics for the detection of diabetic retinopathy". In: *Proceedings of the 2020 CHI conference on human factors in computing systems*. 2020, pp. 1–12.
344. Vaishak Belle and Ioannis Papantonis. "Principles and practice of explainable machine learning". In: *arXiv preprint* arXiv:2009.11698 (2020).
345. Hamid Reza Boveiri, Raouf Khayami, Reza Javidan, and Alireza Mehdizadeh. "Medical image registration using deep neural networks: A comprehensive review". In: *Computers & Electrical Engineering* 87 (Oct. 2020), p. 106767. ISSN: 00457906. https://doi.org/10.1016/j.compeleceng.2020.106767. URL: https://linkinghub.elsevier.com/retrieve/pii/S0045790620306224 (visited on 10/27/2020).
346. Matthew Brett et al. *nipy/nibabel: 3.2.1*. Version 3.2.1. Nov. 2020. https://doi.org/10.5281/zenodo.4295521. URL: https://doi.org/10.5281/zenodo.4295521.
347. Xiaoran Chen, Suhang You, Kerem Can Tezcan, and Ender Konukoglu. *Unsupervised Lesion Detection via Image Restoration with a Normative Prior*. 2020. arXiv: https://arxiv.org/abs/2005.00031.
348. Yunyun Duan, Wei Shan, Liying Liu, Qun Wang, Zhenzhou Wu, Pan Liu, Jiahao Ji, Yaou Liu, Kunlun He, and Yongjun Wang. "Primary Categorizing and Masking Cerebral Small Vessel Disease Based on "Deep Learning System"". In: *Frontiers in Neuroinformatics* 14 (2020), p. 17.
349. Frank Emmert-Streib, Olli Yli-Harja, and Matthias Dehmer. "Explainable artificial intelligence and machine learning: A reality rooted perspective". In: *Wiley Interdisciplinary Reviews: Data Mining and Knowledge Discovery* 10.6 (2020), e1368.
350. Yaxiang Fan, Gongjian Wen, Deren Li, Shaohua Qiu, Martin D Levine, and Fei Xiao. "Video anomaly detection and localization via gaussian mixture fully convolutional variational autoencoder". In: *Computer Vision and Image Understanding* 195 (2020), p. 102920.
351. Frédéric Gretsch, Hendrik Mattern, Daniel Gallichan, and Oliver Speck. "Fat navigators and Moiré phase tracking comparison for motion estimation and retrospective correction". In: *Magnetic Resonance in Medicine* 83.1 (2020), pp. 83–93. https://doi.org/10.1002/mrm.27908. eprint: https://onlinelibrary.wiley.com/doi/pdf/10.1002/mrm.27908. URL: https://onlinelibrary.wiley.com/doi/abs/10.1002/mrm.27908.
352. Ning He, Ruolin Wang, and Yixue Wang. "Dynamic MRI reconstruction exploiting blind compressed sensing combined transform learning regularization". In: *Neurocomputing* 392 (2020), pp. 160–167.
353. Xiuxiu He, Yang Lei, Yabo Fu, Hui Mao, Walter J Curran, Tian Liu, and Xiaofeng Yang. "Super-resolution magnetic resonance imaging reconstruction using deep attention networks". In: *Medical Imaging 2020: Image Processing*. Vol. 11313. International Society for Optics and Photonics. 2020, 113132J.
354. Boah Kim, Dong Hwan Kim, Seong Ho Park, Jieun Kim, June-Goo Lee, and Jong Chul Ye. "CycleMorph: Cycle Consistent Unsupervised Deformable Image Registration". In:

arXiv:2008.05772 *[cs, eess, stat]* (Aug. 2020). arXiv: 2008.05772. URL: http://arxiv.org/abs/2008.05772.
355. Qiegen Liu, Qingxin Yang, Huitao Cheng, Shanshan Wang, Minghui Zhang, and Dong Liang. "Highly undersampled magnetic resonance imaging reconstruction using autoencoding priors". In: *Magnetic resonance in medicine* 83.1 (2020), pp. 322–336.
356. Qinghui Liu, Michael C. Kampffmeyer, Robert Jenssen, and Arnt-Børre Salberg. "Self-Constructing Graph Convolutional Networks for Semantic Labeling". In: *IGARSS 2020–2020 IEEE International Geoscience and Remote Sensing Symposium* (2020), pp. 1801–1804.
357. Qing Lyu, Hongming Shan, Yibin Xie, Debiao Li, and Ge Wang. "Cine Cardiac MRI Motion Artifact Reduction Using a Recurrent Neural Network". In: *arXiv preprint* arXiv:2006.12700 (2020).
358. Ričards Marcinkevičs and Julia E Vogt. "Interpretability and explainability: A machine learning zoo mini-tour". In: *arXiv preprint* arXiv:2012.01805 (2020).
359. Sergio Naval Marimont and Giacomo Tarroni. *Anomaly detection through latent space restoration using vector-quantized variational autoencoders*. 2020. arXiv:2012.06765 [cs.CV].
360. Evan M Masutani, Naeim Bahrami, and Albert Hsiao. "Deep learning single-frame and multiframe super-resolution for cardiac MRI". In: *Radiology* 295.3 (2020), pp. 552–561.
361. Shervin Minaee, Yuri Boykov, Fatih Porikli, Antonio Plaza, Nasser Kehtarnavaz, and Demetri Terzopoulos. *Image Segmentation Using Deep Learning: A Survey*. 2020. arXiv: 2001.05566 [cs.CV].
362. Hiba Mzoughi, Ines Njeh, Ali Wali, Mohamed Ben Slima, Ahmed BenHamida, Chokri Mhiri, and Kharedine Ben Mahfoudhe. "Deep multi-scale 3D convolutional neural network (CNN) for MRI gliomas brain tumor classification". In: *Journal of Digital Imaging* 33.4 (2020), pp. 903–915.
363. Chen Qin, Shuo Wang, Chen Chen, Huaqi Qiu, Wenjia Bai, and Daniel Rueckert. "Biomechanics-informed Neural Networks for Myocardial Motion Tracking in MRI". In: arXiv:2006.04725 *[cs, eess]* (July 2020). arXiv: 2006.04725. URL: http://arxiv.org/abs/2006.04725.
364. Arshia Rehman, Saeeda Naz, Muhammad Imran Razzak, Faiza Akram, and Muhammad Imran. "A deep learning-based framework for automatic brain tumors classification using transfer learning". In: *Circuits, Systems, and Signal Processing* 39.2 (2020), pp. 757–775.
365. Matteo Ronchetti. "TorchRadon: Fast Differentiable Routines for Computed Tomography". In: *arXiv preprint* arXiv:2009.14788 (2020). eprint: arXiv:2009.14788.
366. Ekaterina Sherina, Lisa Krainz, Simon Hubmer, Wolfgang Drexler, and Otmar Scherzer. "Displacement field estimation from OCT images utilizing speckle information with applications in quantitative elastography". In: *ArXiv* abs/2008.07373 (2020).
367. Roberto Souza, Youssef Beauferris, Wallace Loos, Robert Marc Lebel, and Richard Frayne. "Enhanced Deep-Learning-Based Magnetic Resonance Image Reconstruction by Leveraging Prior Subject-Specific Brain Imaging: Proof-of-Concept Using a Cohort of Presumed Normal Subjects". In: *IEEE Journal of Selected Topics in Signal Processing* 14.6 (2020), pp. 1126–1136.
368. Anuroop Sriram, Jure Zbontar, Tullie Murrell, Aaron Defazio, C Lawrence Zitnick, Nafissa Yakubova, Florian Knoll, and Patricia Johnson. "End-to-end variational networks for accelerated MRI reconstruction". In: *International Conference on Medical Image Computing and Computer-Assisted Intervention*. Springer. 2020, pp. 64–73.

369. Anuroop Sriram, Jure Zbontar, Tullie Murrell, C Lawrence Zitnick, Aaron Defazio, and Daniel K Sodickson. "GrappaNet: Combining parallel imaging with deep learning for multi-coil MRI reconstruction". In: *Proceedings of the IEEE/CVF Conference on Computer Vision and Pattern Recognition*. 2020, pp. 14315–14322.
370. Kun Tang, Zhi Li, Lili Tian, Lihui Wang, and Yuemin Zhu. "ADMIR–Affine and Deformable Medical Image Registration for Drug-Addicted Brain Images". In: *IEEE Access* 8 (2020), pp. 70960–70968. ISSN: 2169-3536. https://doi.org/10.1109/ACCESS.2020.2986829.
371. Irem Ulku and Erdem Akagunduz. *A Survey on Deep Learning-based Architectures for Semantic Segmentation on 2D images*. 2020. arXiv: https://arxiv.org/abs/1912.102301912.10230 [cs.CV].
372. Muhammad Usman, Siddique Latif, Muhammad Asim, Byoung-Dai Lee, and Junaid Qadir. "Retrospective motion correction in multishot MRI using generative adversarial network". In: *Scientific reports* 10.1 (2020), pp. 1–11.
373. Zhihao Wang, Jian Chen, and Steven CH Hoi. "Deep learning for image super-resolution: A survey". In: *IEEE transactions on pattern analysis and machine intelligence* (2020).
374. Garrett Wilson and Diane J Cook. "A survey of unsupervised deep domain adaptation". In: *ACM Transactions on Intelligent Systems and Technology (TIST)* 11.5 (2020), pp. 1–46.
375. Alan Yilun Yuan, Yang Gao, Liangliang Peng, Lingxiao Zhou, Jun Liu, Siwei Zhu, and Wei Song. "Hybrid deep learning network for vascular segmentation in photoacoustic imaging". In: *Biomedical Optics Express* 11.11 (2020), pp. 6445–6457.
376. Yibin Zhang, Jie Wang, Jinlong Sun, Bamidele Adebisi, Haris Gacanin, Guan Gui, and Fumiyuki Adachi. "CV-3DCNN: Complex-valued deep learning for CSI prediction in FDD massive MIMO systems". In: *IEEE Wireless Communications Letters* 10.2 (2020), pp. 266–270.
377. Rongjian Zhao, Buyue Qian, Xianli Zhang, Yang Li, Rong Wei, Yang Liu, and Yinggang Pan. "Rethinking dice loss for medical image segmentation". In: *2020 IEEE International Conference on Data Mining (ICDM)*. IEEE. 2020, pp. 851–860.
378. Fuzhen Zhuang, Zhiyuan Qi, Keyu Duan, Dongbo Xi, Yongchun Zhu, Hengshu Zhu, Hui Xiong, and Qing He. "A comprehensive survey on transfer learning". In: *Proceedings of the IEEE* 109.1 (2020), pp. 43–76.
379. Ying Zhuge, Holly Ning, Peter Mathen, Jason Y Cheng, Andra V Krauze, Kevin Camphausen, and Robert W Miller. "Automated glioma grading on conventional MRI images using deep convolutional neural networks". In: *Medical physics* 47.7 (2020), pp. 3044–3053.
380. Wadhah Ayadi, Wajdi Elhamzi, Imen Charfi, and Mohamed Atri. "Deep CNN for brain tumor classification". In: *Neural Processing Letters* 53.1 (2021), pp. 671–700.
381. Fabio Carrara, Giuseppe Amato, Luca Brombin, Fabrizio Falchi, and Claudio Gennaro. "Combining gans and autoencoders for efficient anomaly detection". In: *2020 25th International Conference on Pattern Recognition (ICPR)*. IEEE. 2021, pp. 3939–3946.
382. Elizabeth Cole, Joseph Cheng, John Pauly, and Shreyas Vasanawala. "Analysis of deep complex-valued convolutional neural networks for MRI reconstruction and phase-focused applications". In: *Magnetic resonance in medicine* 86.2 (2021), pp. 1093–1109.
383. Francisco Javier Díaz-Pernas, Mario Martínez-Zarzuela, Míriam Antón-Rodríguez, and David González-Ortega. "A deep learning approach for brain tumor classification and

segmentation using a multiscale convolutional neural network". In: *Healthcare*. Vol. 9. 2. Multidisciplinary Digital Publishing Institute. 2021, p. 153.

384. Tianming Du, Honggang Zhang, Yuemeng Li, Stephen Pickup, Mark Rosen, Rong Zhou, Hee Kwon Song, and Yong Fan. "Adaptive convolutional neural networks for accelerating magnetic resonance imaging via k-space data interpolation". In: *Medical Image Analysis* 72 (2021), p. 102098.
385. Gastón García González, Pedro Casas, Alicia Fernández, and Gabriel Gómez. "On the Usage of Generative Models for Network Anomaly Detection in Multivariate Time-Series". In: *ACM SIGMETRICS Performance Evaluation Review* 48.4 (2021), pp. 49–52.
386. Suhita Ghosh, Andreas Krug, Georg Rose, and Sebastian Stober. "Perception-Aware Losses Facilitate CT Denoising and Artifact Removal". In: *2021 IEEE 2nd International Conference on Human-Machine Systems (ICHMS)*. IEEE. 2021, pp. 1–6.
387. G Gungor, M Michalet, A Lombard, T Roque, B Atalar, B Temur, I Serbez, D Azria, L de Vitry, O Riou, et al. "Human-Level Precision Upper Abdominal OAR Contouring With Anatomically Preserving Deep Learning During Magnetic Resonance Imaging Guided Adaptive Radiotherapy (MRgRT)". In: *International Journal of Radiation Oncology, Biology, Physics* 111.3 (2021), S44–S45.
388. Jongyeon Lee, Byungjai Kim, and HyunWook Park. "MC2-Net: motion correction network for multi-contrast brain MRI". In: *Magnetic Resonance in Medicine* 86.2 (2021), pp. 1077–1092.
389. Johannes Leuschner, Maximilian Schmidt, Poulami Somanya Ganguly, Vladyslav Andriiashen, Sophia Bethany Coban, Alexander Denker, Dominik Bauer, Amir Hadjifaradji, Kees Joost Batenburg, Peter Maass, and Maureen van Eijnatten. "Quantitative Comparison of Deep Learning-Based Image Reconstruction Methods for Low-Dose and Sparse-Angle CT Applications". In: *Journal of Imaging* 7.3 (2021), p. 44. https://doi.org/10.3390/jimaging7030044. URL: https://www.mdpi.com/2313-433X/7/3/44.
390. Grace W Lindsay. "Convolutional neural networks as a model of the visual system: Past, present, and future". In: *Journal of cognitive neuroscience* 33.10 (2021), pp. 2017–2031.
391. Huilin Mu, Yun Zhang, Yicheng Jiang, and Chang Ding. "CV-GMTINet: GMTI Using a Deep Complex-Valued Convolutional Neural Network for Multichannel SAR-GMTI System". In: *IEEE Transactions on Geoscience and Remote Sensing* 60 (2021), pp. 1–15.
392. Changheun Oh, Dongchan Kim, Jun-Young Chung, Yeji Han, and HyunWook Park. "A k-space-to-image reconstruction network for MRI using recurrent neural network". In: *Medical Physics* 48.1 (2021), pp. 193–203.
393. Hajer Ouerghi, Olfa Mourali, and Ezzeddine Zagrouba. "Glioma classification via MR images radiomics analysis". In: *The Visual Computer* (2021), pp. 1–15.
394. Fernando Pérez-García, Rachel Sparks, and Sebastien Ourselin. "TorchIO: a Python library for efficient loading, preprocessing, augmentation and patch-based sampling of medical images in deep learning". In: *Computer Methods and Programs in Biomedicine* (2021), p. 106236.
395. Vit Skvara, Jan Franca, Matej Zorek, Tomas Pevny, and Vaclav Smidl. "Comparison of Anomaly Detectors: Context Matters". In: *IEEE Transactions on Neural Networks and Learning Systems* (2021).
396. Takahiro Ueda, Yoshiharu Ohno, Kaori Yamamoto, Akiyoshi Iwase, Takashi Fukuba, Satomu Hanamatsu, Yuki Obama, Hirotaka Ikeda, Masato Ikedo, Masao Yui, et al.

"Compressed sensing and deep learning reconstruction for women's pelvic MRI denoising: Utility for improving image quality and examination time in routine clinical practice". In: *European journal of radiology* 134 (2021), p. 109430.

397. Yijun Zhao, Jacek Ossowski, Xuming Wang, Shangjin Li, Orrin Devinsky, Samantha P Martin, and Heath R Pardoe. "Localized motion artifact reduction on brain MRI using deep learning with effective data augmentation techniques". In: *2021 International Joint Conference on Neural Networks (IJCNN)*. IEEE. 2021, pp. 1–9.
398. Alexander Ziller, Dmitrii Usynin, Moritz Knolle, Kerstin Hammernik, Daniel Rueckert, and Georgios Kaissis. "Complex-valued deep learning with differential privacy". In: *arXiv preprint* arXiv:2110.03478 (2021).
399. David Zimmerer, Jens Petersen, Gregor Koehler, Paul Jaeger, Peter Full, Tobias Roß;, Tim Adler, Annika Reinke, Lena Maier-Hein, and Klaus Maier-Hein. *Medical Out-of-Distribution Analysis Challenge 2021*. Mar. 2021. https://doi.org/10.5281/zenodo.4573948. URL: https://doi.org/10.5281/zenodo.4573948.
400. Ruiyang Zhao, Burhaneddin Yaman, Yuxin Zhang, Russell Stewart, Austin Dixon, Florian Knoll, Zhengnan Huang, Yvonne W Lui, Michael S Hansen, and Matthew P Lungren. "fastMRI+, Clinical pathology annotations for knee and brain fully sampled magnetic resonance imaging data". In: *Scientific Data* 9.1 (2022), pp. 1–6.
401. Jonathan Tompson, Ross Goroshin, Arjun Jain, Yann LeCun, and Christopher Bregler. *Efficient Object Localization Using Convolutional Networks*. URL: http://arxiv.org/pdf/1411.4280v3.

MIX
Papier aus verantwortungsvollen Quellen
Paper from responsible sources
FSC® C105338

If you have any concerns about our products,
you can contact us on
ProductSafety@springernature.com

In case Publisher is established outside the EU,
the EU authorized representative is:
Springer Nature Customer Service Center GmbH
Europaplatz 3, 69115 Heidelberg, Germany

Printed by Libri Plureos GmbH
in Hamburg, Germany